TOPICS IN STEREOCHEMISTRY

VOLUME 22

ADVISORY BOARD

TOPICS IN

STEREOCHEMISTRY

EDITOR

SCOTT E. DENMARK
Department of Chemistry
University of Illinois, Urbana-Champaign
Urbana, Illinois

Volume 22

AN INTERSCIENCE® PUBLICATION
JOHN WILEY & SONS, INC.
New York / Chichester / Weinheim / Brisbane / Singapore / Toronto

This book is printed on acid-free paper. ∞

Published simultaneously in Canada.

Library of Congress Catalog Card Number: 67-13943

ISBN 0-471-25316-2

Printed in the United States of America.

10 9 8 7 6 5 4 3 2 1

To the memory of Vladimir Prelog and Derek H. R. Barton

Vladimir Prelog, sitting comfortably in his office, Zurich, 1989.

Judy and Derek H.R. Barton, a very happy couple, early 1998.

INTRODUCTION TO THE SERIES

Since its first appearance in 1967, the *Topics in Stereochemistry* series has stood as the standard bearer for advances in the broad field of stereochemistry. Those visionaries who founded the series, anticipated with remarkable foresight, the extraordinary growth and impact that stereochemistry has had in all reaches of the chemical enterprise. Fortunately, there is no surcease in the interest in and importance of stereochemistry as the discipline of chemistry evolves and its borders expand and diffuse into the related fields of biology, medicine, physics, materials science, chemical engineering and environmental science.

The field of stereochemistry serves as a unifying theme for the expanded definition and diversification of chemistry. The consequences of molecular and macromolecular shape and topology are central to issues of chemical reactivity, physical properties and biological function. With that view, the importance of stereochemistry had never been greater and it is hoped that this series will provide forum for documentation of the most significant advances in all of these subdisciplines of chemistry.

The *Topics in Stereochemistry* series has set itself apart by maintaining a remarkable balance of chapters that are both definitive, standing the test of time, and current, addressing the impact of stereochemistry at the most exciting frontiers. As a student and researcher, I have often turned to chapters in *Topics in Stereochemistry* for both the foundations and the state of the art in new areas of interest. It is my hope that the series continues to enjoy that level of confidence in the chemistry community and that it retains, in this second incarnation the esteem which the founders have worked to hard to establish.

I am fortunate in having been able to enlist the help and guidance of an international board of Editorial Advisors who have provided great assistance by suggesting chapter topics and suitable authors for articles both here and in future volumes. While I am grateful for the assistance of the Editorial Advisory Board, it is the Editor and the Authors who are solely responsible for any shortcomings of *Topics in Stereochemistry*.

S. E. Denmark

FOREWORD

Following the publication of *Stereochemistry of Carbon Compounds* (Eliel, 1962) and *Conformational Analysis* (Eliel, Allinger, Angyal and Morrison, 1965) N. L. Allinger and I in 1966 decided to launch a series *Topics in Stereochemistry* to keep readers informed of current and new topics in this area. The subject of stereochemistry, which had been in the doldrums during the first half of the 20th century, was then just experiencing a renaissance, thanks, in no small measure, to the pioneering work of Vladimir Prelog and Derek H. R. Barton (both of whom sadly passed away earlier this year). John Wiley & Sons agreed to publish the series, which, over 27 years, grew into 21 volumes. Samuel H. Wilen joined the editorial team in 1982 and Lou Allinger retired as coeditor in 1986. The series was clearly buoyed by the explosive development of stereochemistry in the 1970's and 1980's and, in turn, may have helped along that development at least in a small way. Among the more influential articles should be mentioned an early one on determination of enantiomeric purity by non-polarimetric means (1967), a comprehensive chapter on resolution and resolving agents (1971), a 1978 article on asymmetric synthesis, chapters on the use of carbon-13 NMR in stereochemistry (1974, 1986), an article on the stereochemistry of hydride reductions (1979) and a chapter on enzymatic resolution (1989)—one of several articles on biochemical aspects of stereochemistry. The stereochemistry of inorganic compounds was also treated in several chapters' culminating in a whole volume on this topic, edited by Gregory Geoffroy in 1981. It is clear that in some instances the series anticipated important developments and brought them to the readers' attention at an early stage.

Stereochemistry continues to be a highly viable and prolific subject. Nonetheless, in 1993 the editors reluctantly decided to retire after publication of Volume 21, in part because for some time prior they had been fully occupied with the publication of *Stereochemistry of Organic Compounds* (Wiley, 1994). Sadly, my close friend, able coauthor and coeditor, and valued colleague Sam Wilen died soon thereafter. In as much as the explosive development in stereochemistry is continuing, I am delighted that Scott Denmark, who is eminently qualified for the task, has taken over editorship of the series with Volume 22. I am particularly pleased to see a chapter by my fellow stereochemist Kurt Mislow in this volume, since Kurt helped us launch the series in 1967 by agreeing to be an editorial advisor and by coauthoring a seminal chapter on stereochemical relationships (heterotopicity) for the first volume. I am also happy to see the inclusion of both inorganic and biochemical aspects of stereochemistry in the present volume, including the very timely and important topic of asymmetric catalysis.

I wish the new editor continued success in this relaunch.

Ernest L. Eliel
Department of Chemistry
University of North Carolina
Chapel Hill, NC 27599-3290

PREFACE

The collection of chapters in this first Volume of the second era of *Topics in Stereochemistry* covers a range of topics that reflects the both the evolution of the field of stereochemistry and the most exciting advances that presage its future. Thus, it was expressly by design that there be conceptually focused chapters on the basic principles of molecular chirality and on asymmetric amplification. In addition, there was a conscious effort to incorporate applied chapters on some of the most exciting advances illustrating the importance of stereochemistry in synthetic organic, physical organic and bioorganic chemistry. It is therefore quite satisfying that the collection in this volume matched the hoped for impact for the re-release of the series.

The first chapter in this volume is a conceptually stimulating feast, fittingly authored by Kurt Mislow who also co-authored the first chapter of Volume 1 in this series. This chapter provides an insightful analysis of the foundations of molecular chirality, topological chirality and definitions of chirality measures. Aficionados of the history of stereochemical concepts will particularly enjoy the account of how various models of molecular and topological models have evolved. Also included is a thorough treatment of the chirality of knots and links. Critical passages from early treatises are quoted in the original language (translated as well) which effectively engages the reader as a participant in this "Chirosopher's" telling of the story of chirality.

The importance of catalysis in biological as well as synthetic organic chemistry cannot be overstated. In Chapter 2, Donald Hilvert examines the scope and utility of asymmetric reactions under catalysis by antibodies. From a stereochemical point of view, this has significant impact not only in the production of important compounds in stereochemically defined form, but also in the ability of the antibody catalysts to alter the stereochemical course of organic reactions in fashions contrary to their natural tendencies. The most important chemical transformations carried out by catalytic antibodies are covered and provide the reader with an excellent snapshot of the state of the art of this emerging subfield in asymmetric catalysis. In addition, a critical appraisal of the limitations and future directions is included which should provide ample stimulation for thought.

Chapter Three by Jonathan White and Christopher Clark is a comprehensive analysis of the stereoelectronic effects of group 14 elements in controlling the course of many reactions in organic chemistry. The scope of coverage is thorough from silicon through lead compounds and with analysis of these effects over many attached atoms. Although the manifestation of these effects have been put to great use by synthetic chemists, this chapter is decidedly more physical organic in focus and provides a clear organizational framework for understanding and correlating the many different substrates and reaction types wherein a stereoelectronic effect has been noted. The presentation of some of the more remote effects should stimulate investigations into possible applications.

The fourth chapter by Masakatsu Shibasaki and Hiroaki Sasai provides an up-to-date coverage of one of the most fascinating areas in asymmetric synthesis, namely, the design, development and application of new systems for enantioselective catalysis. This chapter details the evolution of the fascinating new class of chiral lanthanide-BINOL complexes which have found remarkably broad and powerful application as catalysts for Michael, aldol additions, hydrocyanation, hydrophosphonylation and epoxide openings. In addition, important mechanistic insights have been gleaned from investigations of the structure/reactivity/selectivity relationships of the complexes which is also elaborated in the chapter.

One of the most exciting and recently emerging areas in asymmetric synthesis is asymmetric amplification. This topic has extraordinarily broad implications from mechanistic insights provided by non-linear effects to the enhancement of enantiomeric composition of important compounds to new hypotheses for the origin of biomolecular homochirality. The final chapter by Henri Kagan and David Fenwick provides a thorough and insightful analysis of the basic principles of asymmetric amplification and illustrations of some of the more important applications in synthesis.

It is appropriate here to acknowledge Carla A. Fjerstad, Editor and Patrick J. Kelly, Vice President and General Manager, Physical Sciences, at Wiley for their interest in continuing the series and for providing motivation, advice and assistance.

Finally, in this past year, two of this century's giants of stereochemistry passed away. However, the scientific legacies of Vladimir Prelog and Sir Derek H. R. Barton are firmly embodied in the fundament of our knowledge of stereochemistry and their personal legacies irrevocably woven into the hearts and minds of those co-workers, colleagues, students and friends who were graced by their lives.

Scott E. Denmark
Urbana, Illinois

CONTENTS

TOPICS IN STEREOCHEMISTRY

VOLUME 22

Molecular Chirality

KURT MISLOW

Department of Chemistry, Princeton University, Princeton, NJ 08544

Topics in Stereochemistry, Volume 22, Edited by Scott E. Denmark.
ISBN 0-471-25316-2

I. INTRODUCTION

Chirality, in its many and varied manifestations, is ubiquitous; a concept rooted in mathematics, it permeates all branches of the natural sciences.[1] In 1848, Louis Pasteur announced his epochal discovery of a causal relationship between the handedness of hemihedral sodium ammonium tartrate crystals and the sense of optical rotation of the tartrates in solution.[2] This discovery, which marks the beginning of modern stereochemistry, connected enantiomorphism on the macroscopic scale to enantiomorphism on the molecular scale and thus led to Pasteur's recognition that the optical activity of the tartrates is a manifestation of "dissymétrie moléculaire,"[3] that is, of molecular chirality.

Since Pasteur's time, stereochemistry has experienced an enormous intellectual growth and has also found widespread industrial application. In recent years, a spate of articles, reviews, books, and international conferences and symposia have dealt with the role of chirality in chemistry, and three new journals have been specifically devoted to this topic: *Chirality*, by Wiley–Liss in 1989, *Tetrahedron: Asymmetry*, by Pergamon Press in 1990, and *Enantiomer*, by Gordon and Breach in 1996. Much of the research reported in these media, though motivated to some degree by market forces—notably by the demand of pharmaceutical industry for enantiopure drugs[4]—serves as a reminder that molecular chirality remains the centerpiece of stereochemistry and allied branches of science.

The year 1998, which celebrates the sequicentennial anniversary of Pasteur's discovery, seems appropriate as an occasion on which to examine selected features of molecular chirality that are of interest from a theoretical point of view. We begin by describing the foundations of geometrical chirality within a historical context and go on to analyze the problem of selecting appropriate models for chiral and achiral molecules. We next address questions relating to the topological chirality and achirality of molecular knots and links, and follow this with a discussion of the right–left dichotomy that is common to all chiral objects. Our essay concludes with a critical look at the principles that underlie chirality measures.

II. GEOMETRICAL CHIRALITY

A. Incongruent Counterparts

The paradoxical nature of handedness was first examined in depth by the philosopher Immanuel Kant. Take two objects, such as the right and left hand, that are "völlig gleich und ähnlich" [entirely equal and similar] in the sense that their metric properties are exactly the same. Yet although right and left hands are isometric, they cannot be made to coincide in space. Thus they are at the same time congruent (isometric) and not congruent (not superposable). In the 1768 paper in which he introduced this subject,[5] Kant coined the phrase "incongruent counterparts" to describe such nonsuperposable mirror images:

> Ich nenne einen Körper, der einem andern völlig gleich und ähnlich ist, ob er gleich nicht in eben denselben Grenzen kann beschlossen werden, sein *incongruentes Gegenstück.*
>
> [I call a solid that is entirely equal and similar to another, yet that cannot be included within the same boundaries, its *incongruent counterpart.*]

Kant returned to this troublesome paradox—whose spirit is nicely captured in the phrase "the same and not the same"[6]—in his Inaugural Dissertation, published in 1770, and again, in 1783, in a section of his *Prolegomena zu einer jeden künftigen Metaphysik,*[7] in which he stated the puzzle in the following words:

> Was kann wohl meiner Hand oder meinem Ohr ähnlicher und in allen Stücken gleicher sein als ihr Bild im Spiegel? Und dennoch kann ich eine solche Hand, als im Spiegel gesehen wird, nicht an die Stelle ihres Urbildes setzen; denn wenn dieses eine rechte Hand war, so ist jene im Spiegel eine linke, und das Bild des rechten Ohres ist ein linkes, das nimmermehr die Stelle des ersteren vertreten kann.
>
> [What can be more similar to my hand or my ear, and more equal to it in all of its parts, than its image in the mirror? And yet such a hand as is seen in the mirror cannot take the place of the original; for if the original was a right hand, the hand in the mirror is a left one, and the image of the right ear is a left ear, which could never take the place of the original.]

He went on to elaborate:

> [Die] linke Hand kann mit der rechten, ohnerachtet aller beiderseitigen Gleichheit und Ähnlichkeit, doch nicht zwischen denselben Grenzen eingeschlossen sein (sie können nicht kongruieren); der Handschuh der einen Hand kann nicht auf der anderen gebraucht werden. Was ist nun die Auflösung?
>
> [Despite their mutual equality and similarity, left and right hands can nevertheless not be encompassed within the same boundaries (they cannot be rendered congruent); the glove of one hand cannot be used on the other. Then what is the solution?]

Kant's meditations concerning incongruent counterparts inspired theories regarding the nature of space and laid the groundwork for all subsequent epistemological studies dealing with the problem of handedness.[8] The chief question that arose from Kant's work was the following: Does the property of handedness depend on the space of the embedded object? Before we address this question, however, we need to digress in order to introduce some definitions.

B. Geometrical Chirality

The terms *chiral* and *chirality*, whose literal meanings are "handed" and "handedness," made their first appearance in a footnote of a lecture, entitled "The Molecular Tactics of a Crystal," that Sir William Thomson, who had been elevated to Lord Kelvin in 1892, delivered to the Oxford University Junior Scientific Club on May 16, 1893.[9] The famous footnote reads:

> I call any geometrical figure, or group of points, *chiral*, and say that it has chirality, if its image in a plane mirror, ideally realized, cannot be brought to coincide with itself. Two equal and similar right hands are homochirally similar. Equal and similar right and left hands are heterochirally similar or "allochirally" similar (but heterochirally is better). These are also called "enantiomorphs," after a usage introduced, I believe, by German writers. Any chiral object and its image in a plane mirror are heterochirally similar.

Kelvin's lecture was reprinted verbatim in Appendix H of his Baltimore Lectures,[10a] which were delivered at Johns Hopkins University in October of 1884, nine years before the Oxford lecture. It would therefore be only natural to assume that the terms in Kelvin's definition originated in 1884, and not in 1893, and most authors have accordingly ascribed the origin of Kelvin's coinage of "chiral" and "chirality" to the Baltimore Lectures. In his careful examination of the record, however, Ronald Bentley has shown unambiguously that the term *chiral* was not used in any of the Baltimore Lectures at the time.[11] It is now clear that Appendix H was added subsequently, around the time of publication of the Baltimore Lectures in 1904.

Kelvin's definition is restricted to rigid geometrical bodies. His "similar," like Kant's "gleich und ähnlich," means congruent or isometric. The term *enantiomorph* has the same meaning as Kant's "incongruent counterpart." It is also obvious that the phrase "cannot be brought to coincide with itself" allows for rigid rotations and translations but excludes deformations of the geometrical figure—otherwise an asymmetric tetrahedron, for example, would not be chiral. For the present purpose, let us define a proper congruence as a transformation that is effected by any combination of rigid rotations and translations (i.e., rigid motions) in the object's space, and an improper congruence as a transformation that is effected by rigid motions combined with an odd number of reflections in the object's space. Kelvin's definitions can then be reformulated as follows, with the addition of Kelvin's coinage of *chiroid*[10b] to signify a chiral object:

> An object is chiral and a chiroid if and only if it cannot be superposed on its mirror image by a proper congruence, otherwise it is achiral; two chiroids are heterochiral and enantiomorphs if and only if they are improperly congruent; and two chiroids are homochiral and homomorphs if and only if they are properly congruent.

In these definitions, "object" refers to any rigid array, such as a set of points, or to a geometrical figure, or to a model of a molecule "ideally realized" (of which more below); "properly congruent" can be replaced by "superposable;" and "homomorphs," which may be either chiral or achiral, are "congruent counterparts" in Kant's terminology.

Kelvin's coinage was widely ignored until it was rescued from undeserved obscurity in two letters to *Nature* by Lancelot Law Whyte.[12] In his words,

> Chiral forms are more varied and widespread than is sometimes realized. Some examples are: (mathematical) skew polygon, pair of skew lines, area with sense, various skew polyhedra, for example, an isosceles tetrahedron with scalene faces and Coxeter's "orthoscheme," helix, helical polygon, scalene spherical triangle, single overhand knot, eleven enantiomorphous classes of point group symmetry, ratio of two co-initial vectors, associations of a polar with an axial vector, quaternions, spinors, nonsymmetric three-dimensional tensors, pseudoscalar fields; (natural) most optically active molecules, crystals of the eleven enantiomorphous classes, many electromagnetic fields, screw displacements, body torques, torsions, relaxation processes of chirally deformed structures, amino acid and protein structures, conical helices in plants and shells, many parts of animals (including human hand); (man made) coat with buttoning for man or for woman, shoe, propellor, clock face. They merit a clear designation.

Our discussion would be incomplete without a comment concerning the original spelling of the terms introduced in Kelvin's definition.[13] On February 17, 1873, twenty years before his lecture to the Oxford University Junior Scientific Club, Thomson read a lecture to the Royal Society of Edinburgh entitled "Note on Homocheiral and Heterocheiral Similarity."[14] Because the text of the lecture was never published, all that we are left with is the title. Evidently, after an interval of two decades, Thomson/Kelvin had decided to streamline these terms by dropping the "e" in "cheiral," even though the original spelling was a more faithful transliteration of χειρ, Greek for "hand." As it happens, however, a related term, *amphicheiral,* is still in current use in the mathematical literature, its original spelling intact. This term, which means "topologically achiral in 3-space," was introduced and defined in 1877 by the Scottish physicist Peter Guthrie Tait.[15] Tait, the Professor of Natural Philosophy at the University of Edinburgh, was the foremost pioneer of what might be called empirical knot theory, to distinguish it from the mathematically rigorous discipline that it is today. Thomson was the Professor of Natural Philosophy in the neighboring University of Glasgow, and it was his theory of vortex atoms[16] that had stimulated Tait's work on knots. "Homocheiral" and "heterocheiral," which Kelvin later defined (after purging the "e") with reference to geometrical chirality, and "amphicheiral," which Tait defined with reference to topological chirality, are therefore contemporaneous coinages of Scottish provenance, all dating from the same period, around 1872–1877.

On the basis of this circumstantial evidence, it seems likely that "cheiral" and "cheirality" were the original spellings of these terms.

C. Reflections and Isometries

The relationship between an object and its mirror image is at the heart of the incongruity of counterparts. In particular, the statement that enantiomorphs can be brought into congruence only by rigid motions combined with an odd number of reflections in the object's space requires further elaboration.

A reflection is an operation that changes the sign of an odd number of coordinates. It is discontinuous; that is, it cannot be carried out in a continuous manner without distortion of the object, and it is a purely mathematical operation that cannot be physically performed. For example, consider reflection in Euclidean 3-space (R^3): If the mirror occupies the yz plane, then a reflection sends each point from (x,y,z) in front of the mirror to $(-x,y,z)$ in the magic land behind the looking glass. The relationship between an object and the image in the mirror thus acquires a mysterious quality, for the reflection that we see in the mirror is intangible and has no existence in our world.

An isometry, or congruence, is a transformation that preserves the lengths of all line segments between pairs of points. All isometries are generated by reflections and fall into two classes, depending on the parity of the number of generating reflections. Isometries of the first kind, or proper isometries, are continuous operations generated from an even number of reflections and are physically performable in the object's space. Isometries of the second kind, or improper isometries, are discontinuous operations generated from an odd number of reflections and are not physically performable in the object's space.

An even number of reflections in the same plane send each point from (x,y,z) back to itself, which is equivalent to an identity operation (a proper isometry), while an odd number of reflections in the same plane is equivalent to a single reflection (an improper isometry). The product of two reflections in separate planes is a proper isometry. We distinguish two cases: (1) If the two planes are parallel and spaced by a distance d, the isometry is a translation by $2d$ in a direction perpendicular to the two planes; (2) if the two planes intersect at an angle α, the isometry is a proper rotation (or, simply, rotation) through 2α about the line of intersection (the axis of rotation). Translations and rotations are continuous operations because motion can be controlled by continuous changes in the distance or angle between the two mirrors.

The product of three reflections in separate planes is an improper isometry. We again distinguish two cases: (1) Two reflections in parallel planes combined with a third reflection in a plane perpendicular to the other two results is a translation–reflection or glide reflection; (2) two reflections in intersecting planes combined with a third reflection in a plane perpendicular to the line of intersection results in a rotation–reflection or improper rotation. When the angle of intersection $\alpha = 90°$, the isometry is a central inversion that sends a point from (x,y,z) to $(-x, -y, -z)$. All these transformations are discontinuous.

Finally, the product of two reflections in intersecting planes combined with two reflections in parallel planes that are perpendicular to the line of intersection—that is, the product of altogether four reflections and therefore a proper isometry— is a rotation–translation or screw displacement. This also is a continuous transformation.

In analogy to combinations of even and odd integers under addition, the product of n proper isometries is a proper isometry regardless of whether n is even or odd; the product of n improper isometries is a proper isometry if n is even and an improper isometry if n is odd; and the product of a proper isometry and an improper isometry is an improper isometry.

The elements of point groups constitute a special class of isometries, called symmetry operations, that leave the object unchanged in space and that may be similarly classified as proper, that is, proper rotations (C_n), including the identity ($E = C_1$) as a special case, or improper, that is, improper rotations (S_n), including single reflection ($\sigma = S_1$) and central inversion ($i = S_2$) as special cases. In analogy to isometries, these are also called symmetry operations of the first and second kind, respectively. As an example, consider a model of H_2O with C_{2v} symmetry, and suppose that the model lies in the xz plane that bisects all three atoms. The symmetry operation $\sigma_v(xz)$ then results in the transposition of the two halves of each atom. Similarly, $\sigma_v(yz)$ transposes two halves of the oxygen atom, along with the two hydrogen atoms. Both of these transpositions are improper rotations, and neither one is physically performable. It is only when $\sigma_v(xz)$ and $\sigma_v(yz)$ are combined that we achieve the same effect as $C_2(z)$, a proper rotation and a physically performable operation.

The effect of an improper isometry on a rigid object is to generate a second, isometric object, the mirror image. In the general case, the mirror image is displaced in space relative to the object—unless, that is, the isometry is a symmetry operation of the second kind. In that case, and in that case only, object and mirror image end up occupying the same position in space: object and mirror image are superposable. If the symmetry group of the object lacks improper rotations, however, superposition of an object and its mirror image becomes impossible. The effect of an improper isometry on such an object therefore *invariably* results in a spatial displacement of the mirror image (the enantiomorph) relative to the object: The object is not invariant under improper rotation and is said to be chiral. This leads us to the following definition of a chiroid, which is equivalent to Kelvin's:

> An object is a chiroid if and only if its point group contains no improper rotations.

In his Nobel Lecture, Vladimir Prelog defined geometrical chirality by conjoining the two definitions[17]:

> An object is chiral if it cannot be brought into congruence with its mirror image by translation and rotation. Such objects are devoid of symmetry elements which include reflection: mirror planes, inversion centers, or improper rotational axes.

All objects and their full point groups can be partitioned into two classes: chiral and achiral.[18] Half the elements in an achiral group form a subgroup of proper rotations, the chiral (or rotation) subgroup. The other half constitute a set of improper rotations (the coset) that is not a group because the identity element is contained in the chiral subgroup. The union of a chiral subgroup and its coset is the distinguishing characteristic of all achiral groups. Expressed formally, each chiral group is the subgroup of highest order, H, in at least one achiral group, G_{ach}. Hence $G_{ach} = H \cup aH$, where aH is the coset and a, the coset representative, is an improper rotation. For example, the point group for H_2O is $\{E, C_2, \sigma_v(xz), \sigma_v(yz)\} = \{E, C_2\} \cup \{\sigma_v(xz)E, \sigma_v(xz)C_2\}$, so that $C_{2v} = C_2 \cup \sigma_v(xz)C_2$, where $\sigma_v(xz)$ is a coset representative. Because the sets H and aH have the same number of elements, it follows that the order of all achiral groups must be even and that any group of odd order must be chiral. This does not mean, of course, that the order of a chiral group cannot also be even, as in the example of C_2.

D. Dimension Dependence of Chirality

We now return to answer the question raised by Kant's work on incongruous counterparts: Whether an object is chiral depends on the dimension of the space in which it is embedded. That is, chirality is an extrinsic property.

For example, neither a scalene triangle nor an oriented circle is superposable on its respective mirror image by any rigid motion in the plane (Figure 1); this means that scalene triangles or oriented circles are chiral in Euclidean 2-space, R^2. The enantiomorphs can be superposed, however, by a half-turn in R^3 about the mirror line in R^2 (m in Figure 1). Thus scalene triangles or oriented circles are achiral in R^3. The achirality in R^3 is also evident from the fact that a scalene triangle or an oriented circle can be viewed from either side of the plane and can thus be seen in either of its two mirror-image forms. Similarly, two improperly congruent oriented line segments in R^1 that are related by a mirror point, ○—×·×—○, can be rendered superposable in R^2 by a half-turn about the mirror point. And, by extension, chiroids in R^3, including chiral molecular models, are achiral in higher dimensions because—and this is, of course, hard to visualize—the enantiomorphs can be superposed in four-dimensional space R^4 by a half-turn about a plane that is an element of reflection in R^3. To a being in the fourth dimension, a left hand would look just like a right hand when viewed from a different perspective. In summary,

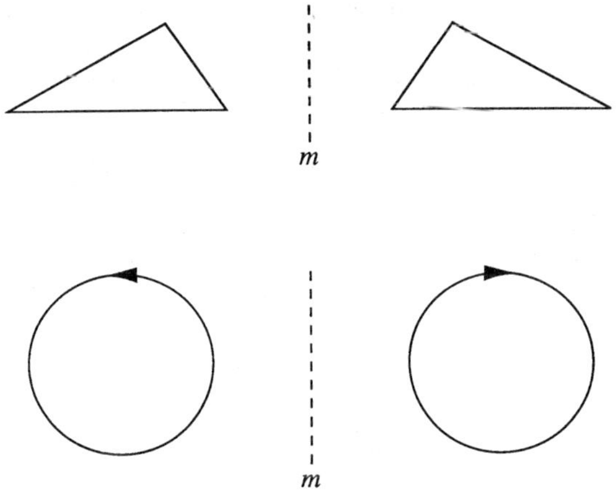

Figure 1. Chiral figures in the plane (m = mirror line). Top: enantiomorphous scalene triangles. Bottom: enantiomorphous oriented circles.

because "reflections in any dimension of space can be considered as rotations in a space one dimension higher,"[19] enantiomorphs in R^n that are related by reflection in an $(n - 1)$-dimensional mirror, m, can be superposed by a half-turn about m in R^{n+1}.

That enantiomorphs of rigid sets in R^n are homomorphs in R^{n+1} was recognized as long ago as 1827. August Ferdinand Möbius, the Professor of Astronomy in Leipzig, observed in *Der barycentrische Calcul*[20] that enantiomorphs in one- or two-dimensional space can be interconverted by half-turns in a higher dimension, and remarked:

> Es scheint sonderbar, dass bei körperlichen Figuren Gleichheit und Aehnlichkeit ohne Coincidenz statt finden kann, da hingegen bei Figuren in Ebenen oder bei Systemen von Punkten in geraden Linien Gleichheit und Aehnlichkeit mit Coincidenz immer verbunden ist. Der Grund davon möchte darin zu suchen seyn, dass es über den körperlichen Raum von drei Dimensionen hinaus keinen andern, keinen von vier Dimensionen giebt ... Zur Coincidenz zweier sich gleichen und ähnlichen Systeme im Raume von drei Dimensionen ... würde also, der Analogie nach zu schliessen, erforderlich seyn, dass man das eine System in einem Raume von vier Dimensionen eine halbe Umdrehung machen lassen könnte. Da aber ein solcher Raum nicht gedacht werden kann, so ist auch die Coincidenz in diesem Falle unmöglich.

> [It seems remarkable that solid figures can have equality and similarity without coincidence, since, in contrast, for figures in the plane or for systems of points on a line equality and similarity are always connected with coincidence. The reason for this may be that beyond the solid space of three dimensions there is no other, none of four dimensions ... For the coincidence of two equal and similar systems in

three-dimensional space... it would be necessary, to reason by analogy, that one should be able to let one of the systems make a half-turn in a space of four dimensions. But because such a space is unimaginable, coincidence in this case also impossible.]

And, almost a century later, the philosopher Ludwig Wittgenstein, in his *Tractatus* (1921),[21] wrote:

> Das Kantsche Problem von der rechten und linken Hand, die man nicht zur Deckung bringen kann, besteht schon in der Ebene, ja im eindimensionalen Raum,
>
> - - - o—x - - x—o - - -
>
> a b
>
> wo die beiden kongruenten Figuren a und b auch nicht zur Deckung gebracht werden können, ohne aus diesem Raum herausbewegt zu werden. Rechte und linke Hand sind tatsächlich vollkommen congruent. Und dass man sie nicht zur Deckung bringen kann, hat damit nichts zu tun. Den rechten Handschuh könnte man an die linke Hand ziehen, wenn man ihn im vierdimensionalen Raum umdrehen könnte.
>
> [Kant's problem concerning the right and the left hand, which cannot be made to coincide, exists even in two dimensions. Indeed, it exists in one-dimensional space,
>
> - - - o—x - - x—o - - -
>
> a b
>
> in which the two congruent figures, a and b, also cannot be made to coincide unless they are moved out of this space. The right hand and left hand are in fact completely congruent. It is quite irrelevant that they cannot be made to coincide. A right-handed glove could be fitted onto the left hand, if it could be turned around in four-dimensional space.]

III. STRUCTURES OF CHIRAL AND ACHIRAL MOLECULES

A. Models of Molecules

Geometrical objects, as we saw, can be neatly partitioned into chiral and achiral classes: Whether such an object is chiral or not merely depends on whether or not it can be brought into proper congruence with its mirror image. But geometrical objects are intangible constructs that exist entirely within our imagination, "beyond the reach of the empirical eye," in Isaiah Berlin's evocative phrase,[22] whereas molecules and their ensembles are concrete entities that exist entirely within the realm of our experience and whose attributes are the concern of physics and chemistry rather than of mathematics. A molecular model is no more a molecule than René Magritte's painting of a briar pipe[23] is the real thing (Figure 2).[24] The construction of a model thus requires a process of abstraction and idealization in which nonessential features are deliberately discarded; as a result, the model summarizes selected aspects of the system and suppresses, or, if necessary, even falsifies others.[25]

The traditional way to represent a molecular structure is illustrated by the space-filling model in Figure 2; the ball-and-stick model is another. By a process of supermagnification, such iconic models[26] bring the unseen molecule into the familiar world of everyday experience: An object with a distinct shape that we can relate to, that we can handle, see, and enjoy. Indeed, the metaphor "molecular shape" works amazingly well: It encapsulates and conveys a great deal of useful chemical information, it is of undeniable heuristic and didactic value, and it serves as a powerful fount of inspiration.[27] In this way the molecular property that we call chirality becomes a vivid reality when we examine molecular models of enantiomers. But certain approximations are needed to yield this familiar picture of a molecule, that is, the classical model, and these need to be discussed next.

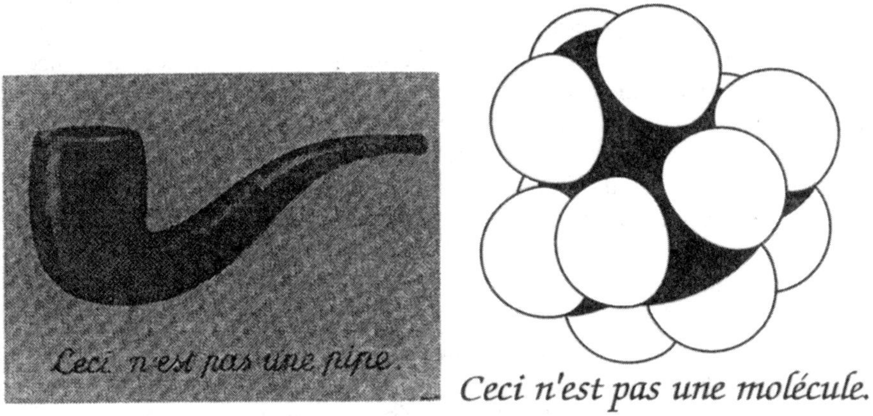

Figure 2. A space-filling molecular representation is no more a molecule than Magritte's pipe is a pipe. Reprinted with permission from B. P. Coppola, S. N. Ege, and R. G. Lawton, *J. Chem. Educ.* **1997**, *74*, 84. Copyright 1997, American Chemical Society.

1. The Classical Model

Consider two enantiomorphous molecules, **R** (right-handed) and **L** (left-handed), isolated from their surroundings and from external fields but not from each other. That is, they are allowed to interact. In the quantum-mechanical treatment of this system, as two particles in a one-dimensional symmetric double-well potential, the two states are degenerate in energy and are related to wavefunctions Ψ_R and Ψ_L localized in the two potential wells. Superposition of these wavefunctions yields ground and first excited states Ψ_+ and Ψ_-:

$$\Psi_+ = \frac{1}{\sqrt{2}}(\Psi_R + \Psi_L); \qquad \Psi_- = \frac{1}{\sqrt{2}}(\Psi_R - \Psi_L)$$

This superposition yields proper eigenstates of the underlying molecular Hamiltonian, that is, solutions of the time-independent Schrödinger equation (Ψ_R and Ψ_L are not eigenfunctions). A crucial difference between Ψ_+/Ψ_- and Ψ_R/Ψ_L is that the former are stationary states while the latter are not. That is, Ψ_+ and Ψ_- remain unchanged under the time evolution given by the time-dependent Schrödinger equation, whereas Ψ_R is transformed to Ψ_L, and vice versa, by a tunneling process whose frequency is inversely proportional to the height of the barrier separating the two wells and to the mass of the particles. For the present purpose, the most important consequence of the superposition (or mixing) of the two nonstationary states is the delocalization of the positions of the nuclei as well as of the electrons. The molecule can thus no longer be said to possess a nuclear framework, or a "chemical structure" in the traditional sense (as represented by iconic models). More specifically, the chirality and enantiomorphism associated with the **R** and **L** structures disappears in the stationary states because the molecule spends equal time in the nonstationary **R** and **L** states. Hence molecular chirality, like molecular structure, is a quantum mechanically undefined concept, at least with respect to an isolated molecule, and it makes no sense to assign a molecule to an **R** or **L** state because the ground state Ψ_+ is neither left-handed nor right-handed. Yet chiral molecules and enantiomers clearly do exist!

Friedrich Hund, who revealed this paradox ("Das Paradoxon der optischen Isomeren") in the early days of quantum mechanics,[28] accounted for the stereochemical stability of chiral biological molecules by assuming an extremely low frequency of tunneling oscillations, corresponding to the observed high-energy barriers for enantiomerization and large molecular masses. A number of alternative explanations have been advanced, among them that parity violation breaks the symmetry of the double-well potential so that Ψ_R and Ψ_L become proper eigenfunctions.[29] Because of its relevance to the definition of "mirror image" and "enantiomer," as discussed below, the last explanation deserves a brief digression.

It was Pasteur who, in 1874, first advanced the notion that our world is chiral ("L'univers est dissymétrique").[30] The discovery, in 1956, that parity is not conserved for the weak interactions governing β decay[31] has led to our present understanding that matter is indeed inherently chiral.[32] Particles and antiparticles possess opposite helicities, as in left-handed electrons and neutrinos versus right-handed positrons and antineutrinos. The unified theory of electromagnetic and weak forces led to the prediction of parity-violating effects in atoms: Thus atoms are chiral and are in principle optically active, although the effect is tiny and has so far been observed only in heavy atoms (Pb, Bi, Tl, Cs). Translated from the atomic to the molecular level, the parity-violating weak neutral current perturbation lifts the degeneracy of the space-inverted enantiomers. Space inversion means central inversion and is represented by the parity operator, $\hat{P}$, which inverts the system through the origin so as to send every point in 3-space from (x,y,z) to $(-x,-y,-z)$. $\hat{P}$-enantiomers, the enantiomers of ordinary stereochemistry, are therefore not strictly degenerate and exhibit a parity-violating energy difference (PVED). In contrast, the combined $\hat{C}\hat{P}\hat{T}$ operation generates the *exact* mirror image: A molecule in the antiworld with the opposite sense of chirality and with exactly the same energy ($\hat{T}$ reverses the direction of motion, and $\hat{C}$, the charge-conjugation operator, interconverts the charges in particles and antiparticles). The PVED, however, is so tiny—approximately 10^{-17} kJ mol^{-1}, which corresponds to an excess of just one molecule of D- or L-alanine in a racemic mixture of 10^{17} alanine molecules in thermodynamic equilibrium at 300 K—that the difference between $\hat{P}$- and $\hat{C}\hat{P}\hat{T}$-enantiomers is normally neglected and comes into play only in questions that deal with the origin and persistence of biomolecular homochirality in the natural world.

Chiral states and enantiomers (strictly: $\hat{C}\hat{P}\hat{T}$-enantiomers) are thus rigorously defined only in some classical limit, as in the limit of infinite nuclear masses or infinitely high enantiomerization barriers. The reason for the existence of enantiomers, as first pointed out by Hund, is then at least partly due to the fact that the molecules we ordinarily associate with chiral structures have substantial nuclear masses and fairly high enantiomerization barriers. Under these conditions, energy-level splitting ($E_+ - E_-$) becomes extremely small and the tunneling process extremely slow. The symmetric ground state Ψ_+ then decays into more stable states Ψ_R and Ψ_L under the influence of small external perturbations, such as collisions with neighboring particles. This leads to the chemist's classical model, in which the molecule has a well-defined nuclear framework.

In short, any discussion of molecular chirality requires the approximation inherent in the classical model, that is, the Born–Oppenheimer approximation. In this model it is assumed that the atomic nuclei in the molecule behave like classical particles whose spatial positions are fixed, and that it is only the electrons that are smeared out as matter waves. This assumption of a molecular structure, which has its origin in traditional chemistry, allows us to account for the existence of enantiomers and to rationalize chirality observables.

2. *Motion-Dependent Chirality*

Time reversal, mentioned above in connection with $\hat{C}\hat{P}\hat{T}$-enantiomers, is also at the heart of a distinction between time-dependent and time-independent enantiomorphism. In a series of publications,[33–36] Laurence Barron has provided the basis for his argument that "The only compelling reason for generalizing Kelvin's definition is to encompass absolute asymmetric synthesis and the concept of chirality in elementary particle physics by including motion-dependent chirality."[37] According to Barron[33]:

> The essential characteristic of a chiral object is that it is found in two distinct enantiomeric states that cannot be interconverted by time reversal combined with any proper spatial rotation.

Barron subsequently reformulated this definition as follows[34]:

> True chirality is exhibited/possessed by systems that exist in two distinct enantiomeric states that are interconverted by space inversion, but not by time reversal combined with any proper spatial rotation.

Note that the original definition has now evolved into a dichotomous classification: "Truly chiral" systems exist in two distinct enantiomorphous states that are interconverted by space inversion *but not* by time reversal combined with any proper spatial rotation, whereas "falsely chiral" systems exist in two distinct enantiomorphous states that are interconverted by space inversion *or* by time reversal combined with any proper spatial rotation.[34, 35] The process of time reversal, represented by the operator $\hat{T}$, "is the same operation as letting a movie film run backward. The act of inversion [i.e., time reversal] is not a physical act, but the study of the opposite chronological order of the same items."[38]

Enantiomorphism and sense of chirality of "truly chiral" systems are invariant under time reversal ($\hat{T}$-invariant); thus an ordinary, immobile chiroid, such as a finite helix or a molecular model, whose sense of chirality is $\hat{T}$-invariant, is "truly chiral." To illustrate the difference between $\hat{T}$-invariant and $\hat{T}$-noninvariant chirality, Barron has described the contrast between a stationary cone spinning about its symmetry axis ($\hat{T}$-noninvariant enantiomorphism and sense of chirality) and a spinning cone, cylinder, or sphere—the shape of the object is immaterial—that is also translating along the axis of spin ($\hat{T}$-invariant enantiomorphism and sense of chirality).[34] This contrast may be expressed in terms of the helicity derived from the combination of polar and axial vectors (Figure 3). The two $\hat{T}$-related translating spinning cylinders in Figure 4 are homomorphs, and time reversal does not affect their sense of helicity (i.e., of chirality). In

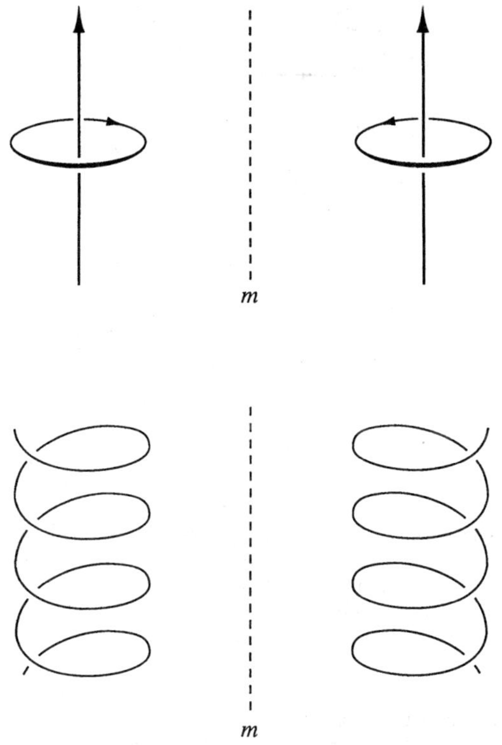

Figure 3. Top: Enantiomorphs (m = projection of mirror plane) of a chiral construction consisting of the product of a polar vector (oriented line segment in a vertical position) combined with an axial vector (oriented circle in a plane perpendicular to the polar vector). Bottom: Left-handed (left) and right-handed (right) cylindrical helices derived from the correponding constructions above. The screw axes coincide with the polar vectors.

contradistinction, the two $\hat{T}$-related stationary spinning cones in Figure 5 are enantiomorphs whose enantiomorphism and sense of chirality, in the absence of translational motion, crucially depend on the polarity of the cone: *Stationary* spinning cylinders or spheres lack polar axes and are thus achiral. Putting it another way, a point on the surface of a translating spinning body describes a helical (chiral) curve, whereas a point on a stationary spinning body—including a cone—describes a circular (achiral) curve.

Barron's classification is based on sound physical principles and is a valuable contribution to our understanding of chirality phenomena. In view of the differences between the symmetry properties of the two classes, as discussed below, it is to be expected that they also differ markedly in the associated physical and chemical observables. In particular, as Barron has repeatedly emphasized, only

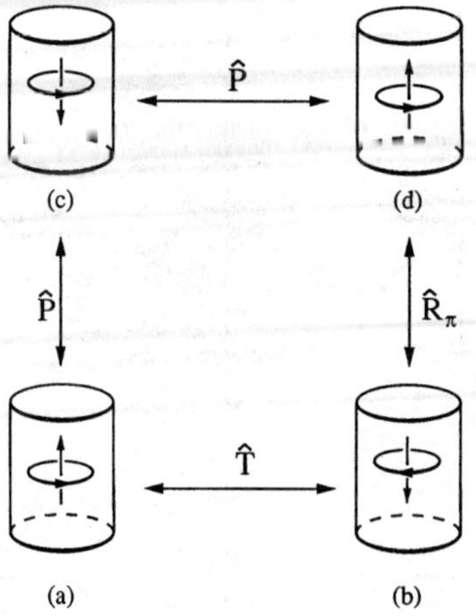

Figure 4. A translating spinning cylinder. The polar vector in the rotation–translation (screw displacement) corresponds to the direction of translation and the axial vector to the direction of spin. Time reversal ($\hat{T}$) does not change the sense of chirality of homomorphous systems (a) and (b); in terms of the helicity generated by the product of the two vectors, (a) and (b) are both right-handed. Space inversion ($\hat{P}$) of (a) yields a left-handed system (c), the enantiomorph of (a). Time reversal of (a), followed by rotation of (b) by 180° ($\hat{R}_\pi$) about an axis perpendicular to the cylindrical axis, yields (d), a homomorph of (a). Space inversion of (d) brings us back to (c).

molecules whose enantiomorphism and sense of chirality are $\hat{T}$-invariant are capable of supporting pseudoscalar observables such as natural optical activity. For further details the reader is referred to the germinal paper by H. Zocher and C. Török[38] and to the extensive discussions in Barron's papers.[33–36] Before we return to a further examination of the two classes, however, we need to digress briefly in order to dispose of some problems of terminology.

One of the reasons given by Barron for introducing the "true" and "false" terminology was to emphasize the distinction between $\hat{T}$-invariant and $\hat{T}$-noninvariant enantiomorphism and sense of chirality; it was not his intention that this should become standard nomenclature.[34, 36] The choice of terms is indeed less than felicitous because "true" and "false" are not only uninformative but carry an inappropriate Manichaean flavor of "good" and "bad," or "right" and "wrong." Additionally, an element of confusion is introduced by Barron's "new definition of the enantiomer of a chiral molecule: It is the molecule with the inverted spatial configuration and composed of antiparticles," the so-called "true enantiomer."[33, 34, 37] Thus $\hat{P}$-enantiomers are not "true enantiomers" because they are not strictly degenerate and exhibit a parity-violating energy difference. In contrast, "the combined $\hat{C}\hat{P}$ operation generates a molecule in the antiworld with the opposite chirality and exactly the same energy [i.e., the $\hat{C}\hat{P}$- or $\hat{C}\hat{P}\hat{T}$-enantiomer]." This means that the space-inverted enantiomer of a "truly

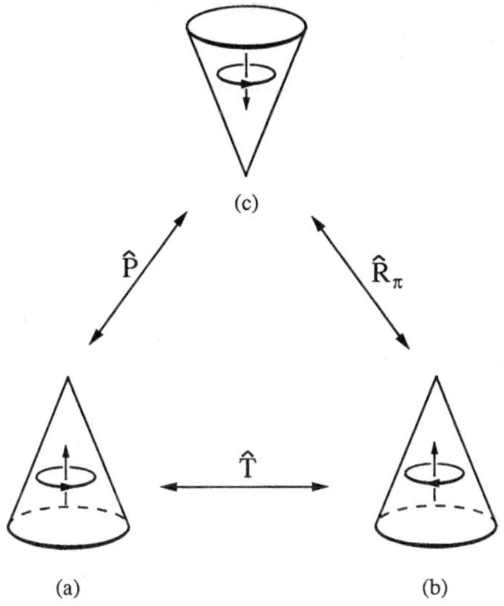

Figure 5. A stationary spinning cone. The polar vector corresponds to the polar axis of the cone, and the axial vector to the direction of spin. Time reversal ($\hat{T}$) changes the sense of chirality of enantiomorphous systems (a) and (b); in terms of the helicity generated by the product of the two vectors, (a) is right-handed and (b) is left-handed. The left-handed system (c) [the enantiomorph of (a) and the homomorph of (b)] is obtained either by space inversion ($\hat{P}$) of (a) or by rotation of (b) by 180° ($\hat{R}_\pi$) about an axis perpendicular to the polar axis.

chiral" molecule is not a "true enantiomer." Is it then a "false enantiomer?" Or should that stigma be reserved for the $\hat{P}$-enantiomer of a "falsely chiral" object? In summary, the "true/false" terminology has little to recommend it.

Barron has also drawn a distinction between dissymmetry and chirality, on the ground that chirality, unlike dissymmetry, "has a physical content: in molecular physics this is the ability to support time-even [i.e., $\hat{T}$-invariant] pseudoscalar observables"; he has gone so far as to suggest that while "dissymmetric" is properly applicable to systems with $\hat{T}$-invariant as well as those with $\hat{T}$-noninvariant enantiomorphism, the word *chiral* should in future be reserved only for the former.[34] It is indeed the case that chirality in the natural sciences has a physical content. In chemistry this consists of observables such as chiroptical properties (natural optical activity), enantioselectivity, etc. In elementary particle physics, chirality is defined as the eigenvalue of the Dirac matrix operator γ^5, with values +1 and −1 associated with right- and left-handed massless leptons (such as neutrinos); it might be added that chirality in this context also coincides with the helicity of the particles, and that for a photon this is just the value ±1 of the circular polarization. Physical content is not, however, a necessary criterion for chirality, any more than it is for dissymmetry. Although "chiral" and "dissymmetric" differ etymologically—the former means, literally, "handed," while the latter means "lacking in symmetry"—both terms are in substance synonymous because both are grounded equally in abstract considerations of symmetry alone, without consideration of physical content. As is clear from Pasteur's papers, the "lack of symmetry" refers to the absence of symmetry operations of the second kind, and it is precisely this absence that characterizes the point groups of "handed" (i.e., chiral) objects. Pasteur's use of "dissymmetric" therefore corresponds exactly to Kelvin's definition of "chiral," which is innocent of physical content. It is, in general, inadvisable to invest a technical term, such as chirality, whose semantic content has been widely accepted by general consensus, with a different meaning. We shall return to the problem of defining chirality at a later point in this chapter.

Accordingly, although the distinction between $\hat{T}$-invariant and $\hat{T}$-noninvariant enantiomorphism is a valid and useful one, the use of "chiral" is, pace Barron, not limited to systems that exhibit the former but is also applicable to those that exhibit the latter. A stationary spinning cone, as exemplified by an achiral molecule in a pure rotational state, is a therefore a chiroid and belongs to a chiral point group, as described below, even though it differs in its physical properties from those of a translating spinning molecule.

3. *The Chiral Point Groups*

Objects that exhibit $\hat{T}$-invariant enantiomorphism are either immobile or undergo a motion that can be reduced to a screw displacement (as exemplified in Figure 4). Objects that exhibit $\hat{T}$-noninvariant enantiomorphism behave effectively like stationary spinning cones (cf. Figure 5). We now show that all these objects belong to chiral groups.

With respect to immobile chiroids, the appropriate symmetries are given by the familiar finite point groups C_1 (nonaxial), C_n (monoaxial), D_n (dihedral), T (tetrahedral), O (octahedral), and I (icosahedral). Molecules that belong to the first three groups are commonplace; molecules with ground-state symmetries T [example: tetrakis(trimethylsilyl)silane],[39] O (example: appoferritin 24-mer),[40] and I (example: human rhinovirus 60-mer),[40] are relatively uncommon.

The symmetry of spinning objects cannot be represented by finite groups. Instead, such objects belong to a subclass of the continuous (or limiting) point groups, that is, groups of infinite order that are characterized by the presence of an infinite-fold rotation axis, C_∞. As A. V. Shubnikov has pointed out,[41] a special characteristic of this subclass of groups is that they cannot be illustrated by the usual geometrical figures. This may account for the fact that they are rarely encountered in the chemical literature and that they are not mentioned in any stereochemistry textbook. They are thus virtually unknown among chemists.

The continuous point groups are developed by systematic desymmetrization of the full group of the sphere (K_h, Figure 6).[41] A symmetric stretch along any one of the C_∞ axes yields a figure with cylindrical symmetry ($D_{\infty h}$), while an asymmetric stretch yields one with conical symmetry ($C_{\infty v}$, the symmetry of a polar vector). Among point groups that cannot be illustrated by geometrical figures, only one, $C_{\infty h}$, is achiral; in addition to the identity, the symmetry elements consist of a C_∞ axis, a σ plane perpendicular to that axis, and a center of symmetry. To visualize this symmetry, imagine a cylinder whose outside is covered with two sets of n (n even) slanted striations that are related by a plane of symmetry, as illustrated at the top of Figure 7. The construction as shown has C_{nh} symmetry. As n approaches infinity, the symmetry of the construction approaches $C_{\infty h}$ in the limit. This is also the symmetry of a stationary cylinder spinning about its C_∞ axis (Figure 7, bottom) and of an axial vector.[41]

To visualize an object with D_∞ symmetry, imagine a cylinder whose outside is covered with n slanted striations, as illustrated at the top of Figure 8. The two constructions shown (D_n symmetry) are enantiomorphs whose sense of chirality is related to the way in which the striations are slanted. As n approaches infinity, the symmetry of the constructions approaches D_∞; in the limit, infinitely many C_2 axes are embedded in a plane perpendicular to the C_∞ axis. This is the symmetry of a stationary cylinder undergoing a twisting motion, as indicated by the arrows on the cylinders at the bottom of Figure 8, and of an axial tensor of the second rank.[41] It is also the helical symmetry of a nonpolar object undergoing a screw displacement, that is, of an object whose enantiomorphism and sense of chirality are $\hat{T}$-invariant.

Similarly, to visualize an object with C_∞ symmetry, imagine a cone whose outside is covered with n slanted striations. The two constructions shown at the top

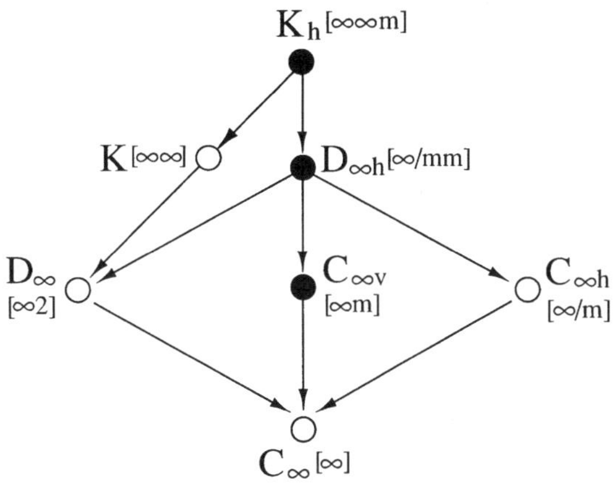

Figure 6. Complete subgroup lattice of continuous point groups. Solid circles represent point goups that can be represented by geometrical figures: K_h (sphere), $D_{\infty h}$ (cylinder), $C_{\infty v}$ (cone). Open circles represent point goups that cannot be represented by geometrical figures. Schönflies notations are accompanied by Hermann–Mauguin (international) notations in brackets.

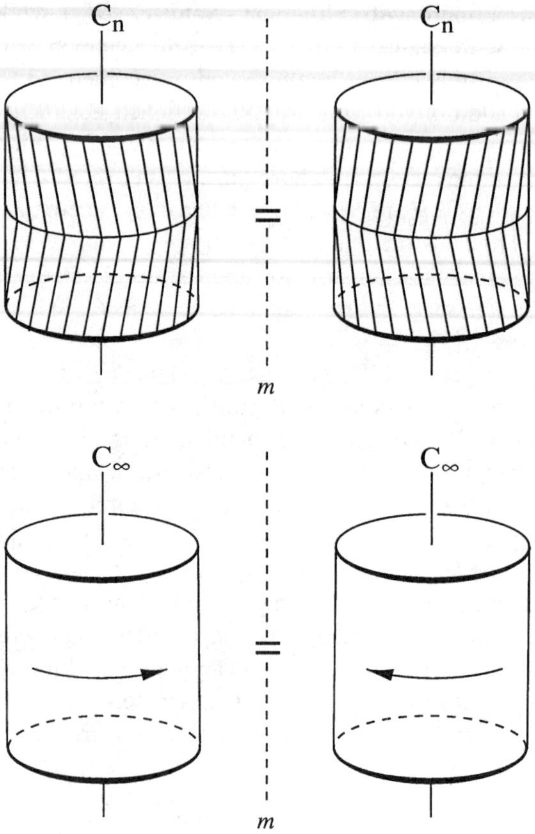

Figure 7. Illustration of $C_{\infty h}$ symmetry. Top: Mirror images of an achiral (C_{nh} symmetry) construction. As the number (n) of striations approaches infinity, the symmetry of the constructions approaches $C_{\infty h}$ in the limit. Bottom: Stationary spinning cylinders with $C_{\infty h}$ symmetry.

of Figure 9 (C_n symmetry) are enantiomorphs whose sense of chirality is related to the way in which the striations are slanted. As n approaches infinity, the symmetry of the constructions approaches C_∞; in the limit, the only symmetry elements are the identity and the C_∞ axis. This is also the symmetry of a cone spinning about its C_∞ axis (Figure 9, bottom), that is, of a polar object whose enantiomorphism and sense of chirality are $\hat{T}$-noninvariant or $\hat{T}$-invariant, depending on whether it is stationary or translating along the axis.

In summary, objects that exhibit enantiomorphism, whether $\hat{T}$-invariant or not, belong to chiral groups. Hence, motion-dependent chirality is encompassed in the group-theoretical equivalent of Kelvin's definition.

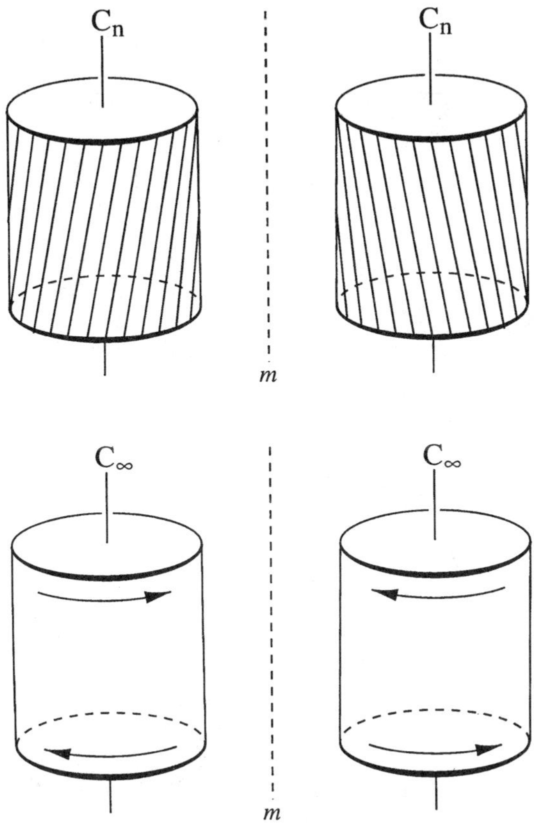

Figure 8. Illustration of D_∞ symmetry. Top: Enantiomorphs of a construction with D_n symmetry. As the number (n) of striations approaches infinity, the symmetry of the constructions approaches D_∞ in the limit. Bottom: Twisted cylinders with D_∞ symmetry.

B. Choosing a Model

The concept of symmetry and chirality in chemistry has a well-defined meaning only in relation to experiment.[18] Consider a system of one or more molecules subject to experimental observation. The properties of any such system are invariant with respect to its symmetry operations.[42] In Pierre Curie's famous dictum, "c'est la dissymétrie qui crée le phénomène."[43] That is, a phenomenon is expected to exist—and can in principle be observed—only because certain elements of symmetry are absent from the system. It follows that all manifestations of chirality flow from a single source: the absence of symmetry elements of the second kind in the group describing the system under observation. Accordingly, if

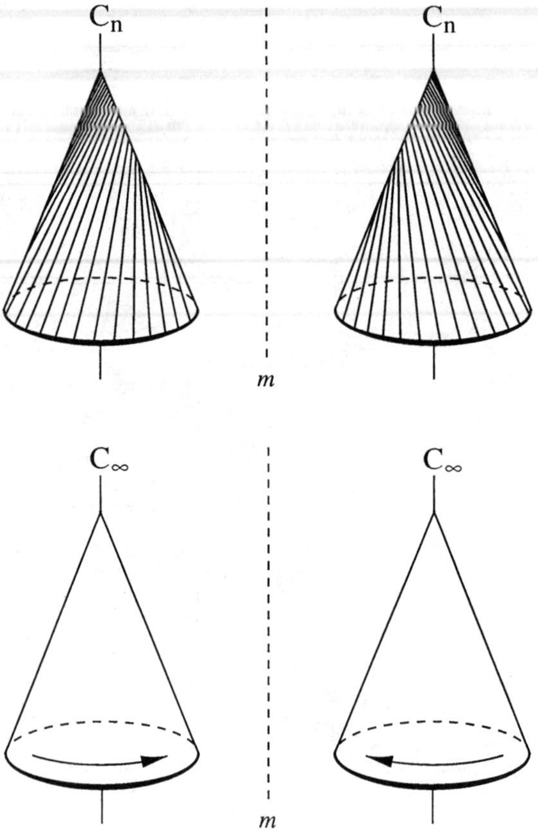

Figure 9. Illustration of C_∞ symmetry. Top: Enantiomorphs of a construction with C_n symmetry. As the number (n) of striations approaches infinity, the symmetry of the constructions approaches C_∞ in the limit. Bottom: Stationary spinning cones with C_∞ symmetry.

the model of the system is appropriate to the conditions of measurement, it may be asserted as a principle, given the property of $\hat{T}$-invariant chirality, that there must exist a set of associated observables (chirality phenomena), as well as pairwise related but distinct species (enantiomers) with indistinguishable scalar (metric) but different pseudoscalar properties.

Each structural model must be chosen so as to match a particular set of observations. That is, the symmetry of the model of a molecule or of a molecular ensemble depends on the conditions of the relevant physical (or chemical) measurement, and may vary for the same system according to time scale of observation and instrumental sensitivity. Hence, whether the model of a chemical system is chiral or achiral depends on the conditions of observation. In what follows, we make the simplifying assumption that the molecule is viewed as existing in an achiral environment; it would take us too far afield to consider the "induced" chirality of achiral molecules embedded in a chiral environment, for example, achiral molecules that are dissolved in a chiral solvent or that crystallize in a chiral space group. Why, it can even be argued, quite reasonably, that molecular achirality is intrinsically impossible, given the inescapable if subtle effect on supposedly achiral molecules of the chirality of the natural environment, including the chirality inherent in matter itself.

In many types of observations, for example, in the determination of X-ray structures, the molecule is approximated as a rigid body and is represented by an iconic (static) model. Of course, even supposedly rigid molecules are far from that: Their atoms vibrate about time-averaged positions. Nevertheless, the choice of a static model under the rigid-body approximation is reasonable because the corresponding molecule occupies only a single minimum on the potential-energy hypersurface.

The idealization inherent in the construction of a static model may be exemplified by a formally achiral molecule that is composed of four or more atoms, such as methane. The motions in such a polyatomic molecule are restricted by the restoring forces imposed by bonding, and achirality is here the result of compensatory internal vibrations. Such a molecule is said to be "chemically achiral" because each momentary conformational geometry can be superposed on its mirror image by a combination of the rotations, translations, *and* intramolecular motions that can occur under the given conditions of observation.[44] Thus, for example, molecular deformations in some vibrational states impart chirality to the methane molecule, with attainable chiral symmetries D_2, C_2, or C_1. Given the infinitely many asymmetric structures that are accessible to the molecule, it is safe to say that individual methane molecules are almost always chiral at any given instant in time, but that all feasible measurements on this system will give the appearance of achirality because of time-averaged cancellations of randomly fluctuating, local chiral effects.[25] Furthermore, cancellation is at best only approximate because it is highly unlikely that two *exactly* enantiomorphous structures will be present in even a large sample—say, a mole—of methane. As this discussion makes clear, the T_d symmetry conventionally ascribed to the methane molecule is a time-averaged property that belongs *exclusively* to the abstract geometrical model: In the real system, in the molecule itself, achievement of *exact* T_d symmetry, though not excluded by logic, is unattainable in practice.

Similar considerations apply to enantiomeric molecules that are represented by static models. Although the time-averaged structures of two such enantiomers (say, D- and L-alanine) will have approximate mirror-image relationships, it must be remembered that we are not dealing with mathematical objects: *Exact* mirror-image relationships belong exclusively to Kelvin's "ideally realized" geometrical model and are impossible to attain in the real system. Nevertheless, all chirality measurements on enantiomeric systems will yield equal but oppositely signed values within experimental error, due to time-averaged cancellations of randomly fluctuating, local effects.

The rigid-body approximation becomes inappropriate in dealing with nonrigid, that is, flexible or fluxional, molecules; the reason is that such molecules occupy more than one minimum on the potential-energy hypersurface, with the various minima separated by low-energy barriers. Under these conditions a dynamic model is required.[25] As an example, consider the proton-decoupled ^{19}F NMR spectrum of *cis*-1,2-difluorocyclohexane. At room temperature, only a single resonance line is observed, and the molecule behaves *as if* it had a plane of symmetry. It is therefore faithfully represented by a model with achiral (C_s) symmetry. One may, if one chooses, refer to a time-averaged structure at the fast-inversion limit that results from the rapid interconversion of enantiomeric structures, but all one really needs to know is that the observation (in the absence of accidental isochrony) does not demand a model of lower symmetry. Because cyclohexane behaves at room temperature *as if* it had D_{6h} instead of D_{3d} symmetry,[45] what is actually observed is the physical manifestation of a symmetry higher than the "true" (i.e., static) one. To the objection that knowledge of structural chemistry demands an asymmetric model for *cis*-1,2-difluorocyclohexane, in which one fluorine atom is axial while the other is equatorial, the response is that the failure to observe more than a single resonance line is merely the result of the particular conditions of observation: Lower the temperature sufficiently, and the multiplet predicted by the static (i.e., asymmetric) model at the slow-exchange limit will be observed. Although the ground-state geometries of these molecules are asymmetric, under the dynamic conditions described above, the achiral C_s model is a proper representation of the system under observation because it is chemically achiral. In that sense the dynamic model differs in no significant way from the model of a conventionally achiral system, such as methane, which is also chemically achiral. We thus have two faithful representations for the *same* chemical system, one chiral and the other achiral, and which one applies depends on the particular conditions of measurement.

In the preceding example, one can at least conceive of a C_s-symmetric structure for *cis*-1,2-difluorocyclohexane and of a D_{6h}-symmetric structure for cyclohexane within the conformational spaces of these two molecules. Yet there are cases in which the symmetry of a nonrigid molecule on the fast-exchange time scale cannot be portrayed by the point-group formalism. The classic example is that of ethane undergoing free internal rotation. In this case the dynamic symmetry is given by a permutation–inversion group[46] of order 36 that is isomorphic to the direct-product group $D_3 \times D_3$. This group is not isomorphic to the point group of any conceivable conformation of ethane, or indeed to any point group at all—even though ethane is obviously chemically achiral. In such a case, establishment of chirality or achirality of the model is far from straightforward.

As an example, consider a compound of the type 4-[(*R*)-Cabc]-4′-[(*S*)-Cabc]-2,2′,6,6′-tetra-R-biphenyl, such as (1*R*)-menthyl (1*S*)-menthyl-2,2′,6,6′-tetranitro-4,4′-diphenate, in which conformational racemization cannot proceed via an achiral intermediate (Figure 10).[47] The four blocking groups (R) in the 2,2′,6,6′ positions are large enough so that the two benzene rings cannot become coplanar. Under these conditions, the only achiral conformation of the biphenyl moiety, taken by itself, is D_{2d}, with the two σ planes intersecting at the central bond axis. At the same time, the only achiral conformations that are available to the two substituent groups –Cabc at the biphenyl 4,4′ positions (taken together as a unit separate from the biphenyl) are either C_s, with a σ plane perpendicular to the central bond axis (as shown in Figure 10), or C_i. Applying Curie's principle of superposition, the identity is the only element common to the point groups of D_{2d} and C_i, and the same is true of D_{2d} and C_s because the σ plane in C_s is perpendicular to both of the σ planes in D_{2d}. It follows that the molecule is asymmetric in all realizable conformations and that the compound exists as a mixture of transient DL pairs. These DL pairs can, however, interconvert by torsion around the bonds to –Cabc; for example, either one of the two enantiomorphous conformations in Figure 10 (related by the mirror line *m*) can convert into the other by a 90° twist of the biphenyl moiety (or, equivalently, of the two –Cabc end groups) in either

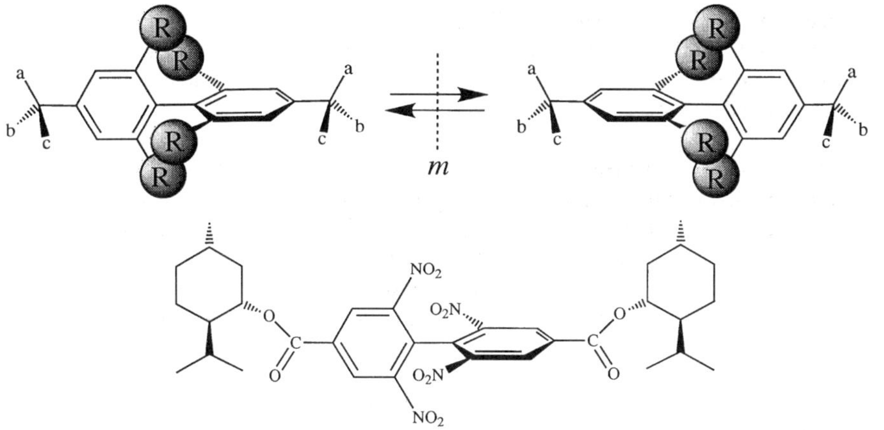

Figure 10. Top: Schematic drawings of two biphenyls with bulky substituents (R) in the 2,2′,6,6′-positions and enantiomorphous *–Cabc* groups in the 4,4′-positions. The two mirror-image-related (*m*) molecules are interconverted by a 90° conrotatory twist about the single bonds to the *–Cabc* groups. Bottom: (1*R*)-Menthyl (1*S*)-menthyl 2,2′,6,6′-tetranitro-4,4′-diphenate.[47] Reprinted with permission from K. Mislow, *Croat. Chem. Acta* **1996**, *69*, 485. Copyright 1996, Croatian Chemical Society.

direction. Like *cis*-1,2-difluorocyclohexane, this molecule is chemically achiral. In contrast to *cis*-1,2-difluorocyclohexane, however, there is no realizable achiral structure within the conformational space of the biphenyl, and conformational enantiomerization must therefore take place exclusively by chiral pathways. The dynamic symmetry of this molecule cannot be expressed by any point group but only by supergroups patterned after the molecular symmetry group of Longuet Higgins,[46] such as Ugi's "chemical identity group"[44] or Günthard's "isometric group."[48] Although the time-averaged symmetries of the biphenyl, C_i, C_s, or S_4, express the chemical achirality of the molecule, no conceivable conformation of the molecule belongs to these point groups.

C. Retournons à Pasteur

Pasteur's discovery, as noted above, connected chirality on the macroscopic scale to chirality on the molecular scale. Unlike the chiral morphology of quartz and tartrate crystals, "dissymétrie moléculaire" is not apparent to the eye, yet Pasteur[3] provided the key insight on the essential requirement for molecular chirality well before the advent of structural theory and the asymmetric carbon atom of Jacobus Henricus van 't Hoff:

> Les atomes de l'acide [tartrique] droit sont-ils groupés suivant les spires d'une hélice dextrorsum, ou placés aux sommets d'un tétraèdre irrégulier, ou disposés suivant tel ou tel assemblage dissymétrique déterminé? Nous ne saurions répondre à ces questions. Mais ce qui ne peut être l'objet d'un doute, c'est qu'il y a groupement des atomes suivant un ordre dissymétrique à image non superposable. Ce qui n'est pas moins certain, c'est que les atomes de l'acide gauche réalisent précisément le groupement dissymétrique inverse de celui-ci.
>
> [Are the atoms of the right [tartaric] acid grouped on the spirals of a right-handed helix, or placed at the summits of an irregular tetrahedron, or disposed according to some particular dissymmetric grouping or other? We cannot answer these questions. But there cannot be any doubt that there exists an arrangement of atoms in a dissymmetric order, with a nonsuperposable [mirror] image. What is no less certain is that the atoms of the left acid precisely realize a dissymmetric grouping that is the inverse of this.]

As forcefully stated by Pasteur, whatever the precise arrangement of atoms in the molecule, "ce qui ne peut être l'objet d'un doute" is that *the chirality of the atomic arrangement is the necessary and sufficient condition for molecular enantiomorphism and for its manifestation in a pseudoscalar property such as optical activity*. It cannot be emphasized too strongly that no recourse to structural theory was needed to arrive at this conclusion, which was based purely on a symmetry argument and which F. M. Jaeger has referred to as "la loi de Pasteur"

(Pasteur's law).[49] In modern terms, Pasteur's law tells us that the location and nature of the bonds between the atoms is immaterial because all that matters is whether the spatial arrangement of the atoms, taken as a whole, is chiral under the conditions of observation. Thus it is futile to search for a local "source of chirality" in the structure of a molecule for the excellent reason that there is no such thing. By the same token, the so-called elements of chirality described in the Cahn–Ingold–Prelog (CIP) system of nomenclature[50] cannot be identified or characterized by any physical or chemical measurements but are defined arbitrarily on the basis of local bonding arrangements; they are in fact stereogenic elements that are not and cannot be symmetry adapted and thus do not depend on local or molecular chirality.[51]

An immediate consequence of Pasteur's law is that the relationship between enantiomers is established by symmetry alone and does not require any knowledge of molecular bonding connectedness (constitution). This is in contrast to diastereomers, the other class of stereoisomers: Diastereomers are not related by symmetry, and their relationship can be defined only by first specifying that their constitutions are the same—otherwise, there would be nothing to distinguish them from constitutional isomers. Thus enantiomers, which have identical scalar properties and differ only in pseudoscalar properties, have more in common with homomers than with diastereomers, while diastereomers, which differ in all scalar properties, have more in common with constitutional isomers than with enantiomers.[51, 52] It therefore makes more sense, in an isomer classification scheme, to give priority to isometry rather than to constitution.[52] In such a scheme there is no need for the concept "stereoisomer;" the concept retains its usefulness only because it normally proves convenient, in chemical reaction schemes, to combine enantiomers and stereoisomers in a common class.

The explosive development of structural theory since Pasteur's day has yielded the tremendous diversity of chiral molecular structures that fill chemical journals and stereochemistry textbooks at the end of this century—yet all of them, without exception, conform to Pasteur's law. In the intervening years, the growing pains of conceptual developments in stereochemistry, and much of the attendant confusion, could to a large extent have been avoided if more attention had been paid to Jaeger's exhortation "Retournons à Pasteur!" [Let's return to Pasteur].[50] Indeed, one might go so far as to argue that, had the present-day collection of these structures been available at the end of Pasteur's life (1895), it would have been possible, with the help of the ball-and-stick models then developed by Friedrich August Kekulé and Johann Friedrich Wilhelm Adolf von Baeyer, to visualize the chirality of the great majority of these organic and inorganic molecules and to predict their optical activity. Admittedly, there are exotic molecules of more than routine interest from the standpoint of structural chirality that could not have been envisioned by Pasteur's contemporaries. Among these are molecules whose chirality depends on differences in isotopic masses, such as chiral acetic acid, CHDTCOOH,[53] as well as the innumerable molecules that owe their chirality to the presence of naturally abundant stable isotopes such as ^{2}H, ^{13}C, ^{15}N, and ^{18}O. The minute percentages of CH_3CHDOH in ethanol and of $(^{12}CH_3)(^{13}CH_3)CHOH$ in 2-propanol are examples in this class. The chirality of these "contaminants" plays a far from trivial role in NMR spectroscopy. For example, consider the isopropyl derivative, $ROCH(CH_3)_2$, of a chiral molecule, ROH. The anisochrony of geminal methyl groups observed in the ^{13}C NMR spectra of such a derivative would normally be considered as being due to the presence of doubly labeled molecules of the type $ROCH(^{13}CH_3)(^{13}CH_3)$. It is highly improbable, however, that two $^{13}CH_3$ groups would reside in the same molecule in a sample of natural isotopic composition. The measurement therefore actually compares $^{13}CH_3$ groups in diastereomers of $ROCH(^{12}CH_3)(^{13}CH_3)$.[54]

Equally beyond the ken of a nineteenth-century chemist would have been the notion that irradiation of an achiral molecule with circularly polarized light can result in an asymmetric distribution of electrons in the photoexcited state and, hence, in molecular chirality. In a classic experiment,[55] P. H. Schippers and H. P. J. M. Dekkers irradiated an achiral 1,7-ketone (Figure 11a) with circularly polarized light and found a large differential polarization. The slowness of the energy transfer between the two carbonyl groups ($k_{ET} \leq 10^7$ s^{-1}) is due to the relatively large distance between them. These observations fully support their claim that local $n \rightarrow \pi^*$ excitation of such achiral (*RS*)-ketones gives rise, in principle, to "two enantiomeric forms, R^*S and RS^*, provided the excitation energy remains localized at the carbonyl group." In a related experiment, E. W. Meijer and co-workers prepared a 1,6-diketone, another example of a molecule that is chiral in the $n \rightarrow \pi^*$ excited state only, by thermal decomposition of enantiopure 1,2-dioxetanes (Figure 11b).[56] The phenomenon that they observed is an exceptionally subtle example of the generation of enantiomers through modification of one of a pair of enantiotopic groups.[57]

Although contemporaries of Pasteur would not have been able to anticipate the dramatic discoveries of modern physics that proved relevant to chirality studies, they would nevertheless have been in a position to appreciate some recent out-of-the-ordinary developments in stereochemistry. For example, a few years after the discovery of C_{60}, Harold Kroto speculated that the most stable 32-atom cluster should be "handed," with D_3 symmetry.[58] In 1991, François Diederich and co-workers succeeded in isolating and characterizing D_2–C_{76}, the first example of a chiral allotrope of an element (Figure 12).[59] The enantiomers of chiral D_2–C_{76}, D_3–C_{78}, and D_2–C_{84} were later separated by kinetic resolution.[60] These and some hypothetical chiral clusters with icosahedral (I: C_{140}), tetrahedral (T: C_{44} and C_{52}), dihedral (D_3: C_{32}, C_{42}, C_{48}, and C_{34}), and monoaxial (C_3: C_{46}) symmetries are considered "textbook examples of interesting point group symmetries."[61]

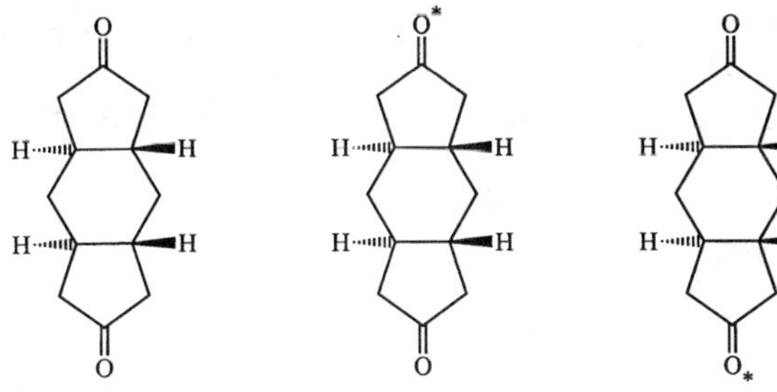

(RS) (RS*) (R*S)

(a)

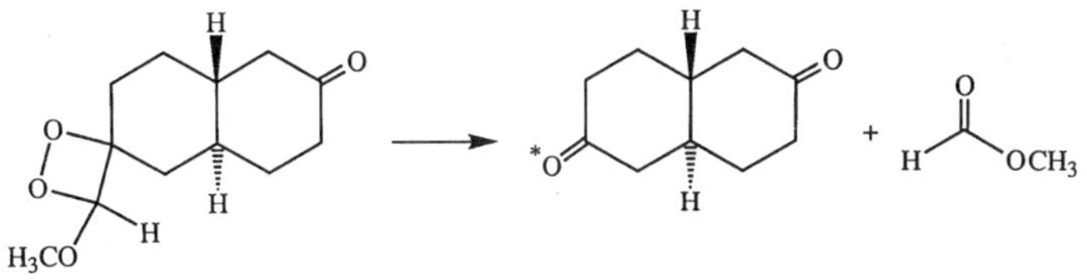

(b)

Figure 11. Molecules that are chiral in the excited state only. (a) (1*S*,3*R*,7*R*,9*S*)-tricyclo-[7.3.0.0^{3,7}]dodecane-5,11-dione in the electronic ground state, (*RS*), and in locally excited $n \rightarrow \pi^*$ states, (RS^*) and (R^*S).[55] (b) 3-($^1n\pi^*$)-(1*S*,6*R*)-bicyclo[4.4.0.]decane-3,8-dione prepared by thermal decomposition of enantiopure 1,2-dioxetanes.[56]

We next turn to another recent and exciting development in stereochemistry: the synthesis and characterization of chiral molecular knots, links, and graphs, and the discovery of these molecules in Nature.

IV. TOPOLOGICAL CHIRALITY

A. Rubber-Sheet Geometry

The models discussed thus far all adhere to what we might call "essential shape conservation." That is, bond distances and angles in such models remain invariant within reasonable limits, while intramolecular deformations (mainly of torsion angles) are limited to those that are energetically feasible. Thus a molecule is chemically achiral if its geometrical structure can be deformed to that of its mirror

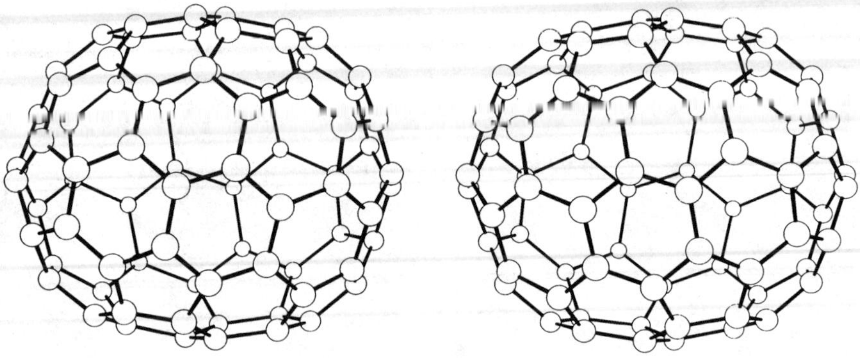

Figure 12. Enantiomers of D_2–C_{76}.[59]

image by way of normally realizable internal molecular motions; otherwise it is chemically chiral.

None of these constraints on shape and energy apply to topological models of molecules. The most common of these is the molecular graph. A graph is a set of vertices together with a set of edges that connect some or all of the vertices; vertices so joined are said to be adjacent. The spatial graph of a molecule is a visualization of the molecule's constitution, with differently labeled vertices representing different kinds of atoms and differently labeled edges representing different types of chemical bonds. As a topological object, a molecular graph is infinitely deformable, provided only that the vertices remain connected in the same way and that no edges are broken and reformed or passed through one another. The various images or presentations that can thus be obtained are topologically equivalent.

This feature of topology is popularly described as "rubber-sheet geometry." What is meant by this phrase is that an object can be changed into a topologically equivalent one by a process of continuous deformation, called ambient isotopy, which may entail unlimited bending, stretching, twisting, squeezing, and the like, but not cutting, breaking, or tearing. Objects that are related by an ambient isotopy are said to be isotopic—the adjective has a different meaning in chemistry—and to belong to the same isotopy type.

The unlimited flexibility of a molecular model as a topological object introduces significant changes in the way that we view molecular chirality and achirality. It is, as we saw, generally a simple matter to establish the chirality or achirality of a molecule modeled as a rigid body or a time-averaged structure. In the case of a topological object, however, one has to deal with infinitely many presentations, all of them interrelated by ambient isotopy. To establish topological chirality or achirality is therefore, in general, a substantially more difficult task.[62–64] The following definition applies not only to molecular models but to any topological construction, whether it represents a molecule or not:

> An object, such as a knot, or a link, or a graph, is topologically chiral if and only if it cannot be converted into its mirror image by continuous deformation (ambient isotopy) in the object's space; otherwise it is topologically achiral.

As an example, consider one of the simplest of topological objects, the trefoil knot. Is it topologically chiral or achiral? Johann Benedict Listing, a student of Carl Friedrich Gauss at Göttingen, who introduced the subject of topology (including the name) into the literature,[65] recognized that trefoil knots exist in two and only two distinct, mirror-image-related types, individual presentations of which are related by a "perversion," the term that Listing used for reflection through a plane. As an aside, the same year, 1848, saw publication both of Listing's work and of Pasteur's momentous discovery: truly an annus mirabilis for stereochemistry!

Listing conjectured that these two knot types cannot be transformed into one another by ambient isotopy. Figure 13 shows the two enantiomorphous isotopy types, arbitrarily designated as **R** and **L**. For each type, individual presentations are shown as projections in the plane, called diagrams. Crossings in these diagrams represent transverse double points, with over- and under-characteristics clearly

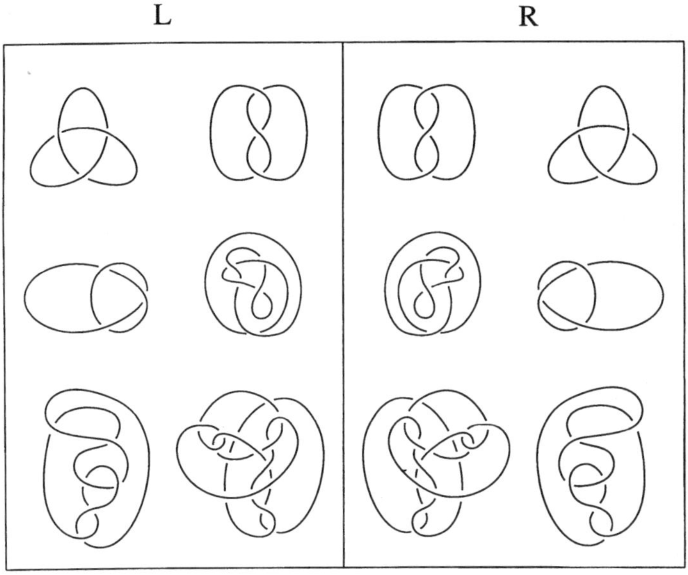

Figure 13. Diagrams of left-handed (**L**) and right-handed (**R**) trefoil knots.

indicated. The diagrams are topologically equivalent because they can all be interconverted by ambient isotopy in R^3, even though they differ in the number of crossings and in other ways. Note that the diagrams in the two isotopy types can be interconverted by reflection in the plane of projection, which changes every overcrossing in one to an undercrossing in the other.

Now, it might seem obvious to the reader, as it did to Listing, that there is no way in which the enantiomorphous knots in Figure 13 can be interconverted by continuous deformation in R^3. This empirical and common-sensical observation does not, however, constitute an acceptable proof of topological chirality because there are infinitely many presentations and infinitely many deformation pathways, and one cannot rule out the possibility, however faint, that interconversion might take place by way of one of them. And so it was not until 1914, after further developments in combinatorial topology, that Listing's empirical observation could be supported by a rigorous mathematical proof.[66]

B. Molecular Graphs

As was remarked above, a molecular (spatial) graph is deformable into infinitely many shapes (presentations or images). A graph that can be embedded in the plane without the crossing of any edges is said to be planar; otherwise it is nonplanar.[67] A planar graph is like the constitutional formulas (line diagrams) that chemists write on blackboards. Such a graph cannot be topologically chiral because any chiral presentation can be isotoped to its mirror image by way of a planar and therefore rigidly, that is, geometrically, achiral presentation. Thus a necessary (but not sufficient) condition for topological chirality in a graph is the absence of planar presentations among its embeddings in R^3.

The vast majority of molecular graphs are planar, no matter how unreasonable or even bizarre the embedding in the plane may appear from a chemical perspective. The molecular graphs of tetrahedrane, cubane, dodecahedrane, and buckminsterfullerene, for example, are all planar because the edges in a molecular graph can be stretched and bent without limit—so long as they are not severed and rejoined. It follows that, with few exceptions, the constitutions of geometrically chiral molecules are represented by topologically achiral graphs. For example, the molecular graph of geometrically chiral CHFClBr can be represented in a stereochemically realistic fashion, with four differently labeled vertices at the corners of a tetrahedron and a fifth at its center, or it can be represented with all five vertices in a plane. The last image, though stereochemically unrealistic, is topologically equivalent to the first. As a further example, depicted in the top row of Figure 14 are geometrically realistic structures for twistane,[68] tritwistane,[69] and [6]chochin,[70] while the bottom row depicts the images of the corresponding planar molecular graphs. Since attainment of planarity is a sufficient (but not necessary)

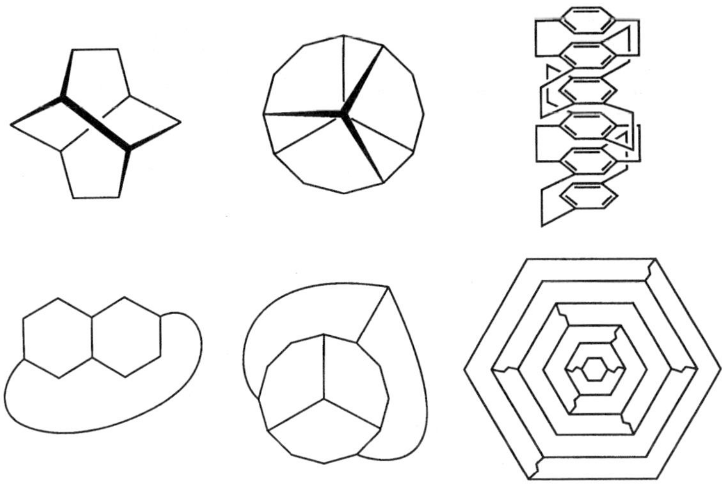

Figure 14. Top row: Molecular structures of (a) twistane, (b) tritwistane, (c) [6]chochin. Bottom row: planar images of the corresponding molecular graphs.[71] Reprinted with permission from K. Mislow, *Croat. Chem. Acta* **1996**, *69*, 485. Copyright 1996, Croatian Chemical Society.

condition for topological achirality, it follows that these three molecules, all of which are geometrically chiral and have been obtained in optically active form, are topologically achiral.[71]

Our asides, that nonplanarity is not a sufficient condition for topological chirality and that planarity is not a necessary condition for topological achirality, now demand an explanation. The reason is that even though planarity might be excluded among the presentations of a given graph, if, among the nonplanar presentations that can be attained by ambient isotopy, there exists at least one that is geometrically achiral—that is, a presentation that belongs to one of the achiral point groups—then any chiral presentation can be isotoped to its mirror image by way of such a rigidly achiral presentation. The corresponding molecular graph is therefore topologically achiral. Figure 15 presents a few examples of this type.[72]

In short, the absence of rigidly achiral presentations, planar or otherwise, is a necessary condition for the topological chirality of a molecular graph. Yet, as we shall later see, it is still not a sufficient one.

Figure 16 depicts some molecules that meet the minimum (i.e., the necessary) requirement for topological chirality: Their graphs cannot attain rigidly achiral presentations. David Walba's insightful conjecture[73] that the graphs of the Simmons–Paquette molecule[74] [Figure 16(a)], of the three-rung Möbius ladder molecule[75] [Figure 16(b)], and of the ferrocenophenone in Figure 16(c)[76] are topologically chiral was subsequently proved by Jonathan Simon and Keith

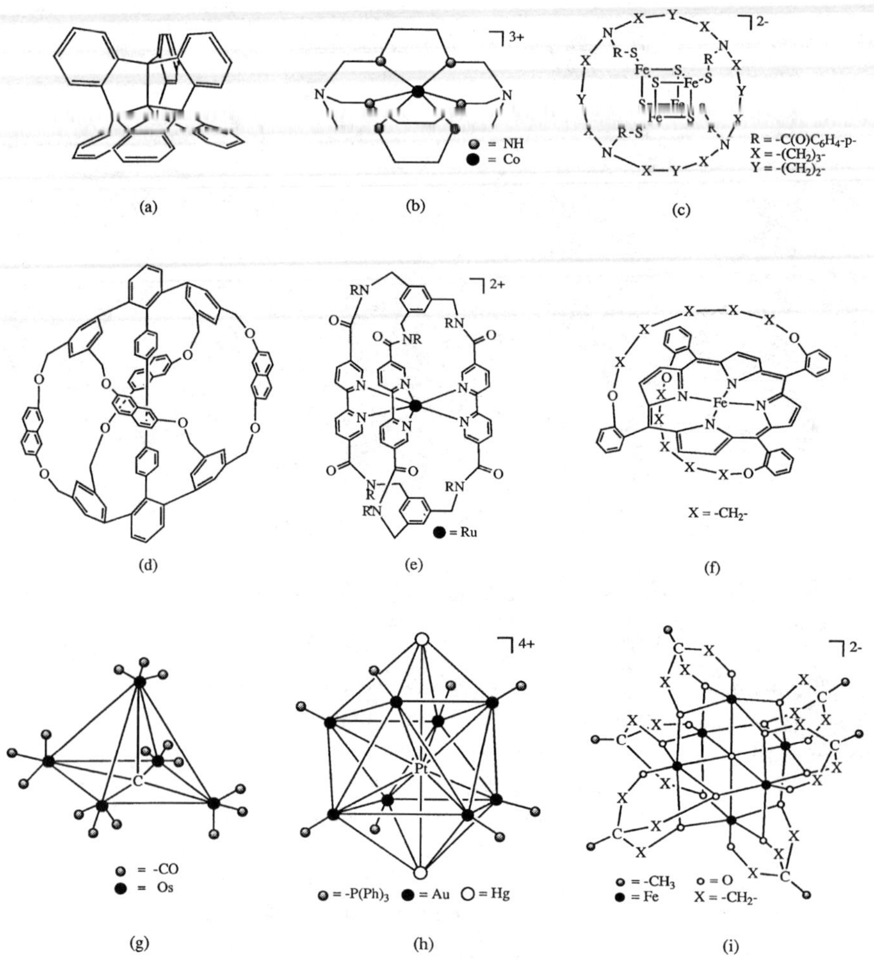

Figure 15. Examples of molecules with nonplanar graphs that have rigidly achiral presentations.[72] Their highest attainable symmetries are: (a) T_d, (b) D_{3h}, (c) D_{2d}, (d) D_{2d}, (e) D_{3h}, (f) D_{2d}, (g) C_s, (h) D_{4d}, and (i) S_6. Reprinted with permission from C. Liang and K. Mislow, *J. Math. Chem.* **1994**, *15*, 245. Copyright 1994, Baltzer Science Publishers.

Wolcott,[77] even though "there is no known [general] algorithm for deciding whether or not a given graph is topologically chiral."[77a] It has also been shown that the C_2-symmetric triple-layered naphthalenophane[78] in Figure 16(d) is topologically chiral.[72] Details of these proofs are beyond the purview of this chapter. In addition, there are numerous topologically chiral proteins,[79] representative examples of which are depicted in Figure 17.

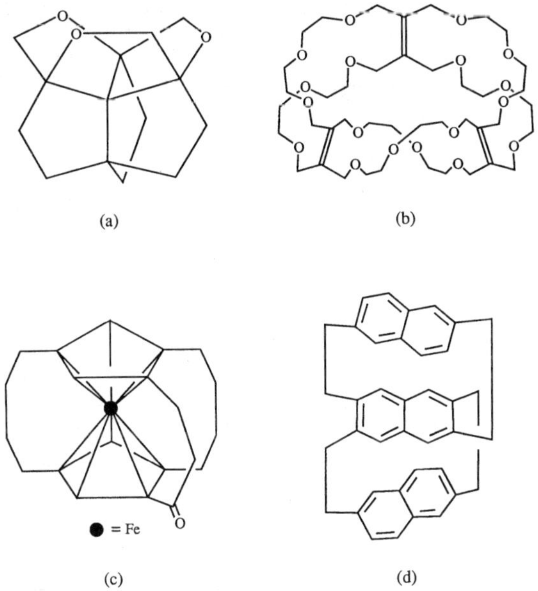

Figure 16. Some topologically chiral molecular graphs. (a) The Simmons–Paquette molecule.[74] (b) Walba's 3-rung Möbius ladder molecule.[75] (c) [4](1,1′)[4](3,3′)[3](4,4′)-Ferrocenophan-16-one.[76] (d) Triple-layered naphthalenophane.[78] Unlabeled vertices represent carbon atoms, and hydrogen atoms are suppressed for clarity. Only one enantiomer of each molecule is shown.

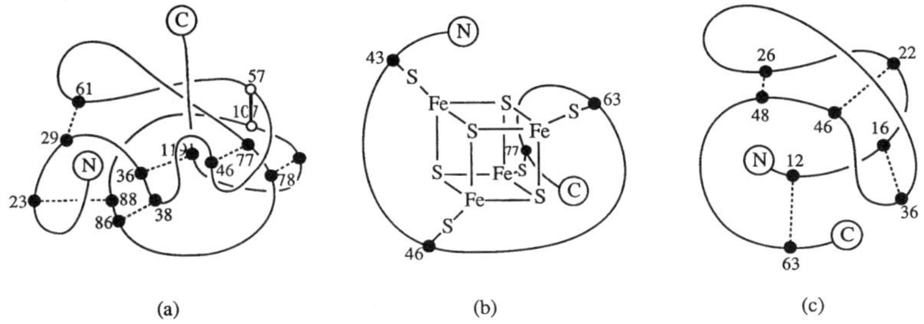

Figure 17. Schematic diagrams of some representative topologically chiral proteins.[79] (a) Condensed schematic drawing of the L subunit of the quinoprotein TV-MADH. The looped line represents the polypeptide backbone with N and C terminals. Cysteine (or half-cystine) residues are numbered, and their α-carbons are indicated by filled circles. Intrachain disulfide bonds are shown as dashed lines joining a pair of filled circles. The heavy line symbolizes an intrachain cofactor link. (b) *Chromatium* high potential iron protein (HiPIP), one of several Fe_4S_4 cluster-containing proteins. (c) Toxin II from the scorpion *Androctonus australis* Hector. Reprinted with permission from C. Liang and K. Mislow, *J. Math. Chem.* **1994**, *15*, 245. Copyright 1994, Baltzer Science Publishers.

C. Molecular Knots

As defined in topology, a nontrivial knot, such as the trefoil knot, is a simple closed polygonal or smooth closed curve in R^3 that does not intersect itself anywhere but that cannot be embedded in the plane without crossings.[62] A circle or a polygon is also a closed curve in R^3, but it *can* be embedded in the plane without crossings. It is therefore called the trivial knot or the unknot. A nontrivial knot thus resembles a nonplanar graph. The word *closed* is crucial to this definition, for a curve that is knotted but not closed can be unknotted, like a tied shoelace, by continuous deformation, whereas a knot as defined above cannot be unknotted without cutting the curve. Thus, from the point of view of topology, only a closed curve can be "truly knotted."[63a] We have defined a molecular knot as a closed or open knot in a molecular graph, where an open knot is a closed knot that has been cut but has not yet been unknotted.[80] Discussion in this chapter will be restricted to closed knots because open knots are not topologically significant.

Trefoil knots are the classic examples of topologically chiral structures. The rational synthesis of molecular trefoil knots, suggested as long ago as 1953,[81] was finally achieved in 1989, by Christiane Dietrich-Buchecker and Jean-Pierre Sauvage[82] (Figure 18); the enantiomers of this knot were subsequently resolved.[83]

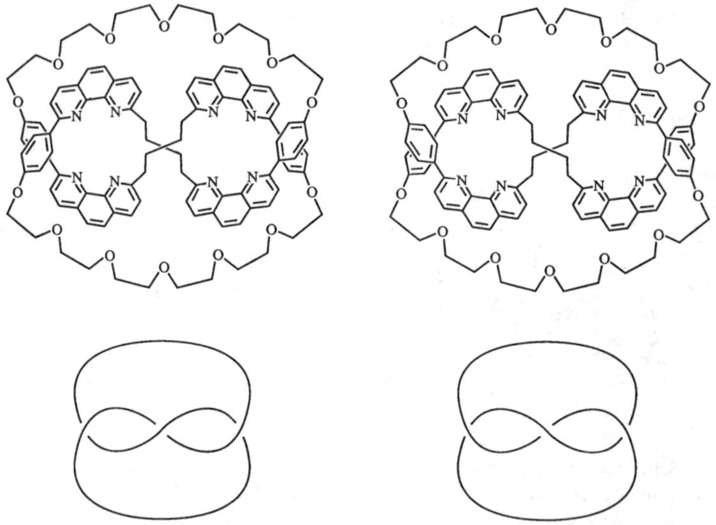

Figure 18. Top: enantiomers of the synthetic trefoil knot K-86.[82] Unlabeled vertices represent carbon atoms, and hydrogen atoms are suppressed for clarity. Bottom: Abstract versions of the same knot. Reprinted with permission from K. Mislow, *Croat. Chem. Acta* **1996**, *69*, 485. Copyright 1996, Croatian Chemical Society.

This milestone in chemical topology[84, 85] and topological stereochemistry[73, 86, 87] was followed in 1991 by the first rational synthesis of trefoil and figure-eight knots made from single-stranded DNA (Figure 19).[88, 89]

Single- and double-stranded DNA knots in Nature were first observed in 1976[90]; a variety of knotted circular DNAs have subsequently been observed in diverse biological systems[91] and by now have become a commonplace in biochemical topology.[92] Knots consisting of polypeptide chain segments combined with cofactors and disulfide intrachain cross-links have recently been observed among some metalloproteins (Figure 20).[93] All these natural products are topologically chiral.

How can we tell whether a knot is in fact topologically chiral? Before we address this question we need to introduce some terminology.[62–64]

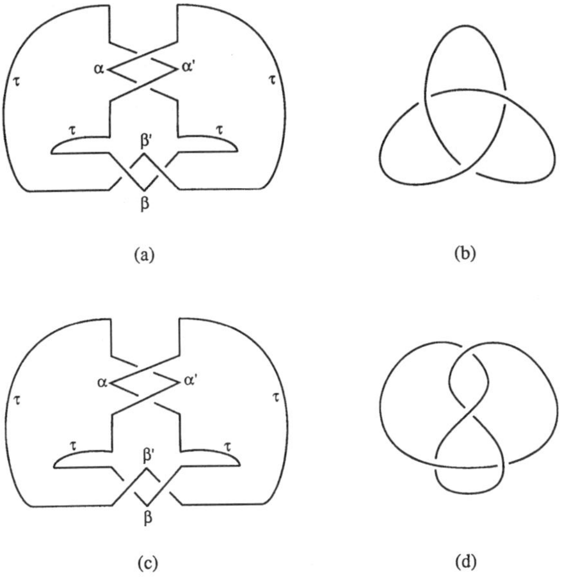

Figure 19. (a) Synthetic single-stranded DNA trefoil knots[88,89] in which α = (A–C–T–G–G–A–C–C–T–C–T), β = (C–G–T–A–G–C–C–G–C–A–T) or $(dCpdGp)_6$, α' and β' refer to sequences that complement α and β by Watson–Crick hydrogen bonding, respectively, and τ = dT_{15} symbolizes a single-stranded linkage between α (α') and β (β'). (b) An abstract version of (a) under denaturing conditions. (c) A synthetic single-stranded DNA figure-eight knot[88, 89] in which β = $(dCpdGp)_6$, and α, α', β', and τ have the same significance as in (a). (d) An abstract version of (c) under denaturing conditions. Reprinted with permission from K. Mislow and C. Liang, *Croat. Chem. Acta* **1996**, *69*, 1385. Copyright 1996, Croatian Chemical Society.

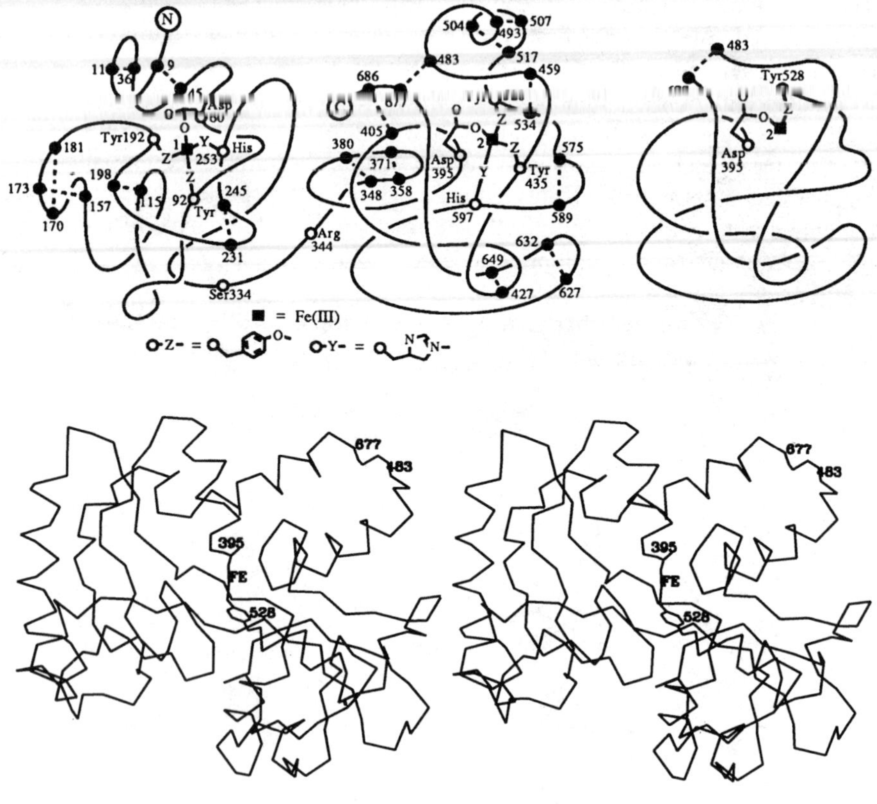

Figure 20. Top left: Condensed schematic diagram of human lactoferrin (hLf), with the C-lobe shown on the right and the N-lobe on the left. The α-carbons of cysteine and selected noncysteine residues are symbolized by filled and open circles, respectively. Cystine cross-links are shown as dashed lines. Unlabeled vertices symbolize carbon atoms, and hydrogen atoms are suppressed for clarity. Top right: The trefoil knot derived from the C-lobe of hLf. Bottom: Stereoview (Cα trace plus cross-links) of the knot shown on the top right.[93] Reprinted with permission from C. Liang and K. Mislow, *J. Am. Chem. Soc.* **1995**, *117*, 4201. Copyright, 1995 American Chemical Society.

1. *Prime Knots and Their Crossing Numbers*

The number of crossings in a knot K may be reduced to a minimum, by an ambient isotopy, to yield a *minimal diagram* with the fewest number of crossings, the crossing number $c(K)$. For example, $c(K) = 3$ for the trefoil knot. The crossing number is a knot invariant, that is, a mathematical object that can be unambiguously associated with an individual knot type, independent of any particular diagram. Knots are characterized by the symbol $c(K)_n$, where n is a numerical index; this is needed because, as shown in Figure 21, two or more

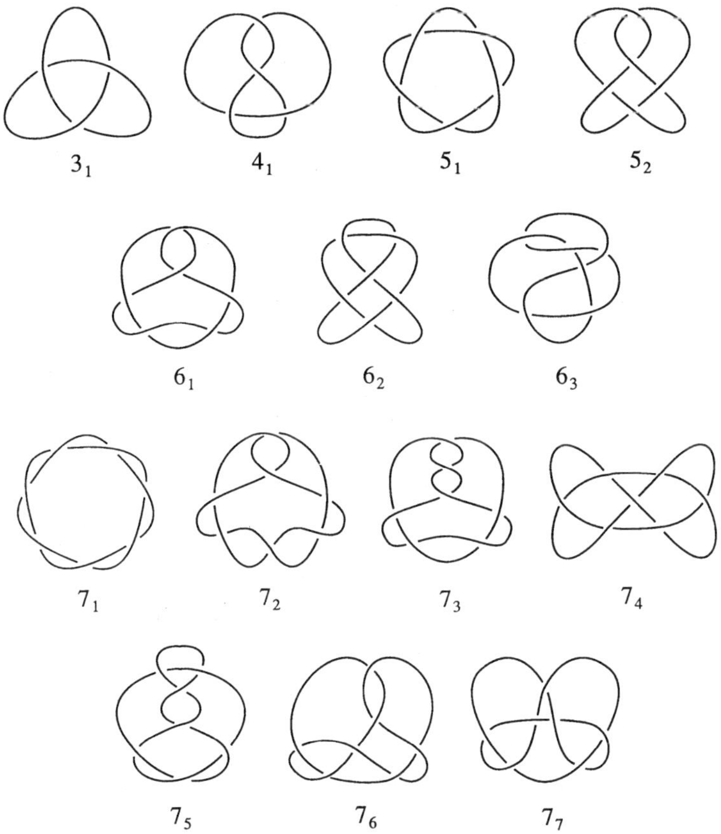

Figure 21. Diagrams of prime knots with up to seven crossings.

nonequivalent knots may share the same crossing number for $c(K) > 4$. The crossing number is therefore a weak invariant. A knot is alternating if overpasses alternate with underpasses all along the curve in the minimal knot diagram; otherwise it is nonalternating. For example, all the knots in Figure 21 are alternating. Incidentally, all the knot types shown in Figure 21 (including both enantiomers of the chiral knots) have been observed in circular DNA.[91]

The knots in Figure 21 are all prime knots because they cannot be divided (factored) into smaller, nontrivial knots. Prime knots are the building blocks of composite knots and of links. Like prime numbers, which yield composite numbers upon multiplication, or like atoms in chemistry, which yield molecules upon combination, prime knots are the elementary units of knot theory. Composite knots are exemplified by the topologically achiral square knot and the topologically chiral granny knot (Figure 22). In each of these knots, a plane perpendicular to the

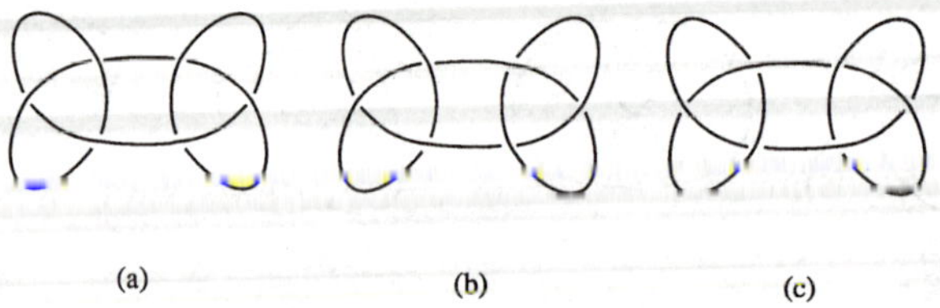

Figure 22. Diagrams of composite knots with $c(K) = 6$. (a) Square, or reef knot. (b) and (c) Enantiomorphs of the granny knot.

plane of projection and pierced in exactly two points cuts the knot in half. If the open ends on both sides of the plane are then joined to form closed curves, two trefoil knots result. Composite knots are symbolized by $K_1 \# K_2$, and their factors, K_1 and K_2, are prime knots other than the unknot. For example, if 3_1^* denotes the enantiomorph of 3_1, then Figure 22(a) = $3_1 \# 3_1^* = 3_1^* \# 3_1$, Figure 22(b) = $3_1 \# 3_1$, and Figure 22(c) = $3_1^* \# 3_1^*$. Three topologically chiral composite knots, $3_1 \# 3_1$, $3_1 \# 4_1$, and $4_1 \# 5_1$, plus their enantiomers, have been identified among the circular DNAs,[91] and a mixture of molecules whose graphs are abstractly symbolized by the three composite knots in Figure 22 has been obtained by dimerization of an open knot structurally related to the molecular knot depicted in Figure 18.[94]

Topologically achiral (amphicheiral) knots are vastly outnumbered by topologically chiral ones: Of the 12,965 prime knots with $c(K) \leq 13$, only 78 are amphicheiral.[13] One reason is that, with a single exception, no amphicheiral knot with an odd crossing number is known up to the present. All alternating amphicheiral knots must have an even crossing number; the lone exception known so far is a nonalternating amphicheiral knot with $c(K) = 15$,[95] though other such exceptions are sure to be lurking among the hordes of knots with $c(K) > 15$. Another reason is that the density of amphicheirals decreases as the crossing number increases. It is instructive, in this connection, to compare the census of topologically chiral and achiral prime knots[13] with a census of chiral and achiral alkanes C_nH_{2n+2} and monosubstituted alkanes $C_nH_{2n+1}X$[96] (Table 1): In alkanes, as in knots, the proportion of chiral members in each class increases with an increase in the size of the class, that is, with an increase in structural complexity.

2. *How to Prove a Knot's Chirality or Achirality*

The existence of a rigidly achiral presentation suffices as proof of the knot's amphicheirality because all chiral presentations can be isotoped to their mirror images by way of the achiral one. The only possible point group for rigidly achiral presentations of prime (but not composite) knots[13] is S_{2n}, $n = 1, 2, \ldots$. Figure 23 depicts diagrams of such presentations for a selected number of amphicheiral prime

Table 1.
Census of Chiral and Achiral Prime Knots and Alkanes[a]

	Prime Knots[b]		C_nH_{2n+2}		$CnH_{2n+1}X$	
n	a_n	c_n	a_n	c_n	a_n	c_n
1			1	0	1	0
2			1	0	1	0
3	0	2	1	0	2	0
4	1	0	2	0	3	2
5	0	4	3	0	5	6
6	1	4	5	0	8	20
7	0	14	7	4	14	60
8	5	32	14	10	23	176
9	0	98	21	34	41	510
10	13	304	40	96	69	1484
11	0	1104	61	284	122	4314
12	58	4236	118	782	208	12624
13	0	19976	186	4226	370	37126
14	274	93396	365	6198	636	109864

[a] Data from references 13, 95, and 96. c_n = number of chiral members. $c_n/2$ = number of enantiomorphous pairs. a_n = number of achiral members.
[b] n = crossing number.

knots.[13] Note that these diagrams are not minimal compare, for example, the minimal diagrams of 4_1 and 6_3, the two amphicheiral knots among the 14 prime knots in Figure 21, with the diagrams of their rigidly achiral presentations in Figure 23. The S_{2n} diagrams of amphicheiral knots can be converted into their mirror images in two ways: either by reflection through the plane of projection, an improper isometry that switches all over- and undercrossings in the diagram, or by a 360°/2n rotation about the C_n axis, a proper isometry. Only the latter qualifies as a continuous deformation because the improper isometry is both discontinuous and physically nonperformable.

Alternatively, proof that a knot is amphicheiral can be supplied by demonstrating that enantiomorphous presentations can be interconverted by continuous deformation, and not necessarily by way of a rigidly achiral presentation. An empirical proof of amphicheirality can therefore be arrived at simply by, say, manipulating a piece of string or wire in the form of a knot. For example, interconversion of the enantiomorphous presentations shown in Figure 24 suffices as proof that 4_1, the figure-eight (or Listing's) knot, is amphicheiral. Note, however, that although there are no rigidly achiral presentations along the indicated interconversion pathway, there are other pathways in which inter-

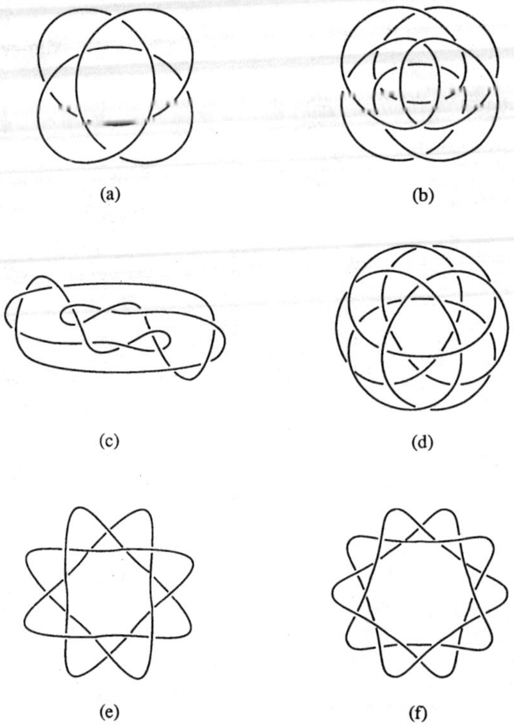

Figure 23. Rigidly achiral presentations of some topologically achiral prime knots. (a) 4_1 (S_4 symmetry); (b) 6_3 (S_4 symmetry); (c) 12_{427} ($S_2 \equiv C_i$ symmetry); 12_{1019} (S_6 symmetry); (e) 8_{18} (S_8 symmetry); (f) 10_{123} (S_{10} symmetry).

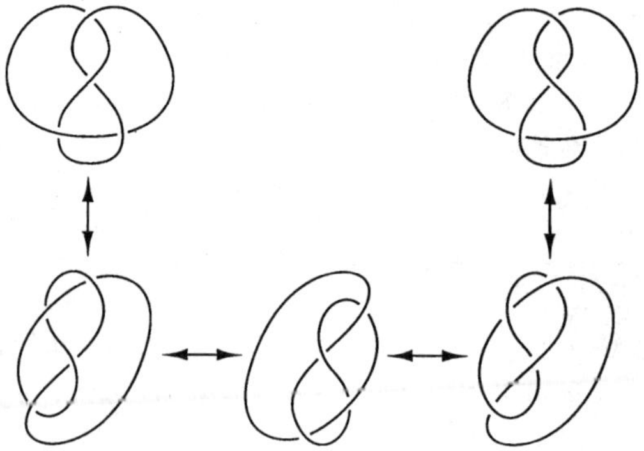

Figure 24. Proof of the amphicheirality of 4_1 by continuous interconversion of enantiomorphous presentations.

conversion takes place by way of the rigidly achiral (S_4) presentation in Figure 23(a). The question then arises whether there exist topological objects whose every presentation is chiral but which are nevertheless topologically achiral; interconversion of enantiomorphous presentations by continuous deformation could in that case proceed *only* by way of rigidly chiral states. The answer, surprisingly, is yes: There are indeed amphicheiral knots that are asymmetric in all their presentations.[97] Objects in this class have been dubbed "topological rubber gloves."[87] The classic example is the knot 8_{17} (Figure 25).[64, 98] A molecular version of a topological rubber glove is single-stranded DNA tied into a figure-eight knot

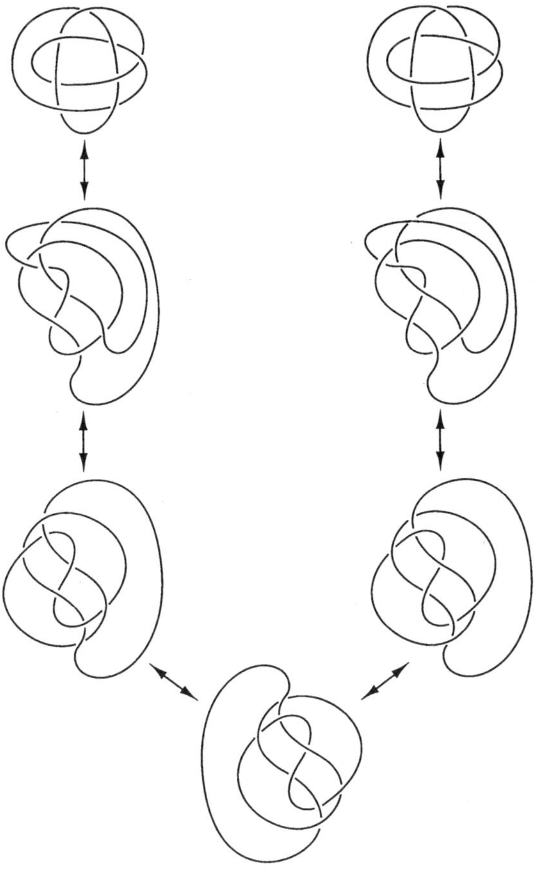

Figure 25. Proof of the amphicheirality of 8_{17} by continuous interconverion of enantiomorphous presentations. Reprinted with permission from K. Mislow, *Croat. Chem. Acta* **1996**, *69*, 485. Copyright 1996, Croatian Chemical Society.

knot [Figure 19(c)][99]; we shall later describe in detail the example of a chemically achiral [2]catenane that is also a topological rubber glove.

The existence of topological rubber gloves proves that the absence of rigidly achiral presentations is a necessary but not a sufficient condition for topological chirality.

One approach to the problem of establishing a knot's chirality or achirality is through the use of knot invariants. The first invariant capable of distinguishing between enantiomorphs, a one-variable Laurent polynomial (a polynomial that has both positive and negative powers), was discovered only as recently as 1985, by Vaughan Jones.[100] More powerful two-variable polynomials have subsequently been developed by others.[101, 102]

As illustrated by the example of the Jones polynomial for the figure-eight knot, the polynomial of an amphicheiral knot must be palindromic with respect to the coefficients[63a]:

$$V_{4_1}(t) = t^{-2} - t^{-1} + 1 - t + t^2$$

In contrast, as illustrated by the example of the Jones polynomial for the trefoil knot, the polynomial of a chiral knot is in principle, and indeed in the majority of cases, not palindromic:

$$V_{3_1}(t) = -t^{-4} + t^{-3} + t^{-1}$$

Nevertheless, even though these polynomials are normally capable of detecting topological chirality, none of them is infallible: In several cases, chiral knots yield palindromic polynomials.[13] The cause for these anomalies is still unknown.

In summary, failure to detect a rigidly achiral presentation does not mean that such a presentation cannot be found among the infinitely many presentations of a knot; failure to interconvert enantiomorphous presentations by ambient isotopy does not exclude the possibility that an interconversion pathway can be found among the infinitely many pathways that are available; and a palindromic knot polynomial does not necessarily mean that the knot is amphicheiral. Consequently, it may be impossible in certain cases to determine with complete certainty whether a knot is topologically chiral or not. The fundamental task of the theory of knots was stated over a hundred years ago by its foremost pioneer: "Given the number of its double points, to find all the essentially different forms which a closed curve can assume."[15] Yet to find invariants that will definitively determine whether or not a knot is chiral remains an unsolved problem to this day.[63a] Vassiliev invariants have been conjectured to be such "perfect" invariants.[63b]

D. Molecular Links

When Edel Wasserman reported the first synthesis of a molecule with two interlocked rings (a [2]catenane) in 1960, he provided "the first demonstrated example of a compound in which the topology of the system must be considered in describing its structure."[84a] At about the same time, in a classic paper with H. L. Frisch,[84b] Wasserman introduced the concept of isomerism between knotted and unknotted and between interlocked (catenated) and noninterlocked rings, and thus launched the subject of topological chemistry. Numerous organic catenanes have since been prepared, thanks largely to the implementation of ingenious synthetic strategies devised by the groups of Jean-Pierre Sauvage, Gottfried Schill, Fraser Stoddart, and Fritz Vögtle. Although these synthetic products are for the most part [2]catenanes, a number of higher catenanes, such as [3]catenanes,[103] a [4]catenane,[104] and the [5]catenane "olympiadane,"[104] have also been prepared. The disjoint rings are held together by what Frisch and Wasserman have called "topological bonds."[84b]

Chemical catenanes are modeled by topological links. A topological link is a finite union of mutually disjoint knots (including the unknot). A knot is therefore the special case of a link with only one component. Links are nontrivial if and only if they cannot be embedded in the plane without crossings. All the links referred to in this chapter are nontrivial, but the components are usually unknots.

Link types are characterized by their crossing numbers according to a convention in which $c(K)$ is superscripted by the number of components in the link and subscripted by a numerical index, needed because two or more nonequivalent links may share the same crossing number for links with $c(K) > 5$. For the case of one-component links, that is, for knots, the superscript 1 is omitted. If directions (denoted by arrows along the curves) are assigned to all the component curves that constitute a link, that link is said to be oriented. Otherwise it is nonoriented. We begin our discussion by a consideration of nonoriented links.

1. Nonoriented Links

The simplest nonoriented, nontrivial link is the 2-crossing Hopf link, 2^2_1 [Figure 26(a)]. This link is the abstract representation of the vast majority of reported chemical [2]catenanes, one of which is depicted in Figure 26(b).[105] The link, and hence the molecule, is topologically achiral because it can attain a rigidly achiral presentation [Figure 26(c)]. A topologically achiral link of exceptional interest is the Borromean link (6^3_2).[106] This link is among the most fascinating of topological constructions: Three mutually disjoint simple closed curves form a link, yet no two curves are linked [Figure 27(a)]. Thus, if any one curve is cut, the other two are free to separate. As shown in Figure 27, the Borromean link can assume rigidly achiral

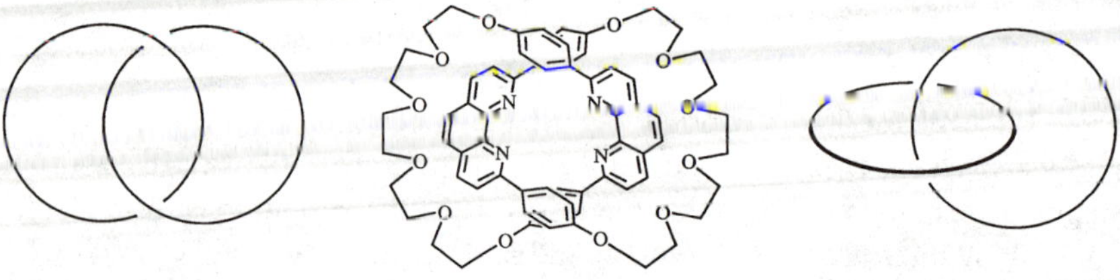

(a) (b) (c)

Figure 26. (a) Minimal diagram of the Hopf link, 2_1^2. (b) A [2]catenane[105] abstractly represented by 2_1^2. (c) A rigidly achiral (D_{2d} symmetry) presentation of 2_1^2.

presentations with a variety of symmetries. It is of interest to note that three golden rectangles in mutually perpendicular planes (whose twelve vertices are the twelve vertices of the regular icosahedron)[107] constitute a T_h-symmetric version of the Borromean link [Figure 27(d)]. In addition to presentations with T_h subsymmetries [including S_6, Figure 27(b)], the Borromean link can also assume D_{2d} symmetry

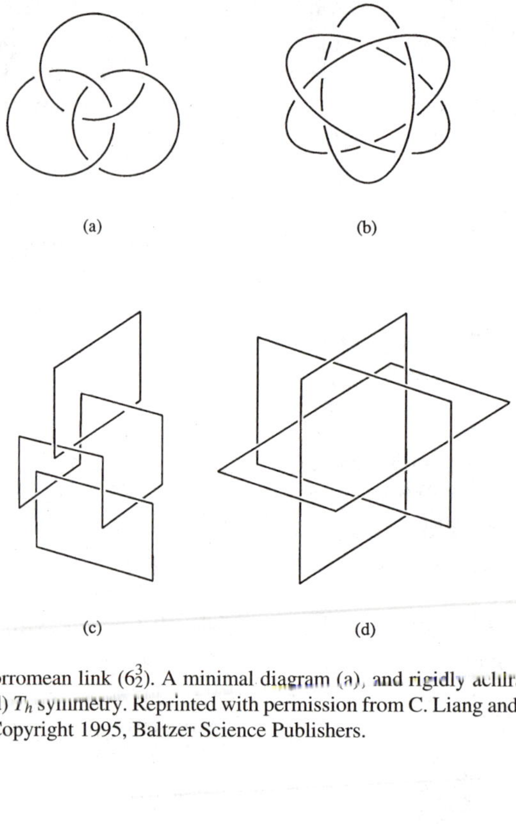

(a) (b) (c) (d)

Figure 27. The Borromean link (6_2^3). A minimal diagram (a), and rigidly achiral presentations with (b) S_6, (c) D_{2d}, and (d) T_h symmetry. Reprinted with permission from C. Liang and K. Mislow, *J. Math. Chem.* **1995**, *18*, 1. Copyright 1995, Baltzer Science Publishers.

[Figure 27(c)]. It remains to be noted that Borromean links composed of metric circles are impossible,[108] that Borromean links composed of three triangles or three squares have inspired sculptures by John Robinson,[109] and that this link, in the form of three interlocked triangles, was known to the ancient Scandinavians as "Odin's triangle" or the "Walknot" (meaning "knot of the slain").[110]

Given its unique construction, it comes as no surprise that realization of the Borromean link in molecular form has long been considered a synthetic goal well worth achieving. As Martin Gardner put it,[111] "Who can guess what outlandish properties a carbon compound might have...if its molecules were joined into triplets, each triplet interlocked like a set of Borromean rings?" More than 30 years ago, Wasserman expressed the view that molecular Borromean links "require a minimum string of 30 carbons" in each of the three rings, and van Gulick discussed the 3-braid approach to the synthesis of such a link.[84b] The synthetic goal was finally reached by an assembly of three single-stranded DNA rings into a Borromean link (Figure 28).[112]

As in the case of knots, a demonstration that enantiomorphous presentations can be interconverted by continuous deformation suffices as proof of a link's amphicheirality. This may be the only option if no rigidly achiral presentations are attainable. An example is the [2]catenane depicted in Figure 29.[113] The ring containing the *p*-tolyl-substituted 1,10-phenanthroline is oriented, thanks to the asymmetric positioning of the substituent. The ring containing the 1,5-dioxynaphthalene, however, is not oriented because it can attain a conformation with C_2 symmetry. All achiral presentations available to the two rings in isolation have C_s symmetry, but the σ planes of the two components cannot become coplanar in the catenated condition. Accordingly, every conceivable presentation of the

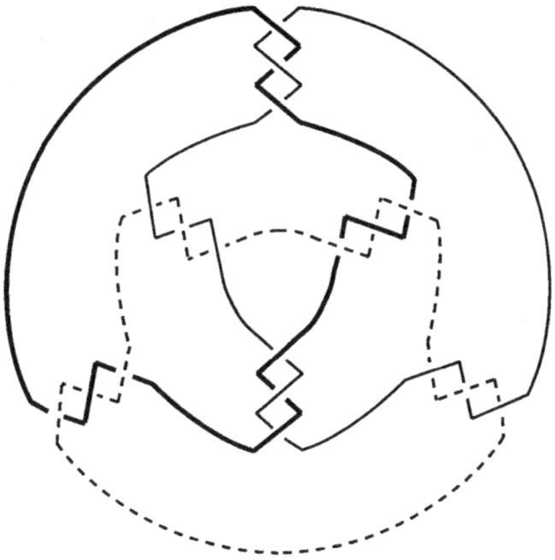

Figure 28. A condensed schematic diagram of a synthetic Borromean link[112] made up of three different circular single-stranded DNA components.

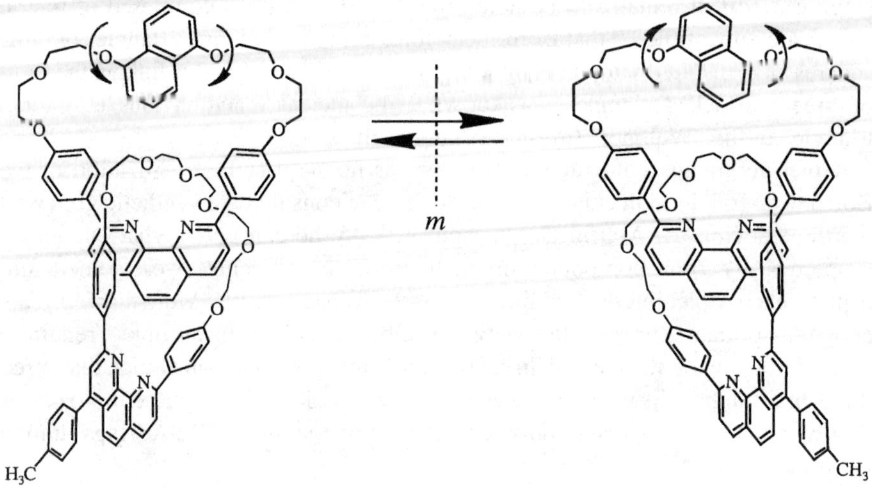

Figure 29. Enantiomorphous conformations of a [2]catenane that is a topological rubber glove.[113] Unlabeled vertices represent carbon atoms, and hydrogen atoms are suppressed for clarity. The enantiomorphs are related by the mirror plane (*m*) and are interconverted by rotation of the 1,5-dioxynaphthalene moiety about the C–O bonds (arrows).

[2]catenane is asymmetric. As shown in Figure 29, and as confirmed by variable-temperature NMR spectroscopy,[113] enantiomorphous conformations are readily interconverted by rotation of the 1,5-dioxynaphthalene plane by 180°. Hence, the [2]catenane is chemically and topologically achiral, even though all chemically accessible conformations encountered along this or any other enantiomerization pathway, as well as all topologically accessible presentations, are asymmetric. The [2]catenane is therefore properly characterized as a topological rubber glove. In contrast to the [2]catenane, the figure-eight knot made from single-stranded DNA [Figure 19(c)], which has been similarly characterized,[99] is incapable of attaining chemical achirality because, in order to convert the DNA strand to its enantiomorph, it is necessary to flatten the sugars so that they are planar, like the purine and pyrimidine bases. There is, of course, no way in which this topological transformation can be achieved by any feasible chemical operation. Furthermore, while it has long been recognized that certain knots and graphs are topologically achiral but not rigidly achiral,[98] the [2]catenane represents the first example of a link in this class.

The time-averaged symmetry of the [2]catenane is C_s, which expresses the chemical achirality of the molecule, even though no individual conformation or presentation of the catenane belongs to this point group. It thus resembles (1*R*)-menthyl (1*S*)-menthyl 2,2′,6,6′-tetranitro-4,4′-diphenate, in which conformational racemization also cannot proceed via an achiral intermediate. The crucial difference between the two molecules is that the biphenyl has presentations that are planar and therefore rigidly achiral, whereas no such presentations exist for the [2]catenane. For all the above reasons, the [2]catenane is sui generis.[113]

Like knots, links may be prime or composite. The Hopf and Borromean links are examples of prime links because they cannot be divided (factored) into smaller, nontrivial links. Figure 30(a) is the minimal diagram of a composite link that is the abstract representative of some [3]catenanes, one of which is depicted in Figure 30(b).[103b] That the three-component link is a composite link is shown by the fact that a plane perpendicular to the plane of projection (dashed line) and pierced in exactly two points cuts the link in half: If the open ends on both sides of the plane are now joined to form closed curves, two Hopf links result. In analogy to composite knots, the three-component link in Figure 30(a) is denoted by $2^2_1 \# 2^2_1$, and the five-component composite link that represents olympiadane by $2^2_1 \# 2^2_1 \# 2^2_1 \# 2^2_1$.

With few exceptions, all nonoriented chemical links reported in the literature are topologically achiral. One of the rare exceptions is the two-component 4-crossing link 4^2_1 (Figure 31).[114] The molecule in the depicted conformation, and indeed in any imaginable conformation, is geometrically chiral. It is therefore reasonable to conjecture that it is also topologically chiral. Proof that 4^2_1 is indeed topologically chiral was provided after the development of suitable polynomials.[115]

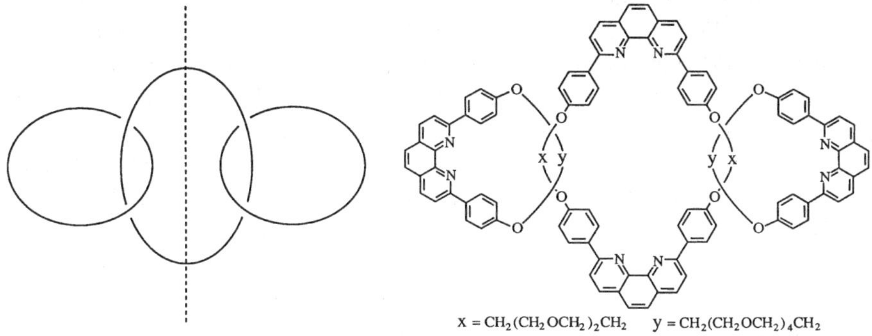

(a) (b)

Figure 30. (a) Minimal diagram of a composite link, $2^2_1 \# 2^2_1$. The dashed line is a projection of the plane that cuts the link into two component Hopf links (after the open ends are joined). (b) A [3]catenane[103b] abstractly represented by $2^2_1 \# 2^2_1$. Reprinted with permission from C. Liang and K. Mislow, *J. Math. Chem.* **1995**, *18*, 1. Copyright 1995, Baltzer Science Publishers.

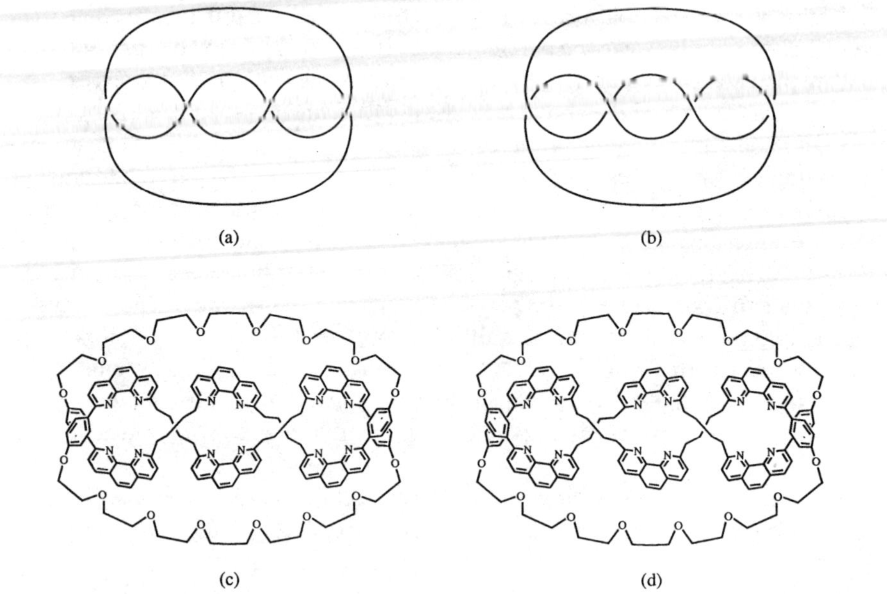

Figure 31. (a) Minimal diagram of the simplest topologically chiral, nonoriented link, 4^2_1. (b) The enantiomorph of (a). (c) A [2]catenane[114] abstractly represented by (a). (d) The enantiomer of (c), abstractly represented by (b). Reprinted with permission from C. Liang and K. Mislow, *J. Math. Chem.* **1995**, *18*, 1. Copyright 1995, Baltzer Science Publishers.

2. *Oriented Links*

Orientation may impart topological chirality to a link that is otherwise amphicheiral. For example, the Hopf link [Figure 26 (a)] becomes topologically chiral when *both* rings are oriented (Figure 32). Orientation of only one ring leaves one plane of symmetry intact, and this construction therefore has a rigidly achiral presentation. For example, the [2]catenane depicted in Figure 29, in which only one of the two rings is oriented, is topologically achiral. Nor can topological chirality in a nonoriented [2]catenane be attained merely through introduction of conformationally chiral features[116] or through attachment of geometrically chiral but topologically planar residues[117]: Orientation of *both* cyclic components remains the unavoidable requirement for topological chirality. This condition is satisfied for the molecule shown in Figure 32[118] because both rings contain 1,10-phenanthrolines asymmetrically substituted by phenyl groups and both are therefore oriented. The direction that is assigned to the rings follows an arbitrary convention, in which the head of the arrow in the phenanthroline moiety points in the direction of the phenyl-bearing ring so that Figure 32(a) matches Figure 32(c), and Figure 32(b) matches Figure 32(d).

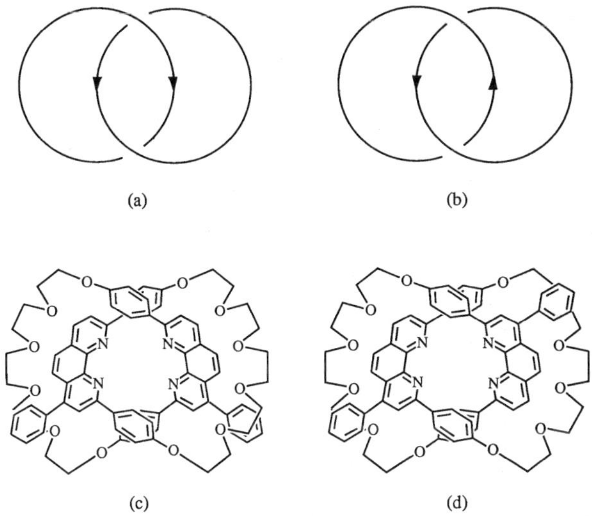

Figure 32. (a) Minimal diagram of an oriented Hopf link. (b) The enantiomorph of (a). (c) A [2]catenane[118] abstractly represented by (a). (d) The enantiomer of (c), abstractly represented by (b). Reprinted with permission from C. Liang and K. Mislow, *J. Math. Chem.* **1995**, *18*, 1. Copyright 1995, Baltzer Science Publishers.

While orientation normally imparts topological chirality, as in the case of the Hopf link, this is not always the case. For example, the Borromean link remains amphicheiral even after orientation.[115, 119] This is easily demonstrated with reference to the S_6 presentation of the nonoriented link [Figure 27(b)]: No matter in which direction the three rings are oriented, the resulting diagram remains centrosymmetric and therefore rigidly achiral.

Polypeptides in native proteins and polynucleotides in nucleic acids are oriented chains with an asymmetric distribution of monomeric components. Molecular links made from circular variants of these chains are therefore topologically chiral.[72] Catenated molecules in Nature were first discovered in the mitochondrial DNA of human cells by Jerome Vinograd and co-workers in 1967.[120] A great variety of catenated DNAs have since been observed in diverse biological systems,[91b] for example, a 5-crossing DNA link composed of two unknots, and a 6-crossing DNA link composed of one oriented unknot and one figure-eight knot.[92, 121] A number of catenated structures have also been recently observed among proteins[93]; an example is shown in Figure 33. Nadrian Seeman and co-workers designed and carried out the synthesis of intricately interlaced structures made up of single-stranded DNA; the most spectacular examples are polyhedra (a [6]catenane in the form of a cube[88, 122] and a [14]catenane in the form of a truncated octahedron[123]) whose faces are made up of cyclic DNAs interlinked with their nearest neighbors.

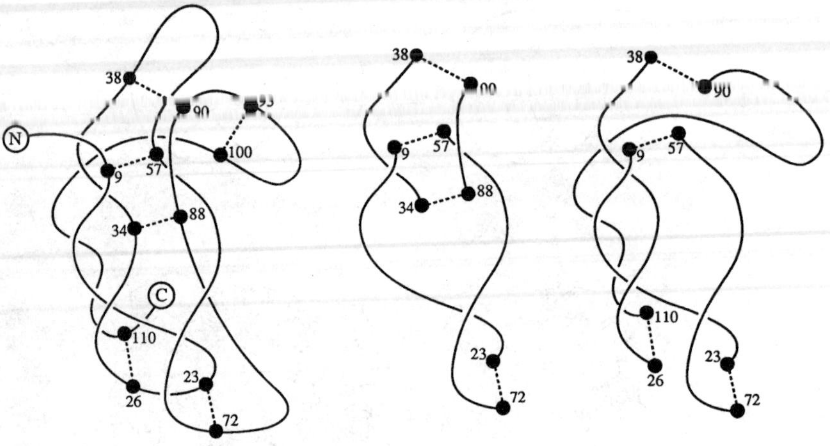

Figure 33. (a) Condensed schematic diagram of the β-subunit of human chorionic gonadotropin (hCG). Cysteine α-carbons are represented by filled circles. Intrachain disulfide linkages are shown as dashed lines. (b) A topological link derived from (a), in which disulfide linkage Cys9–Cys57 penetrates the ring (Cys34–X_3–Cys38–Cys90–X_1–Cys88–Cys34), where X_n represents *n* contiguous amino acid residues. (c) Another topological link, in which the two component rings in (a) penetrate one another through their polypeptide segments instead of their disulfide linkages.[93] Reprinted with permission from C. Liang and K. Mislow, *J. Am. Chem. Soc.* **1995**, *117*, 4201. Copyright 1995, American Chemical Society.

E. Limitations of the Topological Model

1. *On the Choice of Topologically Significant Bonds*

There is no problem in identifying the vertex set in a molecular graph that represents the constitutional formula of a molecule because each vertex bears a one-to-one correspondence to an appropriately labeled atom in the molecule. The relationship of edges in the graph to bonds in the molecule is, however, far less well defined. This point warrants a reiteration of earlier remarks on this theme[52, 71, 72, 79, 93] because it has an important bearing on the subject of topological chirality in molecules.

The crucial question is: Which bonds in the molecule are to be regarded as "topologically significant?"[73, 87] Different authors differ in their answers to this question.[72] The root of the difficulty "lies in the historical sanction of localized valence bond formulas, which, for all their virtues of convenience, imply a sharpness of definition that is physically unsound: For the sake of simplicity, degrees of atomic interaction in a molecule are ignored, and pairs of atoms are regarded as either bonded or not."[52] The problem is in principle solvable by edge-weighting the complete graph, with each atom bonded to every other atom in the molecule and the set of edges mapped one-to-one onto a set of real numbers that express in some manner the nature and extent of bonding interactions between pairs of atoms.[52] In practice, however, the question remains: Which of these bonds are to be included in the edge set of the molecular graph? Inevitably, therefore, considerable fuzziness is built into the definition of a molecular graph. According to Walba, only covalent bonds are to be regarded as topologically significant, whereas "H bonds, ion–ion bonds, ion–dipole bonds, or dipole–dipole bonds are not considered edges of a molecular graph."[73] While this definition has the advantage of being consistent with common usage in organic chemistry, where "molecular graph" carries the same meaning as "constitutional formula" or "localized valence bond diagram," there is no good reason to limit the edge set to covalent bonds. Thus, as Walba remarked, "even the term covalent bond is arbitrary,"[73] since there is, in the general case, a continuum between covalent and ionic bonds. "Indeed, in a trivial sense, crystalline sodium chloride is rich with topological networks, if each coulombic interaction is viewed as a connection."[124]

As an illustration of the sort of quandary that one may have to face in the choice of topologically significant bonds, consider the hypothetical derivative of [1.1.1]propellane shown in Figure 34.[71] The bridgehead atoms of this derivative are different isotopes of carbon, and the remaining carbons are bridged by three polymethylene chains of different lengths. There is no question whatsoever that this molecule is *geometrically* chiral, regardless of whether or not there is a bond between the isotopic bridgehead carbons. The *topological* chirality of this molecule, however, depends crucially on whether or not this bond is represented by an edge in the molecular graph. If the bond is thought to exist, and the edge is therefore included in the graph, the molecular graph [Figure 34(a)] becomes nonplanar. Under these conditions, the graph, and hence the molecule, is topologically chiral because all the vertices are nonequivalent. If the bond is thought not to exist, and the edge is therefore not included in the graph, the molecular graph becomes planar [Figure 34(b)], and the molecule is topologically achiral. The question of whether or not there is a central bond in [1.1.1]propellane has been studied at length,[125] but there is no universal consensus since "all of the arguments put forward for the existence of a central bond in [1.1.1]propellane can be matched with a counterargument except for the heat of formation."[125c] On the one hand, the charge-density distribution points to the existence of a bond between the bridgehead carbons.[125d] On the other hand, this electron density makes no contribution to holding the bridgehead carbons together.[125a] Hence, whether this molecule is considered to be topologically chiral or not depends entirely on what criteria are used to define the central bond.

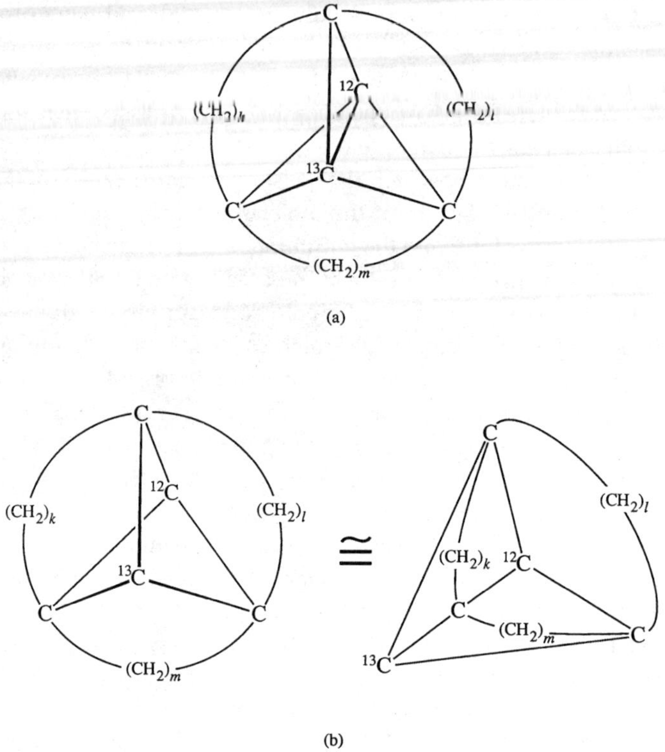

Figure 34. (a) Molecular graph of a hypothetical derivative of [1. 1. 1]propellane. The graph is topologically chiral. (b) The same graph without the central edge is planar and therefore topologically achiral. Reprinted from K. Mislow, *Croat. Chem. Acta* **1996**, *69*, 485. Copyright 1996, Croatian Chemical Society.

2. *On the Relevance of Topological Chirality to Chirality Properties*

While uncertainties in defining membership in the edge set are virtually unavoidable, once the members of the edge set are selected all uncertainty vanishes. The molecular graph is then treated exactly as a topological object[77a] in which considerations of metrics and internal energy play no role. For example, whereas it is physically impossible to flatten the *molecules* in Figure 14 so that all the atoms lie in a plane while all the bonds remain intact, planarization of the corresponding molecular *graphs* is a perfectly unexceptionable topological operation.

The topological chirality or achirality of a molecule thus refers exclusively to its molecular graph, and not necessarily to a physically realistic model. Yet, as remarked above, the concept of chirality in chemistry has a well-defined meaning only in relation to experiment, and it is the geometrical chirality of the molecule (Pasteur's dissymétrie moléculaire), and not the symmetry of the molecular graph, that is responsible for all observed chirality properties. In the world of observables, chirality is a geometrical property,[17] and this property is in no way dependent on molecular constitution.[51] In stark contrast, topological chirality, as we saw above, is critically dependent on a description of bonding connectivity.

Chirality properties (pseudoscalar observables) are therefore tied exclusively to the geometrical model. For example, even though the molecules in Figure 14 are topologically achiral, they exhibit properties (enantiomerism and optical activity) that are manifestations of their geometrical chirality. And any chirality properties observed for topologically chiral molecules, such as the optical activity observed for resolved molecular trefoil knots,[83] is due entirely to the chiral arrangement of the atoms (nuclei and electrons) in space, that is, to the geometrical chirality of the molecule: The molecule's topological chirality is in this respect totally irrelevant.

F. A Note on Definitions of Chirality

We have encountered a number of different definitions of chirality. In the spirit of making "everything as simple as possible, but not simpler,"[126] we suggest a concise and at the same time general definition for a chiroid (and hence of chirality) whose very simplicity makes it broadly applicable in geometry and topology, as well as in the natural sciences:

> An object is chiral if and only if it is not superposable on its mirror image; otherwise it is achiral.

In this definition, "object" is meant to encompass all the "varied and widespread" items listed by Whyte,[12] including abstract mathematical figures, models of molecules or of molecular ensembles, bodies in motion, topological objects such as an overhand knot, and even atoms that, as was mentioned, are inherently chiral. "Superposable" refers to superposition of enantiomorphs by rigid motions or by continuous deformation, and "mirror image" to the product of an isometry of the second kind (including space inversion $\hat{P}$) or to a $\hat{C}\hat{P}\hat{T}$ operation, whichever applies. This comprehensive definition captures the crux of the matter: The single essential characteristic of the property that Kant referred to when meditating on the incongruence of counterparts, that Pasteur called dissymmetry, and that today we call chirality or handedness, is enantiomorphism, that is, the existence of nonsuperposable mirror images.

The breadth in scope of this definition comes at the cost of requiring flexibility and judgment in the interpretation of "object," "superposable," and "mirror image." The model must suit the occasion of its use. Chirality, with respect to an isolated molecule, is a quantum-mechanically undefined concept, but, because we deal with the reality of enantiomeric molecules, chemists resort to the classical model. Matter (or antimatter) is inherently chiral, so that, strictly speaking, there is no such thing as an achiral molecule, either in our world or in the antiworld. Yet chemists normally choose the classical model in which this feature is suppressed, so that individual atoms are regarded as being achiral. By the same token, the difference between $\hat{P}$ and $\hat{C}\hat{P}\hat{T}$ enantiomers is normally ignored, other than in the context of the origin of biomolecular chirality. It is clear that context always needs to be taken into account. For example, we saw that whether a molecular model is chiral or achiral depends on the conditions of observation, that the geometrical model of a molecule such as alanine is chiral but the topological model is achiral, and that the chirality of a spinning molecule is time-noninvariant if it is stationary but time-invariant if it is not. Bearing in mind that the universe of discourse determines the meaning or connotation of the wording, all the various attributes of chirality are subsumed under the all-encompassing umbrella of this definition. That the spareness of this working definition thereby engenders a degree of fuzziness and latitude of interpretation is by no means an undesirable feature in a scientific statement[127] and is in any event inevitable in scientific classifications outside of mathematics.

V. SENSE OF CHIRALITY

The chirality of objects such as scalene triangles and oriented circles in R^2 (Figure 1) and helices in R^3 (Figure 3) is a property shared by both enantiomorphs; the difference between them is their sense of chirality. In what follows, we shall for simplicity use the term *configuration* to stand for "sense of chirality." Two enantiomorphs are thus said to have opposite configurations.

A. Absolute Configuration and the Meaning of Right and Left

The absolute configuration of a chiroid is a representation in the appropriate reference space that differentiates the object uniquely from its enantiomorph. For example, the absolute configuration of what we call a right-handed helix (Figure 3) is given by the descriptions "clockwise as viewed by the observer along the screw axis" with respect to the axial vector and "away from the observer" with respect to the polar vector. The assignment is unambiguous because we all agree on the conventional meaning of "clockwise," since clocks normally run the same way and are viewed in the same way. We are thus able to connect configuration in the abstract world of helices with orientation in the macroscopic world of experience. Similar considerations apply to the description of molecular configuration. The tartrates provide an example of historic significance. Pasteur's success in the manual separation of sodium ammonium tartrate crystals depended on his success in visually identifying the configuration of the hemihedral facets (Figure 35, top). Pasteur showed that the sense of hemihedry was correlated to the sign of rotation of the tartaric acids, but he had no way of establishing a relationship between sign of rotation and molecular configuration. This last step had to await the development of modern crystallographic methods. A little over 100 years after Pasteur's 1848 report, J. M. Bijvoet and co-workers[128] were able to show, using anomalous scattering of the Rb atoms in sodium rubidium (+)-tartrate, that (+)-tartaric acid has the 2*R*,3*R* configuration (Figure 35, bottom). By saying that the absolute configuration of (+)-tartaric acid has the 2*R*,3*R* configuration, we mean no more than to assert that the spatial arrangement of atoms in (+)-tartaric acid shown in Figure 35 is consistent with the configuration of macroscopic chiroids in the same world, a world in which human hearts are located on what we arbitrarily call the left side. The same holds true for the helicities assigned to particles and antiparticles, as in left-handed electrons and right-handed positrons.

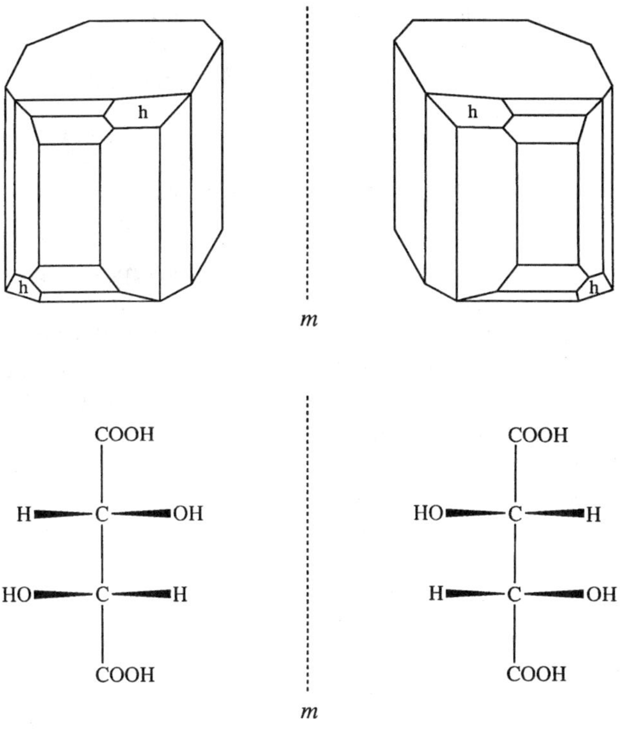

Figure 35. Top: Enantiomorphous crystals of sodium ammonium tartrate. Hemihedral facets are marked by an "h." Bottom: (+)-(2*R*,3*R*)-tartaric acid (left) and (–)-(2*S*,3*S*)-tartaric acid (right).

In discussions dealing with the semantic content of words,[80] it is customary to cite the famous exchange between Alice and Humpty Dumpty, "a philologist and philosopher skilled primarily in linguistic matters"[129]:

> "When *I* use a word," Humpty Dumpty said, in rather a scornful tone, "it means just what I choose it to mean—neither more nor less."
>
> "The question is," said Alice, "whether you *can* make words mean so many different things."
>
> "The question is," said Humpty Dumpty, "which is to be master—that's all."

Humpty Dumpty is right: Words mean "neither more nor less" than what they are intended to mean. According to Martin Gardner, "Lewis Carroll was fully aware of the profundity in Humpty Dumpty's whimsical discourse on semantics. Humpty Dumpty takes the point of view ... that universal terms do not refer to objective existences but are nothing more than *flatus vocis*, verbal utterances. This view ... is now held by almost all contemporary logical empiricists."[129] Thus we are in principle at liberty to call a given chiroid either right-handed or left-handed, whichever we prefer, so long as internal consistency is preserved. In summary, there is no intrinsic meaning to the words *left* and *right,* and all configurational assignments are based on a commonly agreed-upon but arbitrarily chosen ("right-handed") coordinate system.

B. Homochirality Classes

Consider a set of chiroids and their respective enantiomorphs. A sense of chirality can be defined for the members of each heterochiral pair, so that one member within each pair may be designated as "right-handed" (**R**) and the other as "left-handed" (**L**). A homochirality class is one in which all the members have the same configuration, either **R** or **L**. It seems evident that the existence of homochirality classes requires a uniform rule or procedure, some characteristic feature that is shared by the members of the set and that makes it possible to assign a common sense of chirality. Kelvin's definition provides the most stringent example: Two chiroids are homochiral if and only if they are properly congruent. Seen from another perspective, this definition seems unduly restrictive; for example, we saw that the various diagrams of trefoil knots (Figure 13), though clearly not isometric, can nevertheless be assigned to homochiral (**R** or **L**) classes if they can be interconverted by ambient isotopy in R^3. The situation is analogous to one encountered in chemistry, where different conformations (i.e., various geometrical shapes that result from bond angle deformations, bond stretching, and bond torsion) of a given chiral molecule are assigned to a single class, for example, *R* or *S*. As another example, if some standard helix were to be arbitrarily defined as right-handed (Figure 3), we could easily recognize right- or left-handedness in all sorts of helices, regardless of pitch and radius, and we should be able to partition all of them into two (**R** and **L**) homochirality classes. Similarly, all human hands, though clearly never identical, can nevertheless be readily classified as right or left, regardless of differences in size, shape, or color, because they all share an easily recognizable property, appropriately enough called handedness. It is clear from these examples that even though a set of chiroids may not be isometric, they may nevertheless share some characteristic feature that makes it possible to recognize a common sense of chirality.

And so it is with molecules as well, though what we call right-handed or left-handed depends on the necessarily arbitrary choice of a convention. For example, statements such as "amino acid residues in proteins occur only as left-handed molecules" presumably refer to their L-configuration, in which the amino group is on the left of the Fischer projection formula, with the carboxyl group on top. Thus naturally occurring α-amino acids may be regarded as belonging to a single homochirality class characterized by the L configuration. Note, however, that although we are free to call L-amino acids left-handed, there is nothing wrong with calling them right-handed instead (with reference, say, to a different projection formula): As noted above, internal consistency remains the sole requirement.

Homochirality classes provide a unifying overview of objects with a common configuration (e.g., all right hands or all L-amino acids); their usefulness lies in collecting homochirally similar objects without regard to individual differences among the members in each class. But the features that characterize the various classes are not necessarily commensurate: While there is a family resemblance between, say, left hands and left-handed gloves, there is no obvious commonality between left-handed trefoil knot diagrams and left-handed α-amino acids. Each class thus requires its own internal standard.

The members of a homochirality class are chirally connected; that is, they are in principle interconvertible by continuous deformations or by other appropriate transformations along chiral pathways, that is, along pathways that do not contain achiral points. They cannot, however, be chirally connected to the enantiomorphous (heterochiral) set, for enantiomorphs would in that case be homochirally similar, which is a contradiction. In practice, this constraint severely limits the application of the homochirality concept. What may be called the homochirality problem[130] arises whenever there exists a pathway of continuous deformation that connects two enantiomorphous objects but that does not require passage through an achiral point. Under these circumstances, it becomes meaningless to speak of the set of objects anywhere along the deformation path (except at the termini) as right-handed or left-handed; the reason is that no privileged point can be defined, other than arbitrarily, where right switches to left and vice versa. The asymmetric tetrahedron with unlabeled vertices is the simplest example. While conversion of such a tetrahedron to its enantiomorph by way of an achiral tetrahedron is certainly not excluded, achiral pathways are easily circumvented because the set of achiral unlabeled tetrahedra in R^3 does not form a boundary between heterochiral sets. An asymmetric tetrahedron can thus be converted to its enantiomorph by continuous deformation along a chiral pathway without ever having to pass through an achiral intermediate[131]. Enantiomorphous unlabeled tetrahedra are therefore chirally connected. Accordingly, it becomes impossible to isolate a geometrical characteristic that would allow the partition of unlabeled tetrahedra into homochirality classes **R** and **L**: Where, along the chiral pathway that connects two enantiomorphous tetrahedra, would one cross the boundary between **R** and **L**? It therefore makes no sense to speak of chiral unlabeled tetrahedra in general as being either right-handed or left-handed. Indeed, it can be shown that no unlabeled geometrical object in R^3 or higher dimensions can be assigned to homochiral **R** or **L** classes.[132] In contrast to unlabeled tetrahedra, however, enantiomorphous tetrahedra with four differently labeled (indexed) vertices can be partitioned into homochiral R or L classes because any chiral set of four differently labeled points *ABCD* in R^3 cannot be converted to the enantiomorphous set without passage through a planar and therefore achiral state.[130, 132]

In their classic studies on the subject of homochirality in chemistry, Ernst Ruch and co-workers[18, 133] developed a model in which a set of n ligands is partitioned among the n sites of an achiral permutation skeleton. The permutations are transpositions of ligands on sites that are related by mirror planes of the skeleton. The skeletons are of two types[133b]: Those in which every mirror plane contains $n - 2$ sites (category *a*), and those in which there are also mirror planes that contain fewer than $n - 2$ sites (category *b*). Skeletons in category *a* are exemplified by the regular tetrahedron (T_d), the trigonal pyramid (C_{3v}), the trigonal bipyramid (D_{3h}), and the tetragonal disphenoid (D_{2d}), while skeletons in category *b* are exemplified by the square pyramid (C_{4v}), the octahedron (O_h), and the cube (O_h).

Consider now the transposition of two ligands at the vertices of a regular tetrahedron. All six mirror planes of the regular tetrahedron contain $n - 2 = 2$ vertices, and the skeleton therefore belongs to category *a*. Each ligand is associated with a continuously varying, scalar parameter λ. As Ruch has pointed out,[18] ligands may be symbolized by spheres of variable diameter λ that are centered at the vertices of the achiral polyhedron. If the transposition of the two ligands at any two sites is then visualized as a continuous change in the diameters of the corresponding spheres, there inevitably comes a point at which the two diameters are the same and the model becomes achiral. In general, the pathway of continuous deformation that connects two enantiomorphs in models belonging to category *a* necessarily requires passage through an achiral model. The requirement for an acceptable division between right- and left-handed objects is thus satisfied for all skeletons in this category, and the models can be assigned to homochiral **R** and **L** classes.

In contrast to the tetrahedron, two of the four mirror planes of the square pyramid contain only the apical vertex and therefore fewer than $n - 2 = 3$ ligand sites. This skeleton thus belongs to category *b*. If the transposition of the two ligands at sites *a* and *b* in Figure 36 is also visualized as a continuous change in the diameter of the corresponding spheres, there is no point at which the model becomes achiral. The same is true of the subsequent step, the transposition of ligands at sites *c* and *d*. Hence, the enantiomorphs are chirally connected. In general, the pathway of continuous deformation that connects two enantiomorphs in models belonging to category *b* need not entail passage through an achiral model. The requirement for an acceptable division between right- and left-handed objects is therefore not satisfied for skeletons in this category, and the models cannot be assigned to homochiral **R** and **L** classes.

Ruch's classification scheme, though theoretically sound, is severely limited in its application to chemistry. The principal reason is that routine skeletal deformations, which are beyond the scope of Ruch's theory, account for the vast

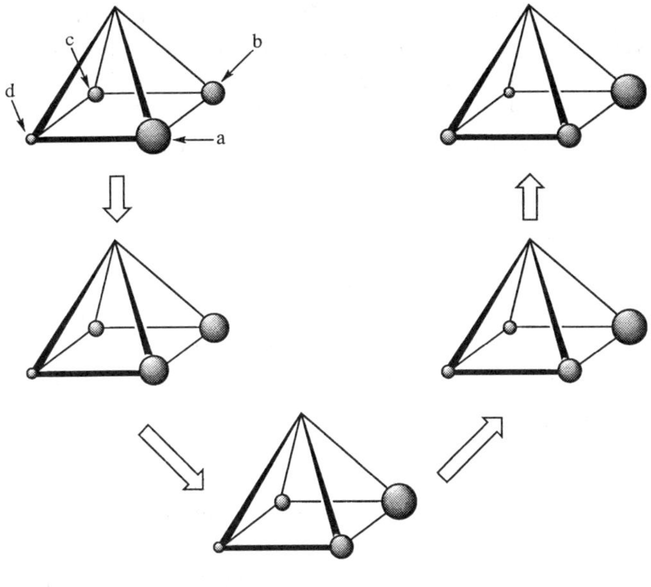

Figure 36. Conversion of an asymmetric square pyramid (top left) into its mirror image (top right) along a chiral pathway. Four spheres with different but variable diameters are centered at the vertices of the square base. The spheres labeled a and b shrink and expand, respectively, until their diameters are switched. This is followed by a similar switch in the diameters of spheres c and d.

majority of molecular enantiomerizations, and it is through the agency of these deformations that enantiomers are most often chirally connected. Thus, while enantiomerization of chiral molecules composed of four different atoms unavoidably requires the intermediacy of achiral structures, as, for example, in the racemization of NHDF by inversion at nitrogen, any chiral molecule composed of five or more different atoms is in principle capable of conversion to its enantiomer by chiral as well as achiral pathways, provided that a feasible route exists on the potential-energy hypersurface.[134, 135] Consider, for example, a conformation of the deuteriated hydroxylamine derivative NHDOH in which the N–O–H plane bisects the H–N–D angle and the O–H hydrogen points at the base of the nitrogen pyramid. By a combination of rotations about the N–O bond and inversions at the nitrogen atom, this conformation can be converted to its enantiomer either by achiral or by chiral pathways.[130] Enantiomerization by an achiral route can take place by way of a transition state in which all five atoms lie in a single plane. Enantiomerization by a chiral route is achieved by a 180° rotation about the N–O bond, followed or preceded by pyramidal inversion.

The existence of chiral pathways in NHDOH is made possible by the existence of the two independent degrees of freedom that govern internal motion: rotation and inversion. As molecular complexity increases, the number of degrees of freedom also increases, and, unless an achiral pathway is energetically much preferred, it becomes more and more likely that enantiomerization proceeds by a chiral pathway. For example, it is extremely improbable that reversal of helicity in a polymeric chain involves an achiral intermediate or transition state. A formal limit is reached when, due to structural constraints, all achiral pathways along the enantiomerization trajectory become energetically inaccessible under normal laboratory conditions. Under these conditions we may be dealing with chemically achiral compounds that are composed exclusively of asymmetric conformations. Examples, in addition to the previously discussed (1*R*)-menthyl (1*S*)-menthyl 2,2′,6,6′-tetranitro-4,4′-diphenate and the [2]catenane characterized as a topological rubber glove, are asymmetric molecular propellers whose enantiomerization by the two-ring flip mechanism involves an exclusively chiral pathway,[136] certain bis(9-triptycyl)methane derivatives, in which the two triptycyl groups behave as highly mobile and tightly meshed bevel gears,[137a] and *cyclo*-Se SSOS which undergoes one-step enantiomerization by pseudorotation via a chiral transition state.[137b]

The absence of an achiral boundary along the conformational enantiomerization trajectory of chemically achiral compounds, such as the ones discussed above, precludes partitioning of conformations along the path into homochirality classes. As noted above, under such circumstances it becomes meaningless to speak of these conformations as right-handed or left-handed because no point can be defined, other than arbitrarily, where right switches to left and vice versa.

Enantiomers, in the world of observables, have oppositely signed pseudoscalar properties—properties that remain invariant under proper rotation but change sign under improper rotation. It is therefore unavoidable that any given pseudoscalar property of a chiral system, such as optical rotation, becomes exactly zero at some point in the conversion of a molecule into its enantiomer along a chiral pathway. That point is called a chiral zero, but it is not the only switching point. The reason is that in principle there are countless pseudoscalar properties, and they all differ in the point along the enantiomerization path at which their value becomes zero. Hence, there is no unique chiral zero, and the switching point remains undefined.

Our discussion of enantiomerization along chiral pathways has been limited to conformational interconversions, but there is no sharp dividing line, in principle, between such interconversions and reaction sequences that connect enantiomers exclusively by chiral intermediates or transition structures. The classic example is the conversion of (+)- into (–)-isopropylmalonamic acid by a reaction sequence reported in 1914 by Emil Fischer and Fritz Brauns.[138] The individual steps are shown in Figure 37, along with configurational descriptors *S* and *R* for each molecule. Although the actual configurations are still unknown, the ones shown in Figure 37 represent an "enlightened guess."[139] Judging by the configurational descriptors, it might appear as though the three compounds at the top of Figure 37 belong to one homochirality class (*S* = **L**) and the three at the bottom to the other (*R* = **R**). This impression would, however, be mistaken. The six compounds shown

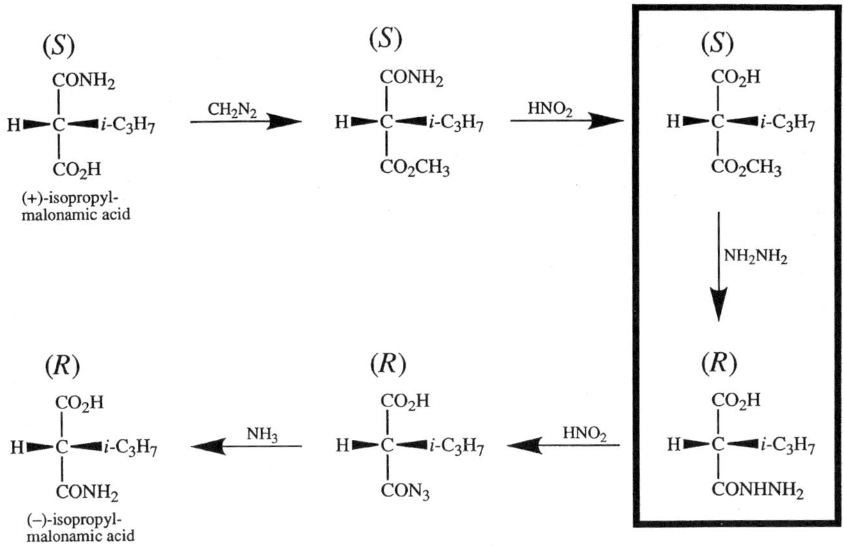

Figure 37. Conversion of (+)- into (–)-isopropylmalonamic acid.[138] The step at which the configuration changes from *S* to *R* is boxed.

in Figure 37 represent particular stationary points on the multidimensional potential-energy surface and are chirally connected through a continuous reaction pathway that involves no achiral intermediates: If this were not the case, the product of the reaction sequence would be racemic. Hence, these molecules cannot be partitioned into **R** and **L** homochirality classes. The *S* and *R* forms of each individual chiral compound—for example, the enantiomeric isopropylmalonamic acids—always belong to heterochiral classes in the restricted sense of Kelvin's definition, but this classification cannot be consistently extended beyond individual enantiomeric pairs and the component conformational sets.

C. Specification of the Sense of Chirality

The sense of chirality of a geometrical model may be specified in a variety of ways.[140] Foremost among these, at least in organic chemistry, is the CIP scheme.[50] This scheme, also known as the "steering wheel" reference system, is based on the combination of polar and axial vectors that, as we saw, is intimately related to the generation of helices (Figure 3). In the simplest version of this scheme, four fiduciary atoms associated with a given stereogenic unit are numbered in a descending order of priority, $1 > 2 > 3 > 4$. The three atoms of highest priority define a plane and effectively represent the oriented circle of an axial vector (the "steering wheel"). The sense of orientation is given by $1 \rightarrow 2 \rightarrow 3$ and defines the direction of this vector. The atom of lowest priority (4) is then placed to one side of the 1-2-3 plane and viewed from the other; the direction of view represents the polar vector or "steering-wheel column." The clockwise or anticlockwise sense of orientation of $1 \rightarrow 2 \rightarrow 3$, as viewed from the side away from 4, then yields the stereochemical descriptor, *R* or *S*.

The steering-wheel concept is physically embodied in the chiral "wheel-and-axle" [2]rotaxane prepared by Ralf Jäger and Fritz Vögtle[141] (Figure 38): the "axle" and the "wheel," whose orientations are given by the unsymmetrical positioning of the amide and sulfonamide groups, symbolize polar and axial vectors, respectively. The enantiomers of this remarkable rotaxane, which might be dubbed the "CIP molecule," have recently been separated.[142]

Another reference system consists of two nonoriented skewed line segments arranged in a D_2 array (Figure 39). The C_2 axis normal to and intersecting both lines serves as an axis of rotation. The chirality of the array is a function of the torsion angle θ subtended by the two lines. Its measure, $\chi(\theta) = \sin 2\theta$, may also be taken as a measure of helicity if one of the pair of lines is the tangent of a cylindrical helix and the other is collinear with the screw axis, as shown in Figure 39. The skew-lines reference system has been applied to assign Δ and Λ configurations to chelate-bridged octahedron edges in bis- and tris(bidentate) complexes, depending on whether $\chi(\theta)$ has positive or negative values.

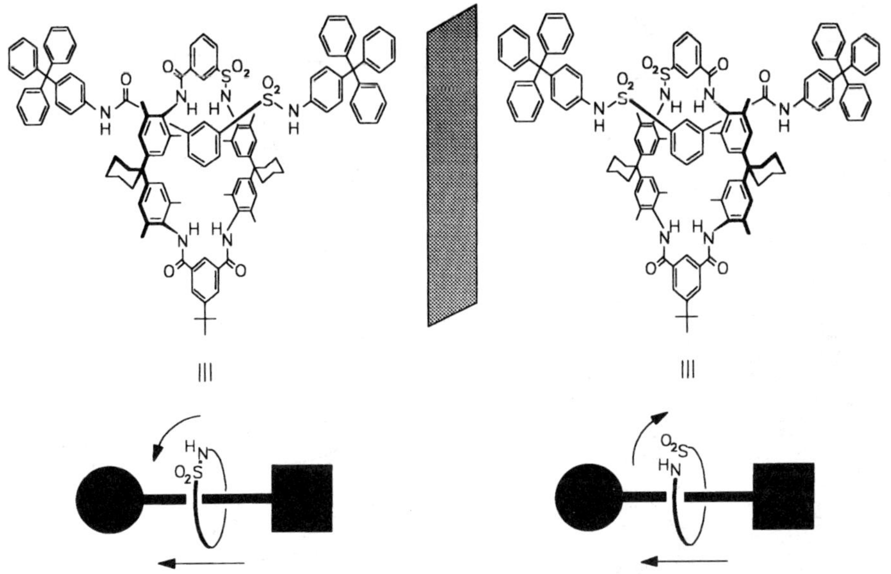

Figure 38. Enantiomers of a [2]rotaxane composed of an oriented "axle" and an oriented "wheel." Arrows indicate the atom sequence (N → S). Adapted from Refs. 141 and 142.

Orientation of the two line segments results in a reference system whose chirality measure is $\chi(\theta) = \sin\theta$. In contrast to the array shown in Figure 39, the chirality for the oriented-skew-lines reference system[140] does not vanish when $\theta = 90°$. This system is used in configurational assignments to knots and links. Crossing points in knot diagrams are associated with a two-valued characteristic, $\varepsilon = +1$ or -1, obtained by assigning a direction to, that is, orienting, the curve—the choice of direction is immaterial—and then labeling the crossing points with the appropriate signs in accordance with the convention in Figure 40. Thus, for example, all three crossings are positive in the minimal diagrams of the trefoil knot with the **R** configuration (Figure 13) and negative in the corresponding **L** diagrams. The writhe $w(K)$ of a knot K is the arithmetic sum of the crossing point characteristics, $w(K) = \Sigma\varepsilon$. For example, the writhes of the diagrams depicted in the top row of Figure 21 are –3, 0, –5, and –5 for 3_1, 4_1, 5_1, and 5_2, respectively. The writhe of a knot can, to a limited extent, be used to assign configuration. In 1963, Stephen J. Tauber[119] proposed a scheme based on the conjecture, subsequently proven to be true,[143] that the minimal diagrams of all amphicheiral alternating knots have an equal number of over- and undercrossings. The writhe of such a knot must always be zero. It follows immediately that if the writhe is not zero, then the alternating knot is topologically chiral. Furthermore, because over- and

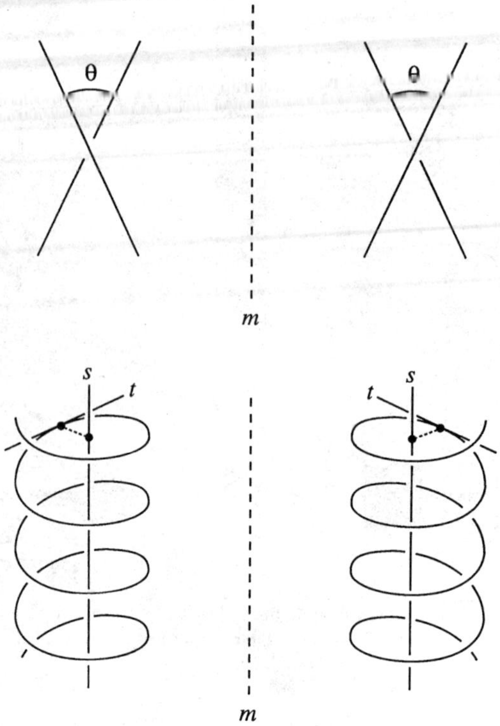

Figure 39. Top: Two nonoriented line segments in a skewed D_2 array. Bottom: One of the pair of lines (t) is the tangent of a cylindrical helix and the other (s) is collinear with the screw axis of the helix.

undercrossings are switched upon reflection in the plane of projection, it also follows that the writhe of a chiral knot and that of its enantiomorph are oppositely signed. Tauber proposed that the absolute configuration of a knot be designated *R* if $\Sigma\varepsilon > 0$ and *S* if $\Sigma\varepsilon < 0$, using the descriptors of the CIP convention. Walba[73a] subsequently proposed a convention for the specification of chirality in knots that was in all essential respects the same as Tauber's, with $\varepsilon = \delta$ and λ replacing +1 and −1, and Δ and Λ replacing *R* and *S*.

In support of his proposal, Tauber pointed out that "For certain knots $\Sigma\varepsilon = 0$. This is exactly as it should be, for precisely these knots are identical with their mirror images." Similarly, Walba asserted that "The number of δs and λs are summed arithmetically. If there are the same number of δ and λ crossings, then the knot must be topologically achiral." Contrary to these assertions, however, alternating knots whose writhe is zero are not necessarily amphicheiral: The simplest example is knot 8_1. Nineteen of the 32 10-crossing prime knots with writhe zero are topologically chiral, and 13 of these are alternating.[144] Two hundred

Figure 40. Convention used to assign characteristics ($\varepsilon = +1$ or -1) to crossings in knot diagrams. Reprinted with permission from C. Liang and K. Mislow, *J. Math. Chem.* **1994**, *15*, 1. Copyright 1994, Baltzer Science Publishers.

sixty-two of the 320 12-crossing prime knots with writhe zero are chiral, and 159 of these are alternating.[95] In short, a writhe of zero is a necessary but not a sufficient condition for the amphicheirality of alternating knots. Furthermore, a writhe of zero is not even a necessary condition for the amphicheirality of nonalternating knots. Because a writhe of zero does not necessarily imply a knot's amphicheirality, the Tauber–Walba scheme is limited in its application to alternating knots whose writhe is not zero. While a scheme has been developed[145] for assigning chirality descriptors to oriented and nonoriented chiral links and knots that overcomes this particular difficulty, the problem of assigning configurations to nonalternating knots remains substantially unsolved. In much the same way, the CIP scheme works well enough for routine organic structures, but significant difficulties are encountered in attempts to assign configurations to unconventional or highly complex structures, such as inorganic or metallorganic compounds, or clusters. In such cases, special schemes have to be designed on an ad hoc basis, as, for example, in the configurational description of chiral fullerenes.[146]

The difficulties described above are symptomatic of a fatal flaw that is common to all schemes designed to specify the configuration of a chiroid: Without exception, they are no more than empirical adaptations to particular structural features of the chiroid. For example, the CIP scheme depends primarily on the identification of bonding arrangements in the molecular graph, which, as noted above, are not always unambiguously defined, and involves an elaborate algorithm for priority orders that requires for its execution reference to arbitrary rules designed purely to meet practical ends. Similarly, the scheme for assigning chirality descriptors to chiral links and knots[145] is limited to alternating constructions: No theory is available to handle nonoriented, nonalternating knots and links.

The unavoidable absence of the theoretical underpinning that would be required for an all-inclusive general application of configurational schemes is akin to the impossibility of finding a characterizing feature common to the diverse homochirality classes. The problem of designing a general scheme for assigning configurations to geometrically or topologically chiral molecules is therefore fundamentally intractable, and we have to be content with the design of empirical "local rules" that can be usefully applied to a restricted set of molecular structures. Pragmatism rules!

VI. QUANTIFICATION OF CHIRALITY

It is natural to attach a quantitative meaning to the chirality of molecular structures and to speak of such structures as being "more" or "less" chiral and as differing in "degree of chirality."[25, 147, 148] That 1-stearoyl-2,3-dipalmitoylglycerin, for example, exhibits no measurable optical activity in the visible and ultraviolet regions, even though this natural triglyceride is enantiomerically pure, is easily understandable because the fatty acid residues in the 1- and 3-positions are "very similar . . . and hence the asymmetry of the molecule is very slight."[149] Similarly, the "very slight asymmetry" of enantiomerically pure $^{14}CH_3(CH_2)_5CH(CH_3)(CH_2)_5CH_3$ leads to the prediction that this compound "would have zero rotation within experimental error on the most sensitive polarimeter now available."[150] By extension, it is possible to rank the degree of chirality of closely related compounds with reference to molecular structures and to chirality properties. It is reasonable to assert, for example, that because the $CH_3(CH_2)_n$ and $^{14}CH_3(CH_2)_n$ groups in $^{14}CH_3(CH_2)_nCH(CH_3)(CH_2)_nCH_3$ resemble each other more closely as n increases, the degree of chirality of $^{14}CH_3CH_2CH(CH_3)CH_2CH_3$ is greater, and that of $^{14}CH_3(CH_2)_{10}CH(CH_3)(CH_2)_{10}CH_3$ is less, than the degree of chirality of $^{14}CH_3(CH_2)_5CH(CH_3)(CH_2)_5CH_3$. Never mind that the chiroptical properties of none of these compounds are likely to be observable with currently available instrumentation: The model still demands different levels of optical activity below the threshold of detection.

How can chirality be quantified? There are two minimum requirements. First, recall that an object X (no matter whether physical or mathematically abstract) is chiral if and only if it is nonsuperposable on its mirror image $\overline{X}$ ($X \neq \overline{X}$). It follows that any chirality measure χ that quantifies this property can equal zero if and only if the object is achiral; any function that does not satisfy this conditio sine qua non fails to qualify as a measure of chirality[147, 148]:

$$\chi(X) = 0 \Leftrightarrow X = \overline{X}$$

Second, the absolute value of $\chi(X)$ must be the same for X and $\overline{X}$.

Additional features are optional, and this allows room for a wide variety of approaches to the problem of quantification.[147, 148]

A. Geometrical Chirality Measures

A particularly transparent example of a chirality measure is the function $\chi(\theta) = \sin 2\theta$, discussed above, of the torsion angle θ subtended by two nonoriented skewed line segments. This real-valued, continuous function is a measure of the chirality of a helix, that is, of its helicity, and its value, the degree of chirality (or degree of helicity), is zero if and only if the tangent line is either parallel or perpendicular to the screw axis. In the first case ($\theta = 0°$), the helix has degenerated to a straight line (infinite pitch), and in the second case ($\theta = 90°$) to a circle (zero pitch): In either event the construction has become achiral. For all intermediate values of θ the degree of chirality is nonzero and has its maximum value [$\chi(\theta) = 1$] at $\theta = 45°$. The function $\chi(\theta) = \sin 2\theta$ thus also meets the objective of similarity invariance, which is desirable because the chirality of helices is measured by shape and not by size.

It has been recognized[147] that chirality measures can be subdivided into two types: those that gauge the extent to which a chiroid differs from an achiral reference object (measures of the first kind) and those that gauge the extent to which two enantiomorphs differ from one another (measures of the second kind). In chirality measures of the first kind, the question to be answered is: How dissimilar are the chiroid and its achiral reference object? In chirality measures of the second kind, the question is: How dissimilar are the two enantiomorphs of a chiroid? In both cases the underlying concept is that of a distance, measured either between a chiral and an achiral object or between two enantiomorphous chiroids. That is, the degree of chirality of a chiroid X is defined in relation to another, chiral or achiral, reference object X_{ref}: The less these two objects match, the more chiral is X.

The chirality measure $\chi(\theta) = \sin 2\theta$ is an example of a measure of the first kind, with X_{ref} the achiral (degenerate) helix. This function is one of a family of chirality measures that take the general form $\chi(\theta) = \sin n\theta$ and that are continuous, normalized in the interval [1,–1], oppositely signed for enantiomorphs, and that vanish if and only if the objects are achiral.[147] As noted above, the measure $\chi(\theta) = \sin \theta$ may be associated with two oriented line segments. The function $\chi(\theta) = \sin 4\theta$, where θ is one of the acute angles in a right triangle, measures the chirality of a right triangle in R^2.

"Overlap" or "common volume" measures are examples of measures of the second kind. Let K and K' denote mirror-image-related solids. When K and K' are overlapped, $K^* = K \cap K'$ represents the volume common to the two solids. Let K^*_{max} denote the intersection for which the ratio of volumes $R = [K^*]/[K]$ is at a maximum (R_{max}). Under this limiting condition, K^*_{max} fills K to its maximum capacity, and K is achiral if and only if $K^*_{max} = K$ and $R_{max} = 1$. The degree of chirality of K is given by $\chi(K) = 1 - R_{max}$. That is, the fraction of nonoverlapping volumes under conditions of maximal overlap becomes a measure of the degree of chirality, and $\chi(K)$ expresses the extent to which K and K' differ from one another.

The variety of functions that can in principle be employed to quantify chirality is potentially limitless,[110] but the very richness of this assortment also means that there is no such thing as a unique scale by which chirality can be measured. Because no one measure has a theoretical advantage over any other, the choice among measures is governed by purely pragmatic considerations, such as, for example, convenience in computation or suitability with respect to some particular experiment.

Consider, for example, the question: What is the shape of the most chiral right triangle in R^2? The answer to this question depends on the particular function that is chosen as the chirality measure: According to measures based on geometric chirality products,[151] on symmetry coordinates,[152] on common volumes,[153] and on Hausdorff distances,[131, 154] the smallest internal angles are 18.8°, 30.0°, 37.5°, and 35.2°, respectively.[147] The function $\chi(\theta) = \sin 4\theta$ mentioned earlier achieves its maximum $[\chi(\theta) = 1]$ at $\theta = 22.5°$. Because there are as many answers to this question as there are functions that can measure the chirality of a triangle, and because there is, in principle, no limit to the number of such functions, the question remains, in a deep sense, unanswerable.

Similar difficulties are encountered in response to the question: What is the shape of the most chiral triangle in R^2? According to the Hausdorff measure, the internal angles of the most chiral triangle are 21.02°, 43.69°, and 115.29°.[154] A dramatically different answer is provided by the other measures mentioned above: According to these measures, the extremal triangles that correspond to the supremum of their respective functions are only approached in the limit, and the most chiral triangles are therefore infinitely flat.[147] Once again, an answer to the question is bound to elude us.

A further consequence of the multiplicity of functions that are available for the measurement of chirality is that relative rankings of chirality become ambiguous. Consider, for example, the question: Which of two right triangles, A and B, with internal angles of 18.8°/71.2°/90° and 30°/60°/90°, respectively, is more chiral in R^2? Let us calculate their degrees of chirality, χ, by two different measures. According to a measure based on geometric chirality products,[151] A ($\chi = 0.074$) is more chiral than B ($\chi = 0.057$). But according to a measure based on symmetry coordinates,[152] B ($\chi = 0.500$) is more chiral than A ($\chi = 0.313$). As shown by this example, the relative ranking of objects according to degree of chirality depends on the particular function that is chosen as a chirality measure.

The same uncertainty attaches to functions designed to measure the chirality of geometrical objects in three dimensions. As a simple example, we saw that the degree of chirality of a helix can be expressed by a function, $\chi(\theta) = \sin 2\theta$, that achieves its maximum value at $\theta = 45°$. Alternatively, the degree of chirality of a helix can be given by the volume of a cylinder on the surface of which is inscribed one complete turn of a helical thread.[155] Let L denote the height of the cylinder, that is, the pitch of the helix, and D the length of the thread. Then the volume of the cylinder is zero if and only if $L = 0$ or $L = D$, that is, under extremal conditions analogous to those that obtain for $\chi(\theta)$. According to this model, the volume is at a maximum when $L = D/\sqrt{3}$. The maximum degree of chirality (helicity) is therefore attained when the angle of rise $\theta = \arcsin 3^{-1/2} \approx 35.3°$. Which one of the two values, 45° or 35.3°, is to be preferred? Both values are correct from the point of view of theory, and the question can therefore be answered only on a pragmatic level: It depends on which of the two models seems more appropriate in connection with some particular observation.

If chirality measures are to have any relevance to the world of observables, a choice must be therefore made among the innumerable functions that can in principle be devised. Such a choice can be justified only on the purely empirical grounds that the measure is successful in correlating the calculated degree of chirality with some experimentally determined chirality property. While there has been no shortage of precedents for this type of correlation, beginning with Guye's "produit d'asymétrie" in 1890[156] and continuing to the present day,[148] any such attempt to bridge the gap between the results of chiral shape analysis and experimental observables is likely to be stymied by two insuperable difficulties.

First, as we saw, different chirality measures give incommensurable values for the shape and the relative degree of chirality of a triangle, the simplex in two-dimensional space, and similar results are obtained for the helix, the quintessential representative of chirality in three dimensions. If these difficulties are unavoidable for such simple geometrical objects, it is blindingly obvious that they cannot be avoided for the vastly more complex models of molecules. The mutual inconsistency among chirality measures, amply illustrated above, places a daunting burden on the "right" choice. Under these circumstances, it is hard to avoid the conclusion that numerical values calculated for degrees of chirality associated with experimental observables are at best of questionable significance.

Second, recall that no unlabeled geometrical object in R^3 or higher dimensions can be assigned to homochiral **R** or **L** classes.[132] This has immediate implications for the choice of functions suitable for use as chirality measures: In principle, no sign-changing continuous functions, and, in particular, no continuous pseudoscalar functions, $\eta(\overline{X}) = -\eta(X)$, can be used as chirality measures in three and higher dimensions because such functions necessarily have chiral zeroes $[\eta(X) = 0 \text{ for } X \neq \overline{X}]$, in violation of the condition that the value of any chirality measure can be zero if and only if the object is achiral.[148] Thus only sign-preserving functions can in principle be used as chirality measures for geometrical figures or sets of points. This requirement can be circumvented only by the imposition of constraints, and such constraints may be severely limiting. As an example, the partitioning of skeletons into categories a and b in Ruch's model[18, 133] becomes possible only if the ligands differ by no more than one continuously varying scalar parameter, or, as mentioned above, if the skeletons remain geometrically rigid; relaxation of either constraint destroys the theoretical basis for the quantification of chirality by the use of this model.

B. Topological Chirality Measures

In stark contrast to the numerous functions that are available to measure geometrical chirality, no measure has yet been reported for the quantification of topological chirality. In analogy to geometrical chirality measures, topological chirality measures $\chi(K)$ must satisfy two minimal conditions: They can be equal to zero if and only if the knot or link is achiral, and they have to have the same absolute value for two topological enantiomorphs.

It would seem that a topological chirality measure should be readily derivable from one of the numerous knot or link invariants that are sensitive to chirality; examples are numerical invariants (e.g., the writhe) or polynomial invariants (e.g., the Jones polynomial). The simplest chirality measure of this kind is the value of the writhe $|w(K)|$, which obeys the double inequality $-N \leq |w(K)| \leq N$, where N is the minimal number of crossings of the knot or link diagram. With this method, the highest degree of chirality of a link with N crossings is attained when all crossings have the same sign, so that $|w(K)| = \pm N$. The trouble is, however, that *all* knot or link invariants devised up to now fail in some instances to detect the chirality of a knot or link[115]; as an example, we saw that a writhe of zero does not necessarily imply a knot's amphicheirality. The ever-present possibility of chiral zeroes, that is, $\chi(K) = 0$ even though the knot or link is topologically chiral, precludes the use of this approach.

An alternative approach that carries some promise of success is a measure of the first kind, in which the degree of topological chirality is gauged by a procedure involving "cutting and pasting" of strings in a string model.[157] These "cut-and-paste operations" are continued until an achiral knot or link is reached. The least number of such operations is then used as a measure of topological chirality. The cut-and-paste (equivalently unknotting or unlinking) operation is represented on a knot or link projection by a switch of an over- to an undercrossing that can be illustrated with reference to Figure 40 by removal of the orientations from the skewed lines, followed by a crossing switch from left to right. A topological chirality measure $\chi(K)$ can then be defined as follows: A knot or link K has a degree of chirality $\chi(K)$ if there exists a diagram of K such that changing $\chi(K)$ crossings in the diagram by the unknotting/unlinking operation turns K into an achiral knot or link, and there are no diagrams of K in which fewer than $\chi(K)$ changes would have turned K into an achiral link. This definition exactly parallels the definition of "unknotting number,"[63a] with "knot" substituted by "knot or link" and "unknot" substituted by "achiral knot or link." This measure, which has the crucial advantage that it satisfies the two minimal conditions mentioned above, remains to be implemented.

While measures of topological chirality have yet to be developed, there is no shortage of hierarchical systems in which topological constructions, including molecular graphs, are ranked as "more" or "less" chiral.[72] Although there are differences among the various systems, all agree on one point: that the "most chiral" topological object is one that is "intrinsically chiral." In mathematics, a property that is independent of local embeddings is called "intrinsic." An object that is chiral in all embeddings is therefore said to be "intrinsically chiral"; otherwise it is "extrinsically chiral" because its chirality depends on a particular embedding. Specifically, a topologically chiral knot can also be embedded in R^3 as a circle, which is topologically achiral; knots are therefore extrinsically chiral. The same is true of chiral links, which can be embedded as disjoint circles in the plane. (In another sense, of course, chirality is *always* an extrinsic property because, as noted above, an object that is chiral in R^n is achiral in R^{n+1}.) There are, however, molecular graphs that are intrinsically chiral: Erica Flapan proved that all embeddings in R^3 of the graphs of the Simmons–Paquette molecule[74] [Figure 16(a)] and of Walba's 3-rung Möbius ladder molecule[75] [Figure 16(b)] are intrinsically chiral.[158] The molecular graphs of the C_2-symmetric triple-layered naphthalenophane in Figure 16(d)[78] and of numerous topologically chiral proteins[79] also belong to this class.[72] It can therefore be claimed, with some justification, that these graphs are "more chiral" than topologically chiral molecular knots or links such as, for example, those shown in Figures 18, 31, and 32.

ACKNOWLEDGMENTS

I am much indebted to Laurence Barron, Brian Coppola, François Diederich, and Fritz Vögtle for helpful correspondence. Special thanks are due to Jaroslav Jonas for a stimulating exchange of ideas and for providing a translation of his article "Chvála Chirality" (In Praise of Chirality)[159] prior to publication. This work was supported by the National Science Foundation.

REFERENCES

1. Gardner, M.; *The Ambidextrous Universe*; Basic Books: New York, 1964; 2nd ed., Charles Scribner: New York, 1979. Gardner, M.; *The New Ambidextrous Universe*; Freeman: New York, 1990.
2. Pasteur, L.; *Ann. Chim. Phys.* **1848,** *24*, 442. See also: Pasteur, L.; *Compt. rend. Acad. Sci. (Paris)* **1848**, *26*, 535; Pasteur, L.; *Ann. Chim. Phys.* **1850**, *28*, 56; *ibid.* **1853,** *38*, 437.

3. Pasteur, L.; *Recherches sur la Dissymétrie Moléculaire des Produits Organiques Naturels*, in *Leçons de Chimie Professées en 1860*; Lib. Hachette: Paris, 1861, pp. 1–48.

4. See, for example: Collins, A. N., Sheldrake, G. N.; Crosby, J.; Eds., *Chirality in Industry*; Wiley & Sons: New York, 1992; Sheldon, R. D.; *Chirotechnology*; Marcel Dekker: New York, 1993; Crossley, R.; *Chirality and the Biological Activity of Drugs*; CRC Press: Boca Raton, 1995.

5. Kant, I.; *Von dem ersten Grunde des Unterschiedes der Gegenden im Raume* (1768), in *Kant's gesammelte Schriften*; Königl. Preuss. Akad. Wissensch., Vol. 2, Verlag Georg Reimer: Berlin, 1905, pp. 375–383.

6. Hoffmann, R.; *The Same and Not the Same*; Columbia University Press: New York, 1995.

7. Kant, I.; *Prolegomena zu einer jeden künftigen Metaphysik*; Vorländer, K., Ed.; Felix Meiner Verlag: Hamburg, 1993, pp. 38–39.

8. Van Cleve, J.; Frederick, R. E.; Eds., *The Philosophy of Right and Left: Incongruent Counterparts and the Nature of Space*; Kluwer Acad. Publ.: Dordrecht, 1991.

9. Kelvin, W. T.; *The Second Robert Boyle Lecture*, in *J. Oxford Univ. Junior Scientific Club*, No. 18, p. 25 (1894).

10. Kelvin, W. T.; *Baltimore Lectures on Molecular Dynamics and the Wave Theory of Light*; C. J. Clay: London, 1904, (a) pp. 602–642, (b) p. 439.

11. Bentley, R.; *Persp. Biol. Med.* **1995**, *38*, 188. Appendix (*On the Derivation and Definition of Chiral*) pp. 221–223.

12. Whyte, L. L.; *Nature* **1957**, *180*, 513; *ibid.* **1958**, *182*, 198. See also: Whyte, L. L.; *Leonardo* **1975**, *8*, 245.

13. Liang, C.; Mislow, K.; *J. Math. Chem.* **1994**, *15*, 1.

14. Thomson, W.; *Proc. Roy. Soc. Edin.* **1872–1875**, *8*, 70.

15. Tait, P. G.; *Trans. Roy. Soc. Edin.* **1876–1877**, *28*, 145.

16. Thomson, W.; *Phil. Mag.* **1867**, *34*, 15; Thomson, W.; *Proc. Roy. Soc. Edin.* **1875–1876**, *9*, 59.

17. Prelog, V.; *Science* **1976**, *193*, 17.

18. Ruch, E.; *Angew. Chem. Int. Ed. Engl.* **1977**, *16*, 65.

19. Stewart, I.; Gulobitsky, M.; *Fearful Symmetry*; Blackwell: Oxford, 1992, p. 37.

20. Möbius, A. F.; *Der barycentrische Calcul*; Georg Olms Verlag: Hildesheim, 1976, Part 2, Chapter 1 (*Von der Gleichheit und Aehnlichkeit*), pp. 181–185. Original text published by Verlag Johann Ambrosius Barth: Leipzig, 1827.

21. Wittgenstein, L.; *Tractatus* 6.36111, from *Tractatus Logico-Philosophicus*; Routledge & Kegan Paul: London, 1961, pp. 140–143. First published as *Logisch-philosophische Abhandlung*, in *Ann. Naturphilosophie*, 1921.

22. Berlin, I.; *The Crooked Timber of Humanity*; Fontana Press: London, 1991, p. 41.

23. Calvocoressi, R.; *Magritte*; Phaedon Press: London, 1984. Plate 30: *La Trahison des images: Ceci n'est pas une pipe*, 1929, oil on canvas.

24. Coppola, B. P.; Ege, S. N.; Lawton, R. G.; *J. Chem. Educ.* **1997**, *74*, 84.

25. Mislow, K.; Bickart, P.; *Israel J. Chem.* **1976/1977**, *15*, 1.

26. Trindle, C.; *Croat. Chem. Acta* **1984**, *57*, 1231.

27. Mislow, K.; *Chemtracts-Org. Chem.* **1989**, *2*, 151.

28. Hund, F.; *Z. Phys.* **1927**, *43*, 805.

29. For example, see: Harris, R. A.; Stodolsky, L.; *Phys. Lett.* **1978**, *78B*, 313. Harris, R. A.; Stodolsky, L.; *J. Chem. Phys.* **1981**, *74*, 2145. Mason, S. F.; Tranter, G. E.; *Chem. Phys. Lett.* **1983**, *94*, 34. Fischer, B. R.; Mittelstaedt, P.; *Phys. Lett. A* **1990**, *147*, 411. Berlin, Y. A.; Burin, A. L.; Goldanskii, V. V.; *Z. Phys. D* **1996**, *37*, 333.

30. Pasteur, L.; *Compt. rend. Acad. Sci.* **1874**, *78*, 1515. See also: Haldane, J. B. S.; *Nature* **1960**, *185*, 87.

31. Lee, T. D.; Yang, C. N.; *Phys. Rev.* **1956**, *104*, 254.

32. Latal, H.; *Parity Violation in Atomic Physics*; in Janoschek, R.; Ed., *Chirality: From Weak Bosons to the α-Helix;* Springer-Verlag: Berlin, 1991, pp. 1–17.

33. Barron, L. D.; *Chem. Phys. Lett.* **1981**, *79*, 392. Barron, L. D.; *Mol. Phys.* **1981**, *43*, 1395.

34. Barron, L. D.; *Chem. Phys. Lett.* **1986**, *123*, 423. Barron, L. D.; *J. Am. Chem. Soc.* **1986**, *108*, 5539. Barron, L. D.; *Chem. Soc. Rev.* **1986**, *15*, 189. Barron, L. D.; *BioSystems* **1987**, *20*, 7. Barron, L. D.; *Fundamental Symmetry Aspects of Molecular Chirality*; in Mezey, P. G.; Ed., *New Developments in Molecular Chirality*; Kluwer Acad. Publ.: Dordrecht, 1991, pp. 1–55. Barron, L. D.; *True and False Chirality, CP Violation, and the Breakdown of Microscopic Reversibility in Chiral Molecular and Elementary Particle Processes*; in Cline, D. B.; Ed., *Physical Origin of Homochirality in Life*; Am. Inst. Phys. Conf. Proceed. 379, Woodbury, NY, 1996, pp. 162–182.

35. Barron, L. D.; *Chem. Phys. Lett.* **1987**, *135*, 1. Barron, L. D.; *ibid.* **1994**, *221*, 311. Barron, L. D.; *Science* **1994**, *266*, 1491.

36. Barron, L. D.; *Molecular Light Scattering and Optical Activity*; Cambridge University Press: Cambridge, 1982, pp. 192–194.

37. Barron, L. D.; *Chem. Eur. J.* **1996**, *2*, 743.

38. Zocher, H.; Török, C.; *Proc. Nat. Acad. Sci. (U.S.A.)* **1953**, *39*, 681.

39. (a) Bartell, L. S.; Clippard, Jr. , F. B.; Boates, T. L.; *Inorg. Chem.* **1970**, *9*, 2436. (b) Iroff, L. D.; Mislow, K.; *J. Am. Chem. Soc.* **1978**, *100*, 2121.

40. Beissel, T.; Powers, R. E.; Raymond, K. N.; *Angew. Chem. Int. Ed. Engl.* **1996**, *35*, 1084.

41. Shubnikov, A. V.; *Symmetry and Antisymmetry in Finite Figures*; in Shubnikov, A. V.; Belov, N. V., *Colored Symmetry*; Pergamon Press: New York, 1964, pp. 34–76. See also: Shubnikov, A. V.; Koptsik, V. A.; *Symmetry in Science and Art*; Plenum Press: New York, 1974, pp. 75–78.

42. This principle seems to have been first recognized by Neumann, F. E. See: Shubnikov, A. V.; Koptsik, V. A.; *Symmetry in Science and Art*; Plenum Press: New York, 1974, p. 334. Donaldson, J. D.; Ross, S. D.; *Symmetry and Stereochemistry;* John Wiley and Sons, Inc.: New York, 1972, p. 132.

43. Curie, P.; *J. Phys. (Paris)* (3) **1894**, *3*, 393.

44. Dugundji, J.; Showell, J.; Kopp, R.; Marquarding, D.; Ugi, I.; *Israel J. Chem.* **1980**, *20*, 20. Ugi, I.; Dugundji, J.; Kopp, R.; Marquarding, D.; *Perspectives in Theoretical Stereochemistry* (*Lect. Notes Chem.* **1984**, *36*, Springer Verlag: Berlin, 1984.

45. Leonard, J. E.; Hammond, G. S.; Simmons, H. E.; *J. Am. Chem. Soc.* **1975**, *97*, 5052.

46. Longuet-Higgins, H. C.; *Mol. Phys.* **1963**, *6*, 445.

47. Mislow, K.; *Science* **1954**, *120*, 232. Mislow, K.; Bolstad, R.; *J. Am. Chem. Soc.* **1955**, *77*, 6712. Mislow, K.; *Trans. N.Y. Acad. Sci.* **1957**, *19*, 298. Mislow, K.; *Introduction to Stereochemistry*; Benjamin: New York, 1965, pp. 91–93.

48. Frei, H.; Günthard, H. H.; *Chem. Phys.* **1976**, *15*, 155. Frei, H.; Bauder, A.; Günthard, H. H.; *Top. Curr. Chem.* **1979**, *81*, 1.

49. Jaeger, F. M.; *Bull. Soc. Chim. Fr.* **1923**, *33*, 853. See also: Jaeger, F. M.; *Spatial Arrangements of Atomic Systems and Optical Activity*; McGraw-Hill Book Co.: New York, 1930.

50. (a) Cahn, R. S.; Ingold, C. K.; Prelog, V.; *Experientia* **1956**, *12*, 81. Cahn, R. S.; Ingold, C. K.; Prelog, V.; *Angew. Chem.* **1966**, *78*, 413; *Angew. Chem. Int. Ed. Engl.* **1966**, *5*, 385. (b) Prelog, V.; Helmchen, G.; *Angew. Chem.* **1982**, *94*, 614; *Angew. Chem. Int. Ed. Engl.* **1982**, *21*, 567.

51. Mislow, K.; Siegel, J.; *J. Am. Chem. Soc.* **1984**, *106*, 3319.

52. Mislow, K.; *Bull. Soc. Chim. Belg.* **1977**, *86*, 595.

53. Cornforth, J. W.; Redmont, J. W.; Eggerer, H.; Buckel, W.; Gutschow, C.; *Nature* (*London*) **1969**, *221*, 1212. Luthy, J.; Retey, J.; Arigoni, D.; *ibid.* **1969**, *221*, 1213.

54. Reisse, J.; Ottinger, R.; Bickart, P.; Mislow, K.; *J. Am. Chem. Soc.* **1978**, *100*, 911.

55. Schippers, P. H.; Dekkers, H. P. J. M.; *J. Am. Chem. Soc.* **1983**, *105*, 145.

56. Miesen, F. W. A. M.; Wollersheim, A. P. P.; Meskers, S. C. J.; Dekkers, H. P. J. M.; Meijer, E. W.; *J. Am. Chem. Soc.* **1994**, *116*, 5129.

57. Mislow, K.; Raban, M.; *Top. Stereochem.* **1967**, *1*, 1.

58. Kroto, H. W.; *Nature (London)* **1987**, *329*, 529. The C_3 symmetry reported in that paper should be changed to D_3 symmetry (H. W. Kroto, personal communication).

59. Ettl, R.; Chao, I.; Diederich, F.; Whetten, R. L.; *Nature (London)* **1991**, *353*, 149.

60. Hawkins, J. M.; Meyer, A.; *Science* **1993**, *260*, 1918. Hawkins, J. M.; Nambu, M.; Meyer, A.; *J. Am. Chem. Soc.* **1994**, *116*, 7642.

61. Boo, W. O. J.; *J. Chem. Educ.* **1992**, *69*, 605.

62. Crowell, R. H.; Fox, R. H.; *Introduction to Knot Theory*; Springer-Verlag: New York, 1963. Livingston, C.; *Knot Theory*; Mathematical Association of America: Washington, DC, 1993.

63. (a) Adams, C. C.; *The Knot Book: An Elementary Introduction to the Mathematical Theory of Knots;* W. H. Freeman and Company: New York, 1994. (b) Vassiliev, V. A.; *Cohomology of Knot Spaces*, in *Theory of Singularities and its Applications*; Arnold, V. I., Ed.; Amer. Math Soc.: Providence, RI, 1990, pp. 23–69.

64. Flapan, E.; *Topological Techniques to Detect Chirality*; in Mezey, P. G.; Ed., *New Developments in Molecular Chirality*; Kluwer Acad. Publ.: Dordrecht, 1991, pp. 209–239.
65. Listing, J. B.; *Vorstudien zur Topologie. Göttinger Studien 1847*; Vandenhoeck and Ruprecht: Göttingen, 1848, pp. 3–68.
66. Dehn, M.; *Math. Ann.* **1914**, *75*, 402.
67. Wilson, R. J.; *Introduction to Graph Theory*; Academic Press: New York, 1979. Harary, F.; *Graph Theory*; Addison-Wesley: Reading, MA, 1972.
68. Adachi, K.; Naemura, K.; Nakazaki, M.; *Tetrahedron Lett.* **1968**, 5467.
69. Nakazaki, M.; Naemura, K.; Chikamatsu, H.; Iwasaki, M.; Hashimoto, M.; *Chem. Lett.* **1980**, 1571.
70. Nakazaki, M.; Yamamoto, K.; Tanaka, S.; Kametani, H.; *J. Org. Chem.* **1977**, *42*, 287.
71. Mislow, K.; *Croat. Chem. Acta* **1996**, *69*, 485.
72. Liang, C.; Mislow, K.; *J. Math. Chem.* **1994**, *15*, 245.
73. (a) Walba, D. M.; *Tetrahedron* **1985**, *41*, 3161, and references therein. (b) Walba, D. M.; *Stereochemical Topology*; in King, R. B.; Ed., *Chemical Applications of Topology and Graph Theory*; Elsevier: Amsterdam, 1983, pp. 17–32.
74. Simmons III, H. E.; Maggio, J. E.; *Tetrahedron Lett.* **1981**, *229*, 287. Paquette, L. A.; Vazeux, M.; *ibid.* **1981**, *229*, 291.
75. Walba, D. M.; Richards, R. M.; Haltiwanger, R. C.; *J. Am. Chem. Soc.* **1982**, *104*, 3219.
76. Hisatome, M.; Watanabe, N.; Sakamoto, T.; Yamakawa, K.; *J. Organomet. Chem.* **1977**, *125*, 79.
77. (a) Simon, J.; *J. Comput. Chem.* **1987**, *8*, 718. (b) Simon, J.; *Topology* **1986**, *25*, 229. (c) Wolcott, K.; Ph.D. thesis, University of Iowa, 1986.
78. Otsubo, T.; Ogura, F.; Misumi, S.; *Tetrahedron Lett.* **1983**, *24*, 4851.
79. Liang, C.; Mislow, K.; *J. Am. Chem. Soc.* **1994**, *116*, 3588.
80. Mislow, K.; Liang, C.; *Croat. Chem. Acta* **1996**, *69*, 1385.
81. Ambs, W. J.; *Mendel Bull.* **1953**, *17*/Spring, 26.
82. Dietrich-Buchecker, C. O.; Sauvage, J.-P.; *Angew. Chem. Int. Ed. Engl.* **1989**, *28*, 189. Dietrich-Buchecker, C. O.; Guilhem, J.; Pascard, C.; Sauvage, J.-P.; *ibid.* **1990**, *29*, 1154. Sauvage, J.-P.; *Acc. Chem. Res.* **1990**, *23*, 319. Dietrich-Buchecker, Ch.; Sauvage, J.-P.; *New J. Chem.* **1992**, *16*, 277. Dietrich-Buchecker, C. O.; Sauvage, J.-P.; Kintzinger, J.-P.; Maltèse, P.; Pascard, C.; Guilhem, J.; *ibid.* **1992**, *16*, 931. Dietrich-Buchecker, C.; Sauvage, J.-P.; *Bull. Soc. Chim. Fr.* **1992**, *129*, 113. Dietrich-Buchecker, C. O.; Sauvage, J.-P.; *Synthetic Molecular Knots*; in Balzani, V.; De Cola, L.; Eds., *Supramolecular Chemistry*; Kluwer Acad. Publ.: Dordrecht, 1992, pp. 259–277.
83. Rapenne, G.; Dietrich-Buchecker, C.; Sauvage, J.-P.; *J. Am. Chem. Soc.* **1996**, *118*, 10932.

84. (a) Wasserman, E.; *J. Am. Chem. Soc.* **1960**, *82*, 4433. Wasserman, E.; *Sci. Amer.* **1962**, *207*/5, 94. (b) Frisch, H. L.; Wasserman, E.; *J. Am. Chem. Soc.* **1961**, *83*, 3789. See also: van Gulick, N.; *New J. Chem.* **1993**, *17*, 619 [manuscript originally submitted to *Tetrahedron*, August 1960].

85. Simon, J.; *Proc. Symp. Appl. Math.* **1992**, *45*, 97. Simon, J.; *A Topological Approach to the Stereochemistry of Nonrigid Molecules*; in King, R. B.; Rouvray, D. H.; Eds., *Graph Theory and Topology in Chemistry*; Elsevier: Amsterdam, 1987, pp. 43–75.

86. Sokolov, V. I.; *Russ. Chem. Revs.* **1973**, *42*, 452.

87. (a) Walba, D. M.; *Topological Stereochemistry: Knot Theory of Molecular Graphs*; in King, R. B.; Rouvray, D. H.; Eds., *Graph Theory and Topology in Chemistry*; Elsevier: Amsterdam, 1987, pp 23–42. (b) Walba, D. M.; *A Topological Hierarchy of Molecular Chirality and Other Tidbits in Topological Stereochemistry*; in Mezey, P. G.; Ed., *New Developments in Molecular Chirality*; Kluwer Acad. Publ.: Dordrecht, 1991, pp. 119–129.

88. Seeman, N. C.; Chen, J.; Du, S. M.; Mueller, J. E.; Zhang, Y.; Fu, T.-J.; Wang, Y.; Wang, H.; Zhang, S.; *New J. Chem.* **1993**, *17*, 739.

89. Mueller, J. E.; Du, S. M.; Seeman, N. C.; *J. Am. Chem. Soc.* **1991**, *113*, 6306. Seeman, N. C.; *Mol. Eng.* **1992**, *2*, 297. Du, S. M.; Seeman, N. C.; *J. Am. Chem. Soc.* **1992**, *114*, 9652. Wang, H.; Du, S. M.; Seeman, N. C.; *J. Biomol. Struct. Dyn.* **1993**, *10*, 853. Du, S. M.; Seeman, N. C.; *Biopolymers* **1994**, *34*, 31. Du, S. M.; Wang, H.; Tse-Dinh, Y.-C.; Seeman, N. C.; *Biochemistry* **1995**, *34*, 673. Du, S. M.; Stollar, B. D.; Seeman, N. C.; *J. Am. Chem. Soc.* **1995**, *117*, 1194.

90. Liu, L. L.; Depew, R. E.; Wang, J. C.; *J. Mol. Biol.* **1976**, *106*, 439.

91. (a) Dean, F. B.; Stasiak, A.; Koller, T.; Cozzarelli, N. R.; *J. Biol. Chem.* **1985**, *260*, 4975. Sumners, D. W.; *The Role of Knot Theory in DNA Research*; in McCrory, C.; Schifrin, T.; Eds., *Geometry and Topology*; Marcel Dekker: New York, 1987, pp. 297–318. (b) Dietrich-Buchecker, C. O.; Sauvage, J.-P.; *Interlocked and Knotted Rings in Biology and Chemistry*; in Dugas H.; Ed., *Bioorganic Chemistry Frontiers*; Springer-Verlag: Berlin, 1991, Vol. 2, pp. 195–248. Bates, A. D.; Maxwell, A.; *DNA Topology*; Oxford University Press: New York, 1993.

92. Wang, J. C.; *Sci. Amer.* **1982**, *247/7*, 94. Krasnow, M. A.; Stasiak, A.; Spengler, S. J.; Dean, F.; Koller, T.; Cozzarelli, N. R.; *Nature* **1983**, *304*, 559. Wasserman, S. E.; Cozzarelli, N. R.; *Proc. Nat. Acad. Sci. U. S. A.* **1985**, *82*, 1079. Wasserman, S. E.; Cozzarelli, N. R.; *Science* **1986**, *232*, 951, and references therein. White, J. H.; Millett, K. C.; Cozzarelli, N. R.; *J. Mol. Biol.* **1987**, *197*, 595.

93. Liang, C.; Mislow, K.; *J. Am. Chem. Soc.* **1995**, *117*, 4201.

94. Carina, R. F.; Dietrich-Buchecker, C.; Sauvage, J.-P.; *J. Am. Chem. Soc.* **1996**, *118*, 9110.

95. Thistlethwaite, M. B.; personal communication.

96. Robinson, R. W.; Harary, F.; Balaban, A. T.; *Tetrahedron* **1976**, *32*, 355.

97. Burde, G.; Zieschang, H.; *Knots*; Walter de Gruyter: Berlin, 1985, Appendix C: Tables, pp. 311–343.

98. Flapan, E.; *Pac. J. Math.* **1987**, *129*, 57.

99. Flapan, E.; Seeman, N. C.; *J. Chem. Soc., Chem. Commun.* **1995**, 2249.

100. Jones, V. F. R.; *Bull. Amer. Math. Soc.* **1985**, *12*, 103.

101. Millett, K. C.; *Croat. Chem. Acta* **1986**, *59*, 669. Millett, K. C.; *Algebraic topological indices of molecular chirality*; in Mezey, P. G.; Ed., *New Developments in Molecular Chirality*; Kluwer Acad. Publ.: Dordrecht, 1991, pp. 165–207. Freyd, P.; Yetter, D.; Hoste, J.; Lickorish, W. B. R.; Millett, K.; Ocneanu, A.; *Bull. Amer. Math. Soc.* **1985**, *12*, 239. Lickorish, W. B. R.; Millett, K. C.; *Topology* **1987**, *26*, 107. Lickorish, W. B. R.; Millett, K. C.; *Math. Mag.* **1988**, *61*, 3.

102. Kauffman, L. H.; *Trans. Amer. Math. Soc.* **1990**, *318*, 417. Kauffman, L. H.; *Amer. Math. Monthly* **1988**, *95*, 195. Kauffman, L. H.; *On Knots*; Princeton University: Princeton, NJ, 1987, pp. 444–473.

103. (a) Schill, G.; Zürcher, C.; *Chem. Ber.* **1977**, *110*, 2046. Schill, G.; Rissler, K.; Fritz, H.; Vetter, W.; *Angew. Chem. Int. Ed. Engl.* **1981**, *20*, 187. (b) Sauvage, J.-P.; Weiss, J.; *J. Am. Chem. Soc.* **1985**, *107*, 6108.

104. Amabilino, D. B.; Ashton, P. R.; Reder, A. S.; Spencer, N.; Stoddart, J. F.; *Angew. Chem. Int. Ed. Engl.* **1994**, *33*, 433, 1286.

105. Dietrich-Buchecker, C. O.; Sauvage, J.-P.; Kern, J.-M.; *J. Am. Chem. Soc.* **1984**, *106*, 3043.

106. Liang, C.; Mislow, K.; *J. Math. Chem.* **1994**, *16*, 27.

107. Coxeter, H. S. M.; *Introduction to Geometry*; Wiley: New York, 1969, pp. 162–163.

108. Lindström, B.; Zetterström, H.-O.; *Amer. Math. Monthly* **1991**, *98*, 340.

109. Brown, R.; Robinson, J.; *Amer. Math. Monthly* **1992**, *99*, 376. See also: Brown, R.; *The Mathematical Intelligencer* **1994**, *16*, 62. Burgiel, H.; Franzblau, D. S.; Gutschera, K. R.; *Math. Mag.* **1996**, *69*, 94.

110. Cromwell, P.; *The Mathematical Intelligencer* **1995**, *17*, 3. See also: Gundarsson, K.; *Teutonic Religion*; Llewellyn Publ.: St. Paul, MN, 1993, pp. 36–37. Wilson, D. M.; Klindt-Jensen, O.; *Viking Art*; University of Minnesota Press: Minneapolis, MN, second ed., 1980, Plate XVII.

111. Gardner, M.; *The Unexpected Hanging and Other Mathematical Diversions;* Simon and Schuster: New York, 1969, Chapter 2 (*Knots and Borromean Rings*), pp. 24–33.

112. Mao, C.; Sun, W.; Seeman, N. C.; *Nature* **1997**, *386*, 136.

113. Chambron, J.-C.; Sauvage, J. -P.; Mislow, K.; *J. Am. Chem. Soc.* **1997**, *119*, 9558.

114. Nierengarten, J.-F.; Dietrich-Buchecker, C. O.; Sauvage, J.-P.; *J. Am. Chem. Soc.* **1994**, *116*, 375.

115. Liang, C.; Mislow, K.; *J. Math. Chem.* **1995**, *18*, 1.

116. Piguet, C.; Bernardinelli, G.; Williams, A. F.; Bocquet, B.; *Angew. Chem. Int. Ed. Engl.* **1995**, *34*, 582.

117. Ashton, P. R.; Iriepa, I.; Reddington, M. V.; Spencer, N.; Slawin, A. M. Z.; Stoddart, J. F.; Williams, D. J.; *Tetrahedron Lett.* **1994**, *35*, 4835.

118. Chambron, J.-C.; Mitchell, D. K.; Sauvage, J.-P.; *J. Am. Chem. Soc.* **1992**, *114*, 4625. See also: Mitchell, D. K.; Sauvage, J.-P.; *Angew. Chem. Int. Ed. Engl.* **1988**, *27*, 930.

Kaida, Y.; Okamoto, Y.; Chambron, J.-C.; Mitchell, D. K.; Sauvage, J.-P.; *Tetrahedron Lett.* **1993**, *34*, 1019.

119. Tauber, S. J.; *J. Res. Nat. Bur. Stan.* **1963**, *67A*, 591.

120. Hudson, B.; Vinograd, J.; *Nature* **1967**, *216*, 647. Clayton, D. A.; Vinograd, J.; *Nature* **1967**, *216*, 652.

121. Wasserman, S. A.; Dungan, J. M.; Cozzarelli, N. R.; *Science* **1985**, *229*, 171.

122. Chen, J.; Seeman, N. C.; *Nature* **1991**, *350*, 631.

123. Zhang, Y.; Seeman, N. C.; *J. Am. Chem. Soc.* **1994**, *116*, 1661.

124. Chen, C.-T.; Gantzel, P.; Siegel, J. S.; Baldridge, K. K.; English, R. B.; Ho, D. M.; *Angew. Chem. Int. Ed. Engl.* **1995**, *34*, 2657.

125. For example, see: (a) Jackson, J. E.; Allen, L. C.; *J. Am. Chem. Soc.* **1984**, *106*, 591. (b) Wiberg, K. B.; Bader, R. F. W.; Lau, C. D. H.; *J. Am. Chem. Soc.* **1987**, *109*, 985. (c) Feller, D.; Davidson, E. R.; *J. Am. Chem. Soc.* **1987**, *109*, 4133. (d) Wiberg, K. B.; *Chem. Rev.* **1989**, *89*, 975. For a recent analysis, see: (e) Adcock, W.; Brunger, M. J.; Clark, C. I.; McCarthy, I. E.; Michalewicz, M. T.; von Niessen, W.; Weigold, E.; Winkler, D. A.; *J. Am. Chem. Soc.* **1997**, *119*, 2896.

126. Einstein, A. ("Alles sollte so einfach wie möglich gemacht werden, aber nicht einfacher" [Everything should be made as simple as possible, but not simpler]), in Quadbeck-Seeger, H.-J., *Zwischen den Zeichen. Aphorismen über und aus Natur und Wissenschaft;* VCH: Weinheim, 1988, p. 71.

127. Rouvray, D. H.; *Are the Concepts of Chemistry All Fuzzy?*; in Rouvray, D. H.; Ed., *Concepts in Chemistry: A Contemporary Challenge*; Wiley & Sons: New York, 1997, Chapter 1.

128. Bijvoet, J. M.; Peerdeman, A. F.; van Bommel, A. J.; *Nature* **1951**, *168*, 271.

129. Carroll, L.; *Alice's Adventures in Wonderland & Through the Looking Glass*; in Gardner, M., *The Annotated Alice*; New American Library: 1960, pp. 268–271.

130. Mislow, K.; *Fuzzy Restrictions and Inherent Uncertainties in Chirality Studies*; in Rouvray, D. H.; Ed., *Fuzzy Logic in Chemistry*; Academic Press: San Diego, CA, 1997, pp. 65–90.

131. Buda, A. B.; Mislow, K.; *J. Am. Chem. Soc.* **1992**, *114*, 6006.

132. Mislow, K.; Poggi-Corradini, P.; *J. Math. Chem.* **1993**, *13*, 209. See also: Weinberg, N.; Mislow, K.; *Theor. Chim. Acta* **1997**, *95*, 63.

133. (a) Ruch, E.; Ugi, I.; *Theor. Chim. Acta* **1966**, *4*, 287. Ruch, E.; Schönhofer, A.; Ugi, I.; *ibid.* **1967**, *7*, 420. Ruch, E.; Schönhofer, A.; *ibid.* **1968**, *10*, 91. Ruch, E.; Ugi, I.; *Top. Stereochem.* **1969**, *4*, 99. Ruch, E.; Schönhofer, A.; *Theor. Chim. Acta* **1970**, *19*, 225. Ruch, E.; *Acc. Chem. Res.* **1972**, *5*, 49. Mead, A.; Ruch, E.; Schönhofer, A.; *Theor. Chim. Acta* **1973**, *29*, 269. (b) Ruch, E.; *Theor. Chim. Acta* **1968**, *11*, 183.

134. Salem, L.; Durup, J.; Bergeron, G.; Cazes, D.; Chapuisat, X.; Kagan, H.; *J. Am. Chem. Soc.* **1970**, *92*, 4472. Salem, L.; *Acc. Chem. Res.* **1971**, *4*, 322.

135. Wolfe, S.; Schlegel, H. B.; Csizmadia, I. G.; Bernardi, F.; *J. Am. Chem. Soc.* **1975**, *97*, 2020.

136. Gust, D.; Mislow, K.; *J. Am. Chem. Soc.* **1973**, *95*, 1535. Mislow, K.; *Acc. Chem. Res.* **1976**, *9*, 26. Glaser, R.; Blount, J. F.; Mislow, K.; *J. Am. Chem. Soc.* **1980**, *102*, 2777.

137. (a) Hounshell, W. D.; Johnson, C. A.; Guenzi, A.; Cozzi, F.; Mislow, K.; *Proc. Natl. Acad Sci. U.S.A.* **1980**, *77*, 6961. Guenzi, A.; Johnson, C. A.; Cozzi, F.; Mislow, K.; *J. Am. Chem. Soc.* **1983**, *105*, 1438; Iwamura, H.; Mislow, K.; *Acc. Chem. Res.* **1988**, *21*, 175. (b) Mauksch, M.; Schleyer, P. v. R.; *Angew. Chem. Int. Ed. Engl.* **1997**, *36*, 1856.

138. Fischer, E.; Brauns, F.; *Ber. d. D. Chem. Gesellsch.* **1914**, *47*, 3181.

139. Eliel, E. L.; *Stereochemistry of Carbon Compounds*; McGraw Hill: New York, 1962, p. 18. Eliel, E. L.; Wilen, S. H.; *Stereochemistry of Carbon Compounds*; John Wiley & Sons: New York, 1994, p. 61.

140. Damhus, T.; Schäffer, C. E.; *Inorg. Chem.* **1983**, *22*, 2406.

141. Jäger, R.; Vögtle, F.; *Angew. Chem. Int. Ed. Engl.* **1997**, *36*, 930.

142. Yamamoto, C.; Okamoto, Y.; Schmidt, T.; Jäger, R.; Vögtle, F.; *J. Am. Chem. Soc.* **1997**, *119*, 10547.

143. Murasugi, K.; *Topology* **1987**, *26*, 187. Thistlethwaite, M. B.; *ibid.* **1988**, *27*, 311.

144. Rolfsen, D.; *Knots and Links*; Publish or Perish: Berkeley, 1976; second printing with corrections: Publish or Perish: Houston, 1990, Appendix C: Table of knots and links, pp. 388–429.

145. Liang, C.; Mislow, K.; *J. Math. Chem.* **1994**, *15*, 35. Liang, C.; Cerf, C.; Mislow, K.; *ibid.* **1996**, *19*, 241.

146. Thilgen, C.; Herrmann, A.; Diederich, F.; *Helv. Chim. Acta* **1997**, *80*, 183.

147. Buda, A. B.; Auf der Heyde, T.; Mislow, K.; *Angew. Chem. Int. Ed. Engl.* **1992**, *31*, 989.

148. Weinberg, N.; Mislow, K.; *J. Math. Chem.* **1995**, *17*, 35.

149. Schlenk, Jr., W.; *Angew. Chem. Int. Ed. Engl.* **1965**, *4*, 139.

150. Morrison, J. D.; Mosher, H. S.; *Asymmetric Organic Reactions*; Prentice Hall: Englewood Cliffs, 1971, p. 5.

151. Buda, A. B.; Auf der Heyde, T. P. E.; Mislow, K.; *J. Math. Chem.* **1991**, *6*, 243.

152. Auf der Heyde, T. P. E.; Buda, A. B.; Mislow, K.; *J. Math. Chem.* **1991**, *6*, 255.

153. Buda, A. B.; Mislow, K.; *J. Mol. Struct. (Theochem)* **1991**, *232*, 1. Buda, A. B.; Mislow, K.; *Elem. Mathematik* **1991**, *46*, 65.

154. Weinberg, N.; Mislow, K.; *J. Math. Chem.* **1993**, *14*, 427.

155. Brewster, J. H.; *Top. Current Chem.* **1974**, *47*, 29.

156. Guye, P.-A.; *C. R. Acad. Sci. (Paris)* **1890**, *110*, 714. See also: Crum Brown, A.; *Proc. Roy. Soc. Edin.* **1890**, *17*, 181. Guye, P.-A.; *C. R. Acad. Sci. (Paris)* **1893**, *116*, 1378, 1451, 1454.

157. Mislow, K.; Cerf, C.; Liang, C. Unpublished results.

158. Flapan, E.; *Chirality of Non-Standardly Embedded Möbius Ladders*; in King, R. B.; Rouvray, D. H.; Eds., *Graph Theory and Topology in Chemistry*; Elsevier: Amsterdam, 1987, pp. 76–81. Flapan, E.; *Math. Ann.* **1989**, *283*, 271. Flapan, E.; *New*

J. Chem. **1993**, *17*, 645. Flapan, E.; *J. Mol. Struct.* (*Theochem*) **1995**, *336*, 157. Flapan, E.; Weaver, N.; *J. Combinat. Theory* **1996**, *B68*, 223.

159. Jonas, J.; *Chem. Listy* **1996**, *90*, 410 (in Czech); *Chem. Papers* **1997**, *51(3)*, 167 (in English)

Stereoselective Reactions with Catalytic Antibodies

DONALD HILVERT

Department of Chemistry, The Swiss Federal Institute of Technology (ETH), Universitätstrasse 16, CH-8092 Zürich, Switzerland

I. INTRODUCTION

Enzymes are, in many ways, ideal asymmetric catalysts. They are inherently chiral and typically provide exacting control over the regio- and stereochemical course of the reactions they promote. Exceptional efficiency and mild and environmentally benign reaction conditions are additional bonuses. It is therefore no surprise that enzymes are being used increasingly in organic synthesis for the production of diverse natural products,[1, 2] and the synthetic challenges posed by

Topics in Stereochemistry, Volume 22, Edited by Scott E. Denmark.
ISBN 0-4711-25316-2

structurally complex and biologically important substances such as carbohydrates, nucleic acids, and proteins offer many exciting opportunities for biocatalysis in the future.

Currently, more than 2000 enzymes are known, and the number continues to grow as the complete genomic sequences of familiar and exotic organisms become available. These catalysts promote a wide range of interesting chemical transformations, from simple hydrolyses to carbon–carbon bond-forming reactions. Several hundred of these proteins are commercially available, and many others are accessible with the tools of molecular biology. Nevertheless, natural enzymes may not exist for every transformation of interest, and the narrow substrate specificity and limited stability of many existing enzymes represent significant disadvantages for the chemist seeking general synthetic tools.

Engineered protein catalysts have the potential to circumvent some of these limitations. Approaches for tailoring the properties of existing enzymes or for designing new catalysts from first principles are now being developed in many laboratories. One promising strategy for catalyst design exploits the specificity and diversity of the mammalian immune system to create antibody molecules with catalytic properties.[3,4] The microevolutionary principles that govern the maturation of the immune response to a foreign antigen—together with modern hybridoma methods—enable the routine production of antibodies capable of high-affinity recognition of virtually any molecule.[5–7] When the foreign antigen is a small molecule (more properly called a hapten) that mimics the transition state of a particular chemical transformation, antibodies configured for catalysis can result.[3,4] In practice, this has proved to be a powerful and versatile approach for creating catalysts for a broad array of chemical reactions on a laboratory time scale.

Catalytic antibodies, like natural enzymes, are well suited for asymmetric synthesis. The ability of antibodies to discriminate between closely related configurational or stereochemical isomers has been recognized since the pioneering work of Landsteiner.[8] Extensive shape and chemical complementarity between the binding pocket and its preferred ligand [9, 10] means that enormous binding energies (between 10 and 20 kcal/mol) are potentially available for molecular recognition, only a fraction of which would be necessary to achieve preferential stabilization of a single enantiomeric or diastereomeric transition state. Furthermore, the programmable nature of the antibody binding site allows both mechanism and selectivity to be specified a priori through appropriate design of the hapten. As a consequence, transformations that cannot be achieved efficiently or selectively with available chemical methods are particularly attractive targets of investigation.[11]

This review surveys the progress toward tailoring the selectivity of diverse chemical transformations that has been made in the decade since catalytic antibodies were first described. General considerations relevant to practical applications of this technology—including limitations and strategies for overcoming them—are discussed in the concluding section.

II. ASYMMETRIC REACTIONS

A. Acyl Transfer

1. *Ester Hydrolysis*

Hydrolytic enzymes are widely used in organic synthesis as a consequence of their ability to form ester and amide bonds with high stereoselectivity under mild conditions. Esterases, lipases, and amidases, for example, are frequently exploited as catalysts for stereoselective acylations and kinetic resolutions of alcohols and amines.[1] Enzymatic coupling of unprotected peptide fragments, on the other hand, is being developed for the assembly of large synthetic proteins and other oligopeptides.[12] Because hydrolytic antibodies with tailored specificities could significantly extend the scope of these naturally occurring enzymes, their development has been the focus of much research.[4]

Ester hydrolysis (acyl transfer to water) has been investigated in particular detail. This reaction is energetically less demanding than amide cleavage, and its mechanism is well understood. In addition, many potent inhibitors of natural esterases are known. Tetrahedral transition-state analogs in which a phosphorous or a secondary alcohol is used to mimic the tetrahedral geometry of the transition state for hydrolysis have proved especially reliable as haptens for preparing catalytic antibodies. Well over 50 esterolytic antibodies have been generated to date using phosphonate or phosphonamidate haptens alone.[4] They typically display classical Michaelis–Menten kinetics, rate accelerations of up to 10^6-fold, and substrate specificity mirroring the structure of the original hapten. When the hapten contains stereogenic centers, high levels of stereospecificity can also be attained.

The development of lipaselike antibodies illustrates the potential of antibody catalysis for promoting reactions that require stereochemical control. Lerner and co-workers employed the racemic phosphonate hapten **1** to generate antibodies that catalyze the cleavage of unactivated esters like **2** with rate enhancements in the range 10^3–10^5 over background.[13] Even though the catalytic efficiencies of the antibodies do not approach those of natural lipases (which can be as high as 10^{10} over background), the reactions are highly selective. Of particular note is the fact that the racemic hapten yielded antibodies that exclusively hydrolyze either the (*R*) or (*S*) isomer of **2**. For the two catalysts studied in detail—the (*R*)-specific antibody 2H6 and the (*S*)-specific antibody 21H3—hydrolysis of the preferred stereoisomer was catalyzed at a rate >50 times that of its antipode. These high

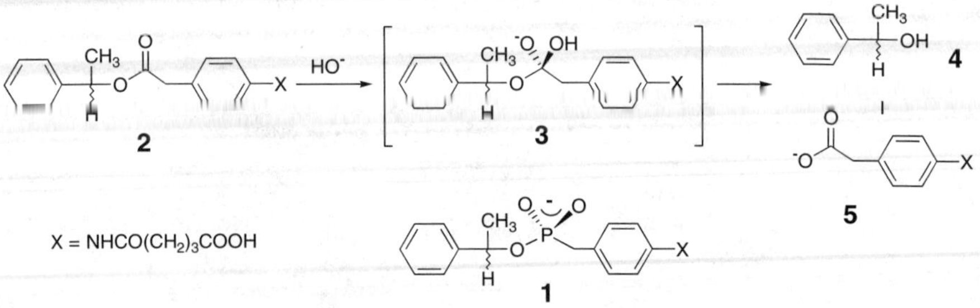

enantioselectivities are more than sufficient for practical resolutions of chiral alcohols. Conceivable applications of such catalysts, provided they could be prepared inexpensively on a large scale, might include deprotection of orthogonal protecting groups differing only in their chirality.

Racemic haptens have the advantage of synthetic expedience, but they do not always yield (*R*)- and (*S*)-specific catalysts in a single experiment. The eighteen catalytic antibodies obtained with the tripeptide phosphonate **6**,[14] for example, only catalyzed the hydrolysis of substrates (**7**) containing D-phenylalanine at the cleavage site. Depsipeptides containing leucine or tryptophan in place of phenylalanine were not hydrolyzed. The selectivity for D- over L-phenylalanine-containing substrates was at least 40 to 1 and, for three of the five catalysts characterized in detail, the D isomer was cleaved more than 200 times faster than the corresponding L isomer. To account for this preference, Schultz and co-workers postulated that peptides containing analogs of noncoded D-amino acid analogs might be more immunogenic than those containing the L-isomers. However, broad extrapolations from this conclusion are probably premature, given

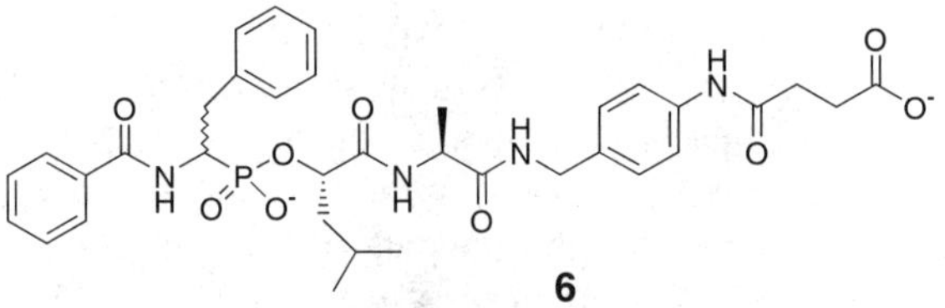

6

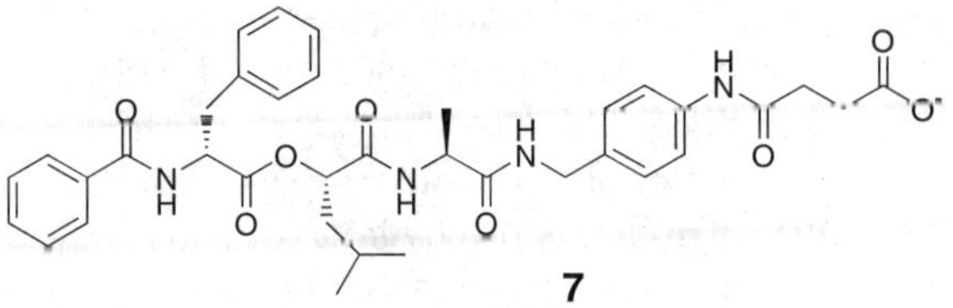

7

that only a single immunization was performed and a relatively small panel of antibodies tested.

To ensure that antibodies with the desired selectivity are obtained, enantiomerically pure haptens can be used to elicit the immune response. This was the approach taken by Kitazume and colleagues to prepare antibody catalysts for the stereoselective synthesis of fluorinated esters.[15] Because antibodies raised against racemic **8** were insufficiently discriminating to be useful for asymmetric synthesis, the *R* and *S* enantiomers were used individually as haptens. The resulting antibodies were highly stereoselective, promoting the hydrolysis of racemic fluoromethylated esters like **9** to give 1-(fluoroalkyl)-alkanols in > 90% yield and 98% ee. Resolution of diastereomeric mixtures by antibodies prepared individually with the four possible stereoisomers of **10** and **11** was equally successful.[16] Moreover, these antibody-mediated reactions were found to be superior to alternative methods using commercially available enzymes.

Enantioselection with catalytic antibodies is also possible, as demonstrated by the enantioselective hydrolysis of meso diester **12** to (1*R*,4*S*)-(+)-4-hydroxy-2-cyclopentenyl acetate **13** (> 98% ee) in the presence of an antibody raised against phosphonate **14**.[17] In other applications, anti-phosphonate antibodies have been developed for the resolution of glycerol derivatives,[18] the degradation of cocaine to the inactive metabolite ecgonine methyl ester,[19, 20] and for the conversion of biologically inactive esters of chloramphenical and 5-fluorodeoxyuridine into their bioactive forms.[21–24] Extrapolating from the foregoing examples, high enantioselectivity (diastereoselectivity) is likely to be a general feature of these and other antibody-catalyzed hydrolytic reactions.

A flurry of recent structural studies on esterolytic antibodies is providing valuable insight into the structural origins of the activity and selectivity of the induced binding pockets.[25–29] The catalyst 17E8, prepared by Scanlan and co-workers using the norleucine-based arylphosphonate transition-state analog **15** as hapten,[30] is a case in point (Figure 1). Antibody 17E8 accelerates the

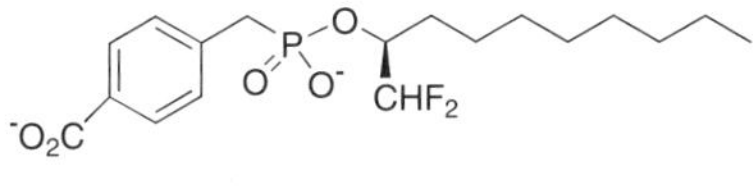

8

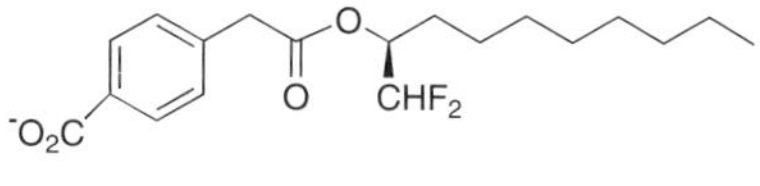

9

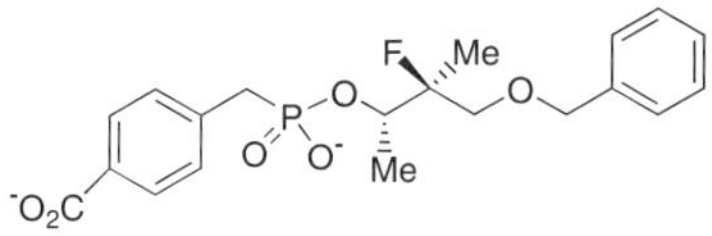

10

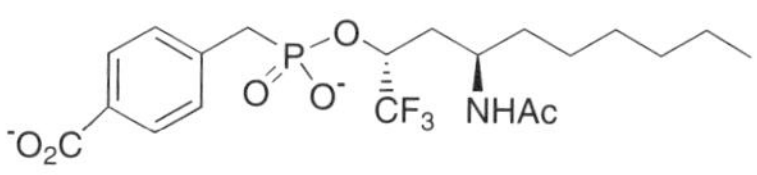

11

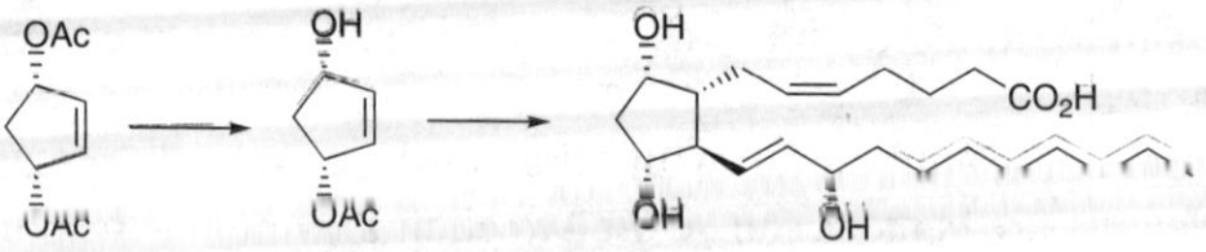

12 13 Prostaglandin $F_{2\alpha}$

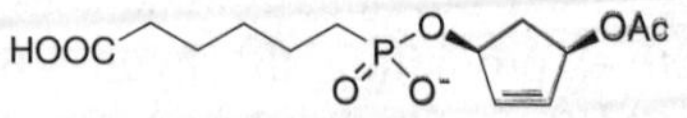

14

enantioselective hydrolysis of unactivated phenyl esters of *N*-formyl-L-norleucine (**16a**, X = CH_2) and N-formyl-L-methionine (**16b**, X = S) with rate accelerations of $\approx 10^4$ over background; hydrolysis of the corresponding D enantiomers was not detectably catalyzed. Not surprisingly, the active site of the antibody, as revealed by X-ray crystallography,[26] is complementary to the L-enantiomer of phosphonate **15** but not to the D-enantiomer. The carboxylate-containing linker arm used as an attachment site to the carrier protein lies at the entrance of the binding cleft, while the aryl group and the hydrophobic *n*-butyl side chain of **15** bind deeply in adjacent apolar pockets formed at the interface of the antibody heavy and light chains. This binding mode positions the phosphonate moiety for favorable electrostatic and hydrogen bonding interactions with several polar residues in the active site, and suggests a simple mechanism in which the antibody facilitates direct hydroxide attack on the substrate by electrostatically stabilizing the oxyanionic intermediate and flanking transition states.[31] Hydrophobic interactions with the pro-*S* Cα

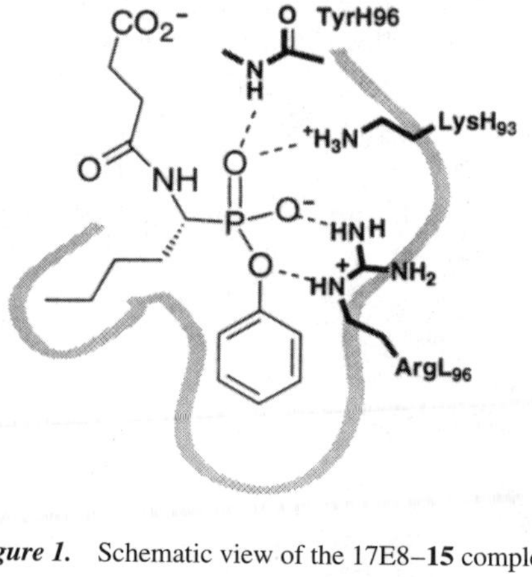

Figure 1. Schematic view of the 17E8–**15** complex

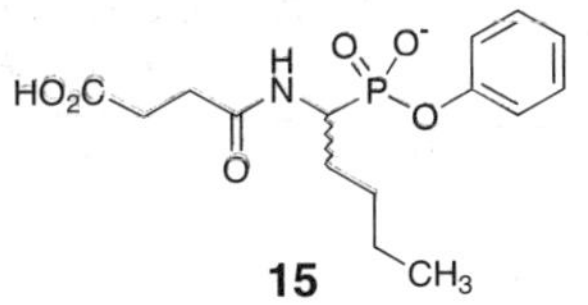

15

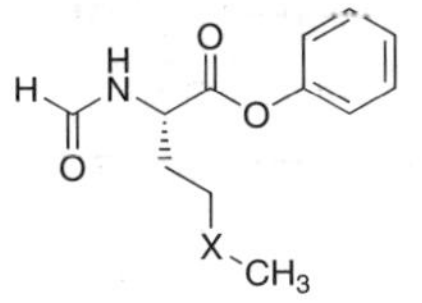

16a, X=CH_2
16b, X=S

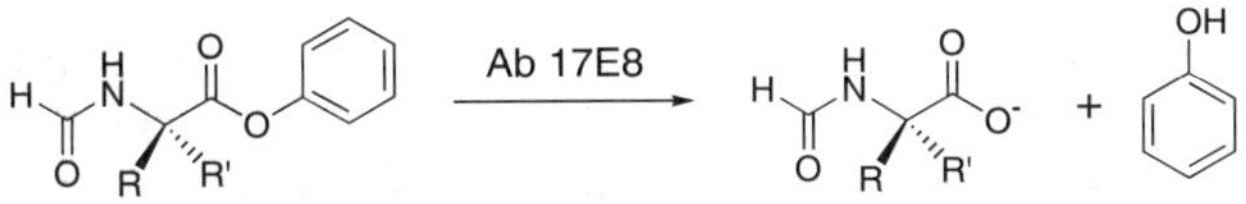

16a, R=H, R'=*n*-Bu
17, R=R'=H
18, R=R'=CH_3

substituent are apparently important for catalysis, as shown by the inability of 17E8 to cleave the glycine ester **17**, but the pro-*R* substituent can be no larger than hydrogen, since the aminoisobutyric acid ester **18** is not a substrate either.[32] Unfavorable steric interactions apparently block binding of the "incorrect" substrate enantiomer, while productive van der Waals interactions with the hydrophobic portion of the acyl group are used to position the "correct" isomer properly within the active site. Thus, like natural enzymes, antibodies apparently exploit binding interactions remote from the reaction center to control enantioselectivity and stabilize the hydrolytic transition state. The remarkable similarities between 17E8 and six other independently-derived esterolytic antibodies[31] suggest that these conclusions are likely to be general.

Although high selectivity is the hallmark of biological catalysis, narrow substrate specificity is disadvantageous for general applications in organic synthesis unless the product of interest is of unusually high value or the transformation cannot be accomplished by standard chemical methods. In general, catalysts that couple high enantio- or diastereoselectivity with broad substrate specificity have the greatest utility, since they obviate the need to develop new agents for each new application.

Fujii and co-workers have exploited the emerging structural understanding of antibody-antigen interactions to devise a clever strategy for eliciting highly selective antibody catalysts that also exhibit broad substrate specificity.[33, 34] They reasoned that two sufficiently large hydrophobic groups linked to a stereogenic center in the hapten—for example, to the alcohol leaving group and to the amine protecting group common to a diverse panel of α-amino acid esters—would suffice to induce a deep, chiral pocket capable of providing the desired stereochemical control over the target reaction. The nature and point of attachment of the linker through which the hapten is coupled to carrier proteins could then be chosen to accommodate broad structural modifications elsewhere in the molecule.

To apply this approach to the preparation of catalytic antibodies that enantioselectively hydrolyze amino acid esters possessing different α-substituents, the Fujii group designed racemic hapten **19**.[33] The relatively bulky and hydrophobic benzyl leaving group and the carbobenzyloxy protecting group were envisioned as the primary epitopes for immune recognition, leaving the Cα position for attachment of the linker moiety. Anti-**19** antibodies 3G2 and 7G12 were found to accelerate the hydrolysis of D-**20** and L-**20**, respectively, and preparative scale kinetic resolutions of racemic **20** yielded the chiral acid in greater than 94% ee. In addition to having high enantioselectivity, these two antibodies accept a broad range of substrates (Table 1). Variation in the values of the steady state parameters, k_{cat} (0.03 to 0.12 min^{-1}) and K_m (5 to 70 μM) is relatively modest, showing that different α-substituents have little effect on activity or enantioselectivity. The nonselective hydrolysis of lysine derivative **21i** by antibody 3G2 (the ε-amino group presumably interferes with substrate recognition by the catalyst) and the failure of either antibody to catalyze the cleavage of proline ester derivatives are exceptions to this general statement which can be sensibly interpreted in terms of unfavorable electrostatic or steric interactions within the antibody–substrate complex. Overall, however, the results show convincingly how ingenious design strategies can be exploited to afford antibody catalysts with broad recognition properties.

The Fujii group has also applied their strategy to the regio- and stereoselective deprotection of acylated sugars.[34] They designed the phosphonylated glucose derivative **22** in which the two 4-(*N*-acetylamino)phenylacetyl groups at C-3 and

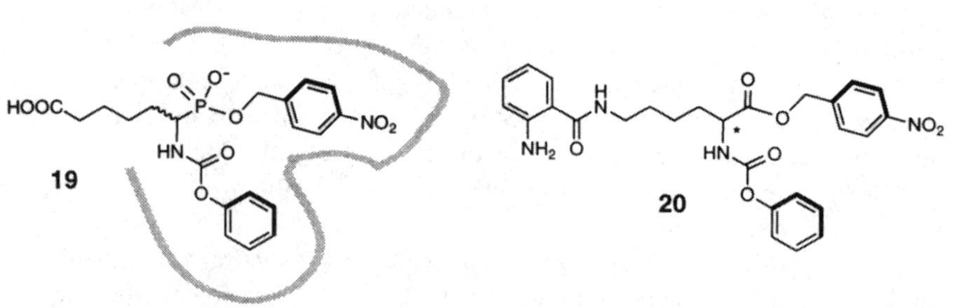

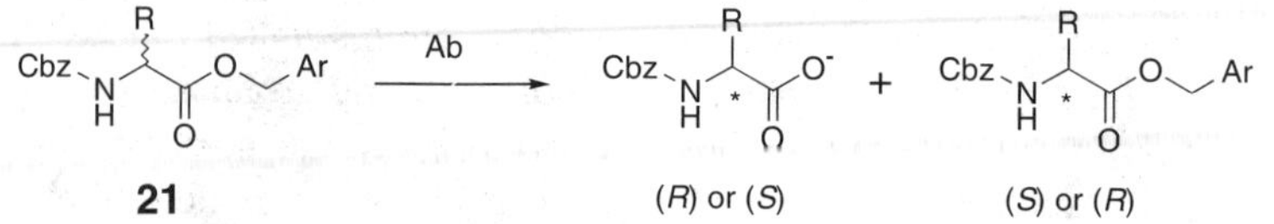

Table 1
Ratio of Initial Velocities for the Enantioselective Hydrolysis of Esters **21** with Antibodies 3G2 and 7G1[2a]

21	Substrate R–	3G2 L-**21**:D-**21**	7G12 L-**21**:D-**21**
21a	CH_3–	5:95	88:12
21b	$(CH_3)_2CHCH_2$–	5:95	98:2
21c	$CH_3(CH_2)_3$–	4:96	98:2
21d	$CH_3SCH_2CH_2$–	6:94	95:5
21e	$PhCH_2$–	7:93	98:2
21f	$(CH_3)_2CH$–	23:77	91:9
21g	Ph–	:99	> 99:< 1
21h	4-HOPh–	31:69	98:2
21i	$H_2N(CH_2)_4$–	48:52	97:3
21j	$H_2NCH_2CONH(CH_2)_4$–	2:98	

[a]Reaction conditions, reference 33: (L- or D-substrate) = 100 μM for **21a,c,f,g,i** and 200 μM for **21b,d,e,f**, [antibody] = 10 μM (active site), 10% DMSO/50 mM Tris-HCl (pH 8.0), 25°C.

C-4 were expected to serve as the anchoring recognition elements. Consistent with this expectation, antibody pockets induced in response to **22** exhibit high selectivity at C-3 and C-4 but have greater tolerance to substitutions elsewhere in the sugar ring. For example, antibody 17E11 hydrolyzes the glucopyranoside derivative **23** but not its galactopyranoside counterpart **24**. Moreover, the antibody is completely regioselective, cleaving only the ester at C-4, even in the presence of chemically identical groups at C-2 and C-3. Glucopyranosides with a bulky ester group at C-2 or the β-configuration at the anomeric center were also substrates for 17E11, in accord with the experimental design.

Although the antibodies described here are still too inefficient to be useful as general reagents in organic synthesis (k_{cat}/k_{uncat} = 10^3 to 10^4), the results demonstrate that broad substrate specificity and high enantioselectivity can be

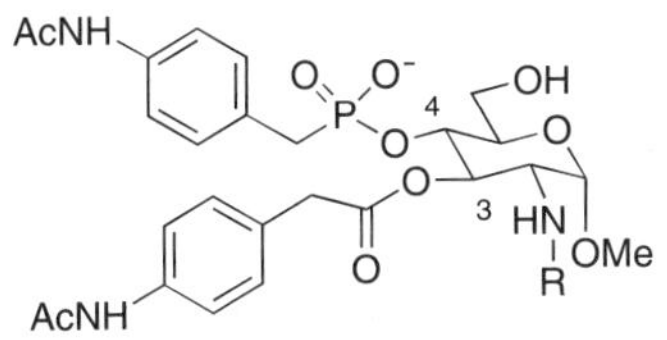

22, R=C(O)CH_2CH_2SH

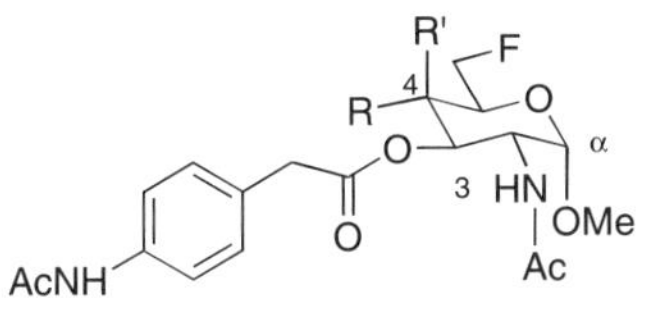

23, R=AcNHC_6H_4COO, R'=H
24, R=H, R'=AcNHC_6H_4COO

successfully combined. If improvements in catalytic efficiency can be achieved, families of catalysts for the preparation of a wide variety of natural and unnatural α-amino acids or complex oligosaccharides in enantiomerically and diastereomerically pure form could conceivably be produced.

2. *Transacylation Reactions*

Acyl transfer to alcohols and amines is related mechanistically to ester hydrolysis but yields a complementary set of products that are useful in their own right and as chiral synthons for the preparation of more complex materials. Such transformations can be difficult to achieve in water, however, because the solvent, which is present in vast excess, can participate directly in the reaction as a reactant. Enzyme-like specificity is thus required to favor the bimolecular reaction between alcohol and ester and prevent spontaneous hydrolysis of the acyl donor.

Although the phosphonate transition-state analog suggests a simple hydrolytic mechanism in which hydroxide attacks the scissile carbonyl group of the ester substrate, individual antibodies raised against such compounds occasionally exploit more complex mechanistic pathways. For example, the *S*-specific lipase antibody 21H3 described above, which was generated against hapten **1**, apparently accelerates ester hydrolysis by a two-step mechanism involving transient acylation of an amino acid within the binding pocket.[35] This mode of reactivity has been successfully exploited by Lerner and co-workers[35] to achieve stereoselective transesterifications. Vinyl ester **25** and alcohol (*S*)-**26** react efficiently in the presence of 21H3 in water to give **27**. The acyl–antibody intermediate formed from **25** has also been trapped by a variety of alcohols other than **26** to yield a series of optically active esters.[36] This transesterification reaction, which is not observed in the absence of antibody catalyst, can also be performed in mixed aqueous–organic media. Moreover, the efficiency of 21H3 in water is comparable to that of the lipase SAM II in methylene chloride.[36] The lipase does not catalyze transesterifications in water because of rapid, competing substrate hydrolysis.

The serendipitous transesterification chemistry and mechanistic sophistication of 21H3 reveal the chemical potential of the immune system. Nevertheless, induction of essential catalytic groups within an antibody pocket is unpredictable and generally expected to have a low probability. The requirement for a specific and properly oriented nucleophile or general base can be circumvented if the

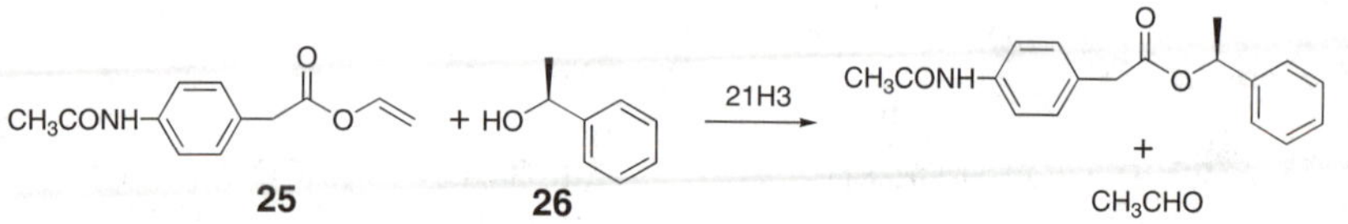

25 26 27

substrate carries the necessary functionality into the active site every time it binds. "Substrate assisted catalysis" was first demonstrated with an engineered variant of the serine protease subtilisin,[37] but this strategy has also been used for antibody catalysis by Napper et al.[38] They designed the racemic cyclic phosphonate ester **30** and used it to generate antibodies for the lactonization of 6-acetamido-5-hydroxyhexanoate **28**. Of the 24 immunoglobulins that bound the hapten, one enhanced the rate of the intramolecular cyclization by a factor of 170. In this case, the reduction of the conformational degrees of freedom available to the flexible substrate upon binding at the active site of the antibody is believed to be a significant factor in catalysis. Not surprisingly, only 50% of the racemic δ-hydroxy acid is converted to product in the presence of the antibody. ^{1}H-NMR spectroscopy with a chiral lanthanide shift reagent confirmed that product δ-lactone **29** was predominantly a single enantiomer, obtained with an enantiomeric excess of about 94%.

Phosphonate ester **30** can also be considered as a mimic of the transition state for subsequent esterolysis and aminolysis of the δ-lactone. In fact, the antibody that promotes ring formation was shown to catalyze the stereoselective reaction between **29** and 1,4-phenylenediamine.[39] The kinetic mechanism of the bimolecular process involves random equilibrium binding of lactone and amine, and the observed turnover rate could be approximated from the measured difference between the binding of reactants and the TSA. Again, entropic factors are presumed largely responsible for the observed rate acceleration, with minimal contributions derived from specific catalytic groups at the active site.

Neutral phosphonate diesters, such as **30**, are potentially better surrogates for the transition state for transesterification (and the reaction of esters with amines) than simple phosphonates since they contain structural equivalents of both the nucleophile and the leaving group. One might therefore expect that such compounds would be generally useful for sculpting active sites tailored to the steric demands of bimolecular reactions. Among the experiments designed to test this prediction, the production of an unusually effective artificial aminoacyl tRNA-synthetase by Schultz and co-workers[40] stands out. Phosphonate **31** was used as a mixture of diastereomers to elicit antibodies that aminoacylate the 3′-hydroxyl group of thymidine (**32**) with an activated alanyl ester (**33**). Four of the antibodies preferentially accelerated the reaction of L-alanine ester **33**, three preferred the D-isomer, and one was unselective. Interestingly, the L-selective

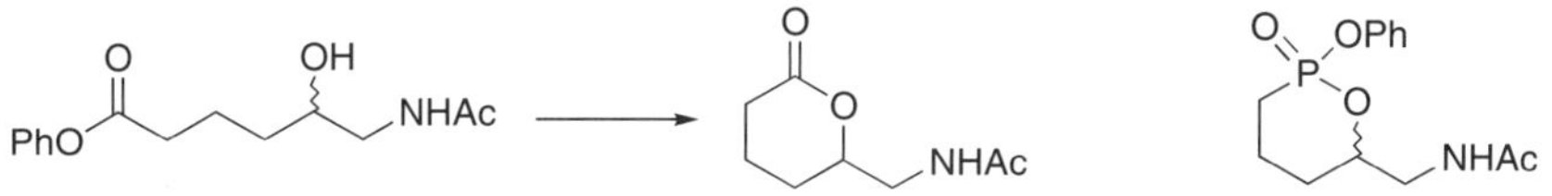

28 **29** **30**

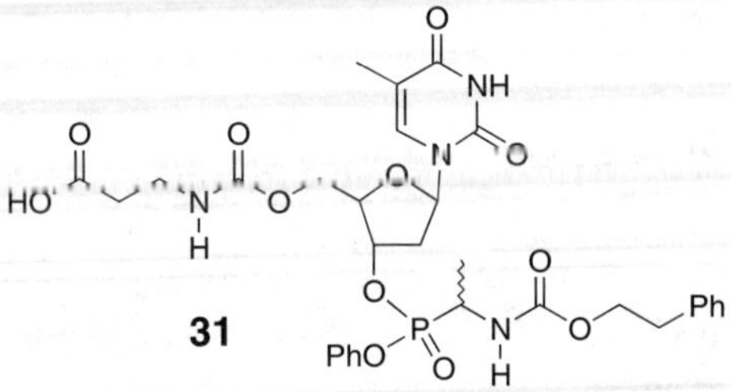

31

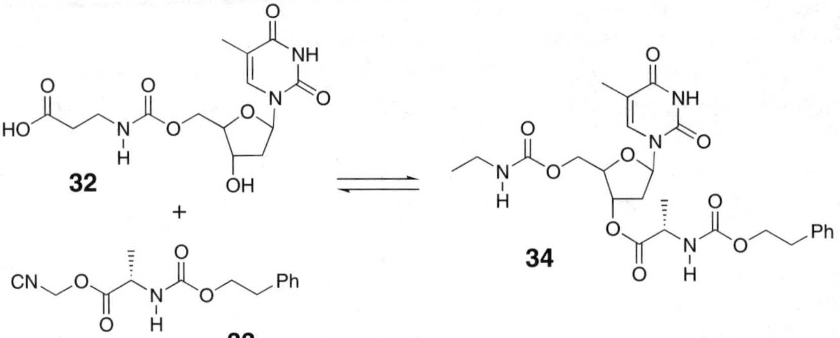

32 + **33** ⇌ **34**

antibody 18R.136.1 binds the L- and D-amino acid esters with comparable affinity, as judged by their similar K_m values, but it discriminates between the diastereomeric transition states for transesterification, achieving a 22-fold difference in k_{cat} values. Furthermore, the effective molarity for acyl transfer in the induced pocket is 50,000 M, indicating highly successful preorganization of the two substrates. Unfortunately, severe product inhibition by **34** (K_i = 18 nM) probably limits the synthetic potential of this first-generation catalyst.

Product inhibition is not unique to 18R.136.1; it limits the effectiveness of many antibody-catalyzed reactions, and special strategies are often needed to overcome this complication. Extensive screening of the immune response often succeeds in identifying catalysts that are not crippled by inhibition. Product inhibition was not observed, for example, in the peptide bond-forming reactions catalyzed by one of 28 antibodies raised against the neutral phosphonate diester **35**.[41] This antibody accelerates the condensation of phenylalanine derivatives with an acyl azide derived from L-alanine (the corresponding phenyl ester is not sufficiently reactive) with modest selectivity for L- versus D-alanine, but it is relatively inefficient, as judged by an effective molarity of 0.16 M and a rate acceleration $[(k_{cat}/K_m)/k_{uncat}]$ of only 10^2.

In special cases, product inhibition can be avoided by taking advantage of chemical or conformational changes in the initially formed product. An example

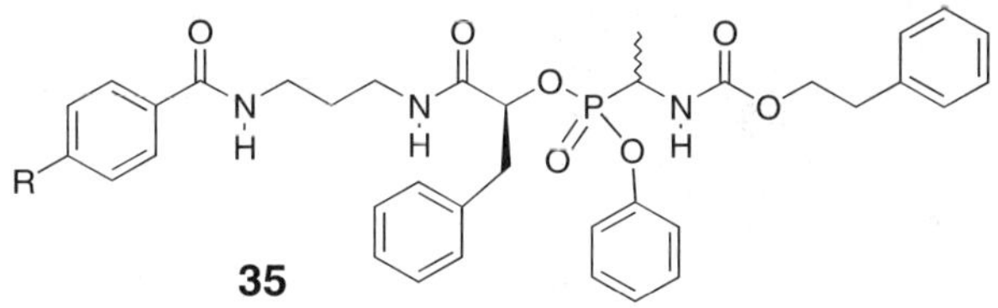

35

of the latter strategy was recently reported by Oda and co-workers.[42] The esterolytic antibody 1C7 was raised against phosphonate **36**. Although it enantioselectively hydrolyzes the *R*-isomer of ester **37**, it does so in a nearly stoichiometric fashion because the resulting carboxylic acid binds tightly at the active site. In contrast, the carbonate ester **38** is an excellent substrate for 1C7. Upon hydrolysis, the initially formed carboxylate undergoes rapid decarboxylation to give a neutral alcohol that is a poor ligand for the antibody. Consequently, more than 100 turnovers per active site are observed with no loss of catalytic activity after 72 hours. Furthermore, the stereoselectivity of 1C7 towards **38** was even higher than towards **37**: (*R*)-**38** is hydrolyzed 120 times faster than (*S*)-**38**. An analogous mechanism likely accounts for the efficiency of one of the first catalytic antibodies described,[43] although this point was not explicitly investigated.

Large structural units in the hapten that do not find a direct match in either the product or the leaving group can also be used to facilitate the dissociation of sterically less demanding products from antibody catalysts. This approach was adopted by Hirschmann and colleagues[44] to catalyze the formation of peptides. They designed compound **39** with the expectation that the cyclohexyl group would induce a generic, hydrophobic pocket that would accommodate a variety of acyl donors. In accord with this design, the anti-**39** antibody 16G3 catalyzes the coupling of 4-nitrophenyl esters (**40**) of *N*-acetyl valine, leucine, and phenylalanine with tryptophan amide (**41**) to form the corresponding dipeptides (**42**). It achieves high turnover rates and also accelerates the reaction of activated amino acids and

36

37

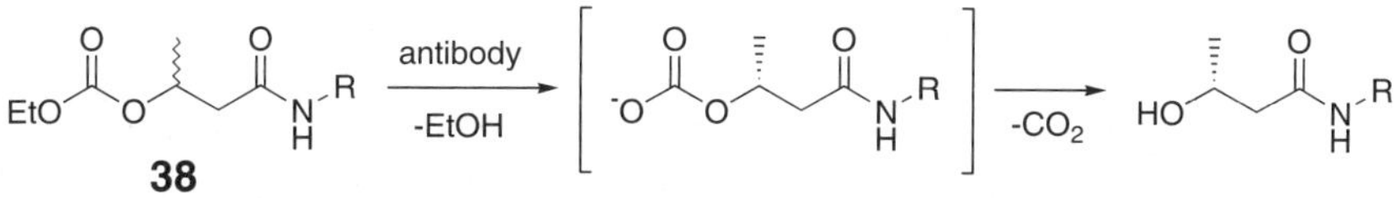

38

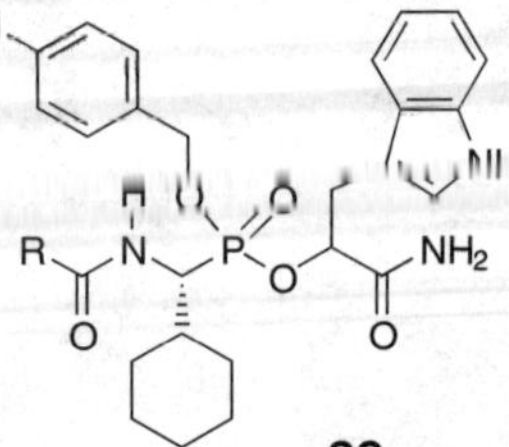

39

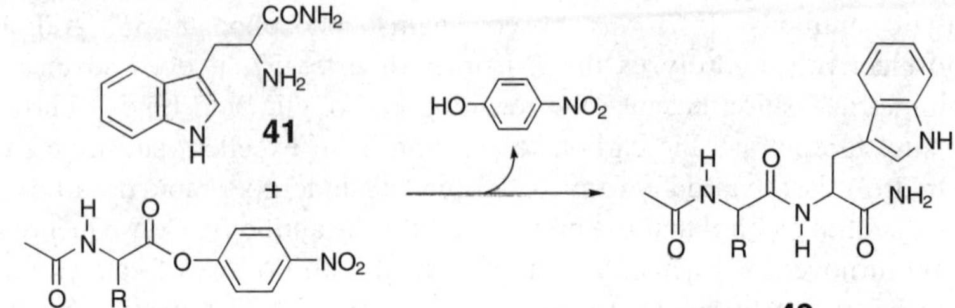

42

40 R = $CHMe_2$, CH_2CHMe_2, CH_2Ph

dipeptides with another dipeptide to give tri- and tetrapeptides, respectively.[45] Neither racemization of the activated esters nor hydrolysis of the final products was observed, but all possible stereoisomeric combinations of ester and amide substrates were coupled at comparable rates. The lack of selectivity in the coupling reactions was rationalized in terms of the small size of the substrate compared to the hapten. Although fortuitous, weak chiral recognition on the part of the antibody is potentially advantageous with respect to the development of general reagents for the synthesis of polypeptides insofar as it expands the repertoire of possible reaction partners. The relative cost of broad selectivity appears small in this case: The effective molarities for these reactions are consistently in the range 10^2 to 10^3 M, and product yields of 80% were typically observed. The synthesis of L-leucine–D-tryptophan amide on a 1 mM scale underscores the relatively low level of product inhibition observed with the antibody.[44] These results thus establish an encouraging precedent for the development of tailored peptide ligases for the condensation of unprotected peptide segments.

3. *Amide Hydrolysis*

Amide hydrolysis is energetically more demanding than ester hydrolysis or transacylation, and its catalysis by antibodies represents a formidable challenge. Although phosphonamidates would appear to be excellent transition-state analogs, numerous attempts in many laboratories to use these compounds to produce amidases have had only limited success. The difficulty encountered in making amidases presumably reflects the intrinsic stability of the amide bond and the low probability of eliciting an effective constellation of catalytic groups within the antibody pocket.

The most efficient and best characterized antibody amidase is 43C9.[46] It was generated with the achiral phosphonamidate transition-state analog **43** and accelerates the hydrolysis of an activated 4-nitroanilide (**44**) by a factor of 2.5×10^5.[47] More recently, Martin and co-workers have identified an antibody that catalyzes the enantioselective hydrolysis of an unactivated primary amide.[48] A rapid throughput assay that allowed large numbers of immobilized antibodies to be screened directly for catalysis undoubtedly contributed to their success. Antibody 13D11, one of 68 antibodies raised against the dialkylphosphinate hapten **45**, was found to promote the cleavage of a dansyl derivative of (*R*)-phenylalaninamide.[46] At pH 9.0, the rate enhancement for this substrate (k_{cat}/k_{uncat}) was only 132. However, neither the *S* isomer of **46** nor a shorter acylated derivative **47** was hydrolyzed by 13D11, demonstrating that specific binding interactions are essential for catalysis. The preferred substrate enantiomer is opposite that of most natural amino acids, making it unlikely that the observed activity is due to contaminating proteases. This conclusion was strengthened by several additional control experiments. These included studies showing that two independently purified batches of the full-length 13D11 immunoglobulin as well as its *Fab*' fragment had identical specific activities. Furthermore, the amidase activity was found to be resistant to nine inhibitors of common proteolytic enzymes but was effectively and selectively blocked by a single hapten enantiomer ($K_i = 14\ \mu M$).

These results establish the feasibility of cleaving unactivated amide bonds with antibodies. Nevertheless, given the low activity of 13D11, the often-stated goal of

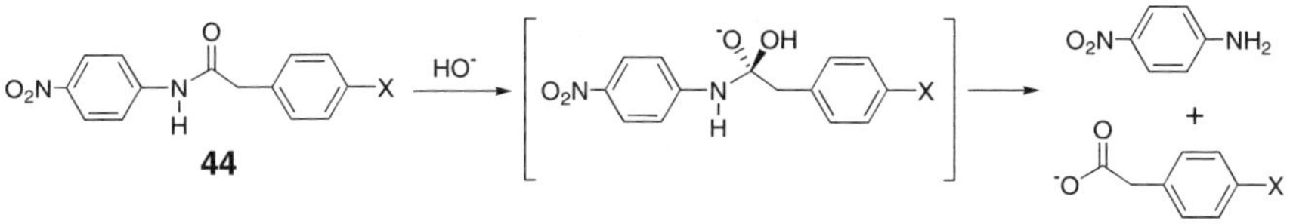

44

X = NHCO(CH2)3COOH

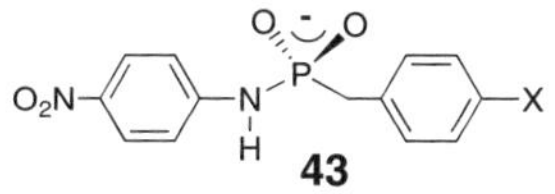

43

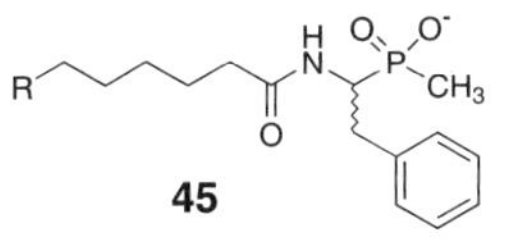

45

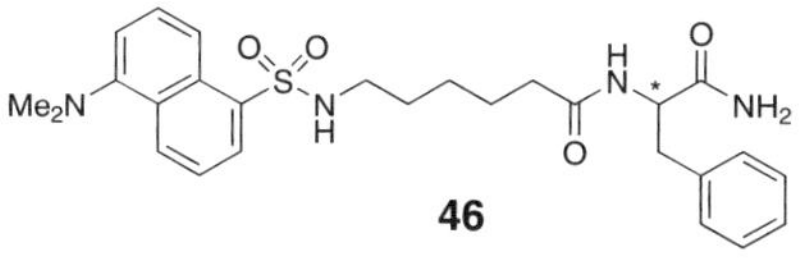

46

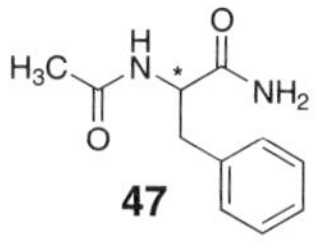

47

creating site-specific antibody proteases remains more hope than reality. Realization of this goal will require more sophisticated approaches. Introduction of catalytic functionality into the active site, either through more clever hapten design or by site-specific mutagenesis, will likely be a key feature of such strategies. Incorporation of external cofactors, particularly metal ions,[49] into the antibody active site can also be expected to significantly extend the capabilities of these catalysts.

As in the lactonization reaction described above, the peptide itself can be recruited to assist in amide cleavage. Deamidation of asparagine residues in proteins and the rearrangement of an asparaginyl–glycine peptide bond are biologically relevant examples of substrate-assisted catalysis that have been implicated in the denaturation and degradation of proteins in vivo. Benkovic and co-workers[50] designed the bifunctional cyclic phosphinate **48** to mimic the transition state for the formation and subsequent hydrolysis of the key succinimide intermediate in this process; by virtue of its two tetrahedral moieties—the phosphinate and the secondary alcohol—this hapten potentially programs for both hydrolytic pathways available to the succinimide leading, respectively, to aspartyl–glycinamide and isoaspartyl–glycinamide. In fact, **48** yielded several antibodies that accelerate the rate-limiting step of succinimide formation (≈200-fold) and also direct the course of imide ring opening to give various isoasparate-to-aspartate ratios (Scheme 1, Table 2). These catalysts exhibited considerable stereochemical diversity as well, selectively processing either the D- or L-substrate stereoisomers (14A8 and 39F3, respectively) or catalyzing the

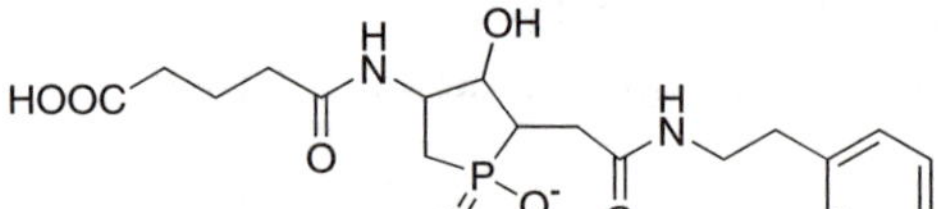

48

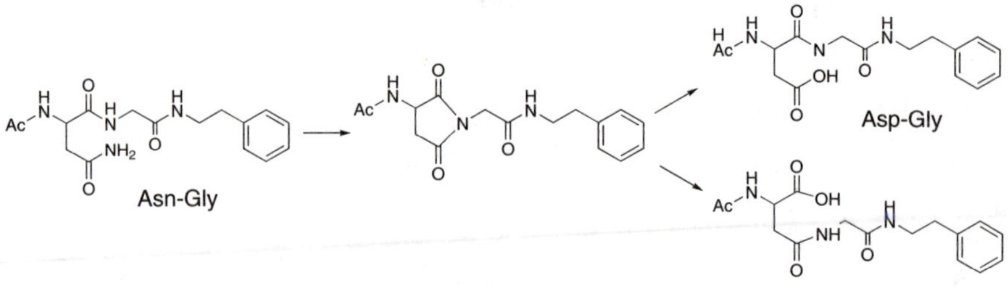

Scheme 1.

Table 2
Isoaspartate versus Aspartate Product Ratios of RG2 Antibodies[a]

	IsoAsp/Asp product ratios	
Antibody	L-isomer	D-isomer
39F3	8.3	3.6[b]
14A8	3.4[b]	1.4
23C7	16.4	1.2
40H4	1.9	5.7
2E4	2.4	4.7
24C3	2.1	4.8

[a]Gibbs, R.A.; Taylor, S.; Benkovic, S. J. *Science* **1992**, *258*, 803–805.
[b]Spontaneous reaction.

hydrolysis of both isomers. The stereoselectivities and activity patterns of the promiscuous antibody 23C7 have been explained in terms of a simple model involving a bifunctional binding site.[51] Perhaps the most significant outcome of these experiments, however, is the demonstration that bifunctional transition-state analogs can be used to induce antibodies that effectively deal with the problem of multiple transition states.

B. Glycosyl and Phosphoryl Group Transfer

In comparison with acyl group transfer, catalysis of glycosyl or phosphoryl group transfer reactions has received relatively little attention. In part, this reflects the high activation barriers for these transformations and the paucity of good transition-state analogs. The difficulty of generating high-affinity antibodies to very polar molecules like sugars is an additional practical complication. Nevertheless, given the potential utility of tailor-made enzymelike catalysts for the synthesis of complex carbohydrates, sequence-selective manipulations of genes, and targeted post-translational modification of proteins, we can expect much more activity in this area in the future.

1. Glycosyl Transfer

Several groups have investigated the hydrolysis of aryloxytetrahydropyrans as a model for the cleavage of glycosidic bonds (e.g., **49** → **51**). This is a two-step reaction limited by the rate of acid-catalyzed expulsion of the leaving group to give

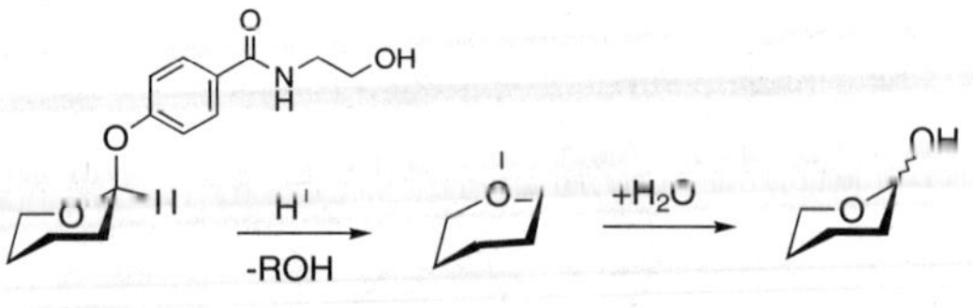

49 **50** **51**

an oxocarbenium intermediate (**50**). Lerner[52] and Schultz[53] have independently shown that haptens containing a positive charge corresponding to the anomeric center of the acetal, like the piperidinium derivatives **52** and **53**, are able to induce antibodies with weak hydrolytic activity. Rate accelerations 60–70-fold over the hydrolysis rate in acetate buffer were observed. The acid dependence of these reactions, together with chemical modification data, indicate the probable involvement of an ionizable carboxylic acid at the antibody active sites. Although data on the stereoselectivity of catalysis were not reported, inhibition studies of antibody 14D9, generated to **52**, with the four possible diastereomers of a disubstituted tetrahydropyran provided circumstantial evidence for hydrolysis occurring via the more labile substrate conformer, in which the phenolate leaving group is axial.[52] If correct, these results implicate stereochemistry as a critical feature in hapten design for this class of reactions.

Consistent with this notion, the positively charged amidine-containing hapten **54** and the neutral D-galactal derivative **55**, which mimic the half-chair conformation of the oxocarbenium intermediate, failed to yield glycolytic antibodies.[53] It was suggested that the planar structure of these haptens does not permit the leaving group in the corresponding substrates to adopt a favorable axial geometry in the induced active site. Another transition-state analog, azetidine **56**, was designed by Masamune and co-workers to mimic the desired half-chair geometry *and* an appropriately disposed leaving group.[54] Antibody ST-8B1 derived from this compound catalyzes the hydrolysis of the enantiomerically and diastereomerically pure substrates (1*R*,4*R*)-**57** and (1*R*,4*S*)-**57** with rate

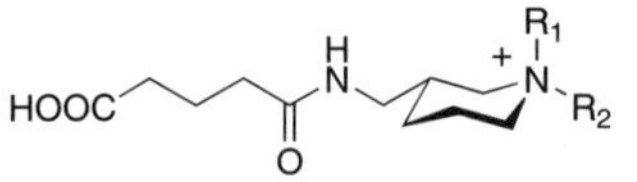

52 **a** $R_1 = CH_2Ar$, $R_2 = CH_3$
b $R_1 = CH_3$, $R_2 = CH_2Ar$

Ar =

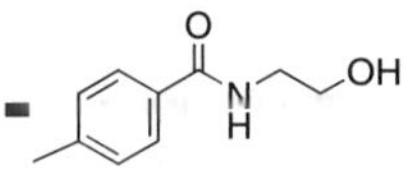

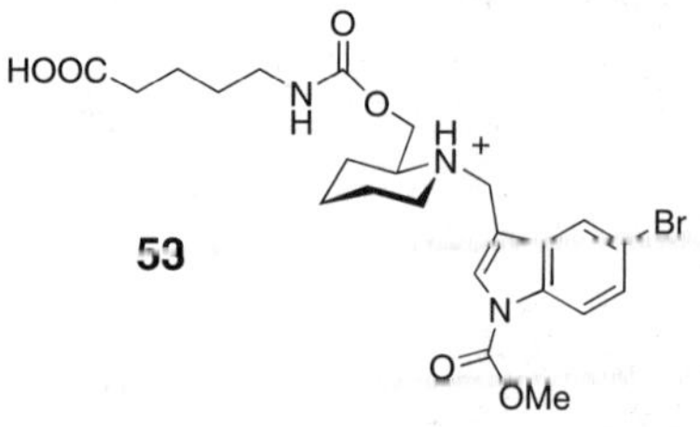

53

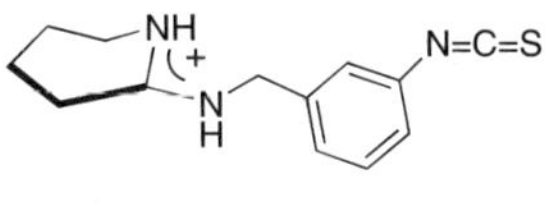

54

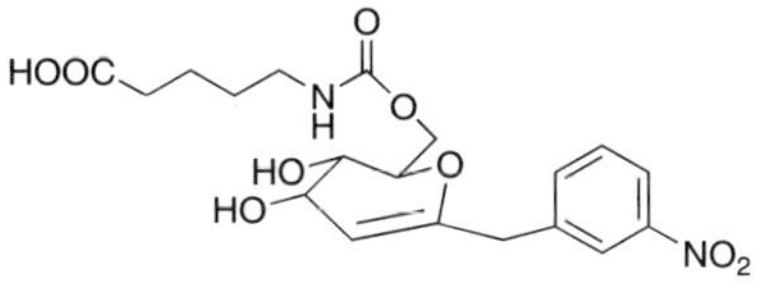

55

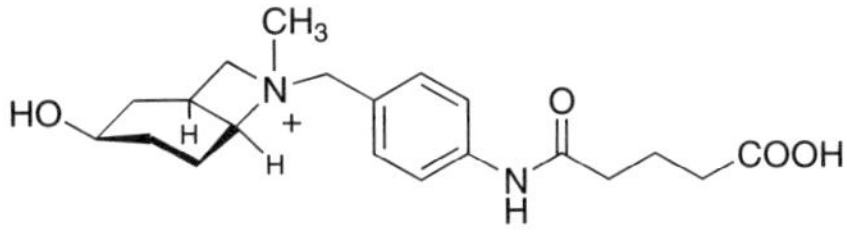

56

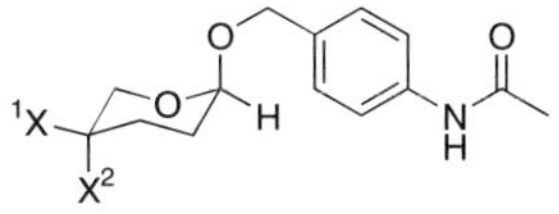

(1*R*,4*R*) or (1*S*,4*S*)-**57a** X^1=OH, X^2=H
(1*R*,4*S*) or (1*S*,4*R*)-**57b** X^1=H, X^2=OH

enhancements of 100–200-fold, but not the other stereoisomers. In contrast to the antibodies prepared against **52** and **53**, ST-8B1 is independent of pH between 5.4 and 7.0, and no evidence could be adduced for the participation of an acidic residue at the active site.

Lerner and co-workers used a fundamentally different approach to prepare antibodies with bone fide glycosidase activity.[55] They employed the mechanism-based inhibitor **58** to trap antibodies displayed on filamentous phage, which promoted glycosidic bond cleavage covalently (Scheme 2). The phage library was constructed from the genes of approximately 100 monoclonal antibodies, prepared with conventional hybridoma techniques, which bind **59**, an iminocyclitol inhibitor of β-galactosidase. One of the successful catalysts, *Fab* fragment 1B, was shown to catalyze the hydrolysis of 4-nitrophenyl-β-galactosidase with a rate enhancement (k_{cat}/k_{uncat}) of 7×10^4 and some turnover. This catalyst is substantially better—by two orders of magnitude—than the antibodies identified by screening hapten binders. Although the stereoselectivity of the catalyst was not investigated, it seems likely that the antibody will exhibit at least modest preferences for certain sugar stereoisomers.

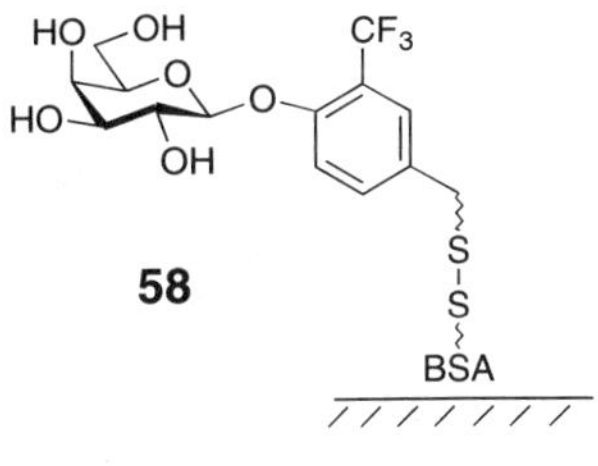

58

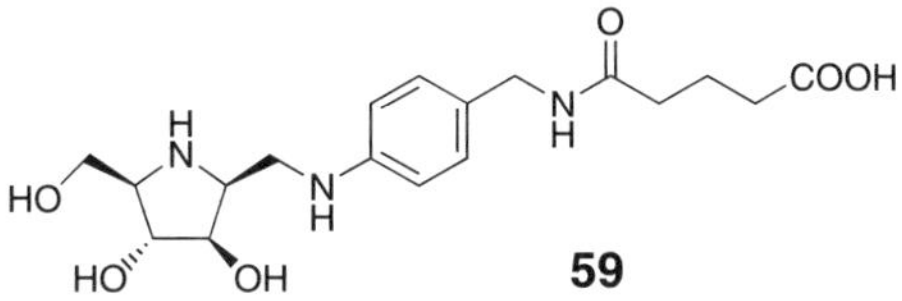

59

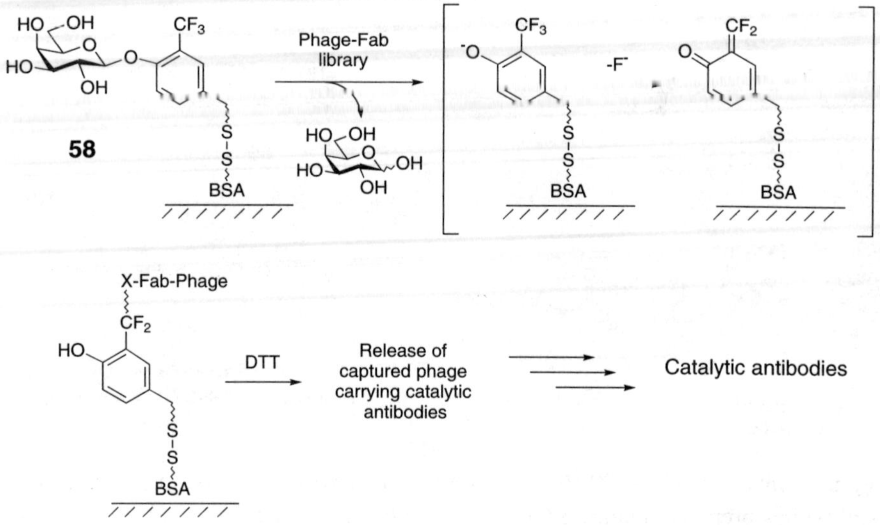

Scheme 2.

2. *Phosphoryl Group Transfers*

The hydrolysis of phosphate esters is an energetically demanding process that proceeds via a pentacoordinate transition state. The half-life for the hydrolysis of dimethylphosphate at pH 7, for example, is 3×10^{10} years,[56] which can be compared with a half-life of only (!) 168 years for the hydrolysis of a typical peptide.[57] Because of the difficulty of fashioning stable analogs that correctly mimic the steric and electronic characteristics of the pentacoordinate transition state, a variety of compromises have been necessary in designing templates for generating antibodies capable of hydrolyzing phosphate mono-, di- and triesters. Among the haptens that have been investigated are nonhydrolyzable substrate analogs,[58] amine oxides,[59, 60] and a pentacoordinate oxorhenium(V) complex.[61] The low activities of the resulting catalytic antibodies ($k_{cat}/k_{uncat} < 10^3$) may reflect the poor resemblance of the hapten and the targeted transition state. In evaluating the claims for phosphataselike antibodies, a note of caution is also warranted. Extremely small amounts of contaminating natural phosphatases could easily account for the low activities observed. In our experience, highly active phosphatases are ubiquitous and extremely difficult to remove from immunoglobulin samples (C. Lewis, unpublished results), so an extraordinary level of rigor with regard to sample purification and characterization is mandatory.

In any case, the reactions reported in the literature to date have exploited achiral substrates and products; so no information is available on the degree of stereochemical control possible with phosphorolytic antibodies. It would be remarkable, though, if these catalysts proved to be less discriminating than other hydrolytic antibodies.

In analogy with early studies by an Israeli group in which antibodies raised against steroid esters were found to be covalently modified by their hapten,[62] Lerner and co-workers have recently identified a set of antibodies that reacts covalently with haptenic phosphonate diesters (e.g., **60**).[63] The immunoglobulins apparently contain reactive nucleophiles within their active sites that can be exploited for ester hydrolysis ($k_{cat}/k_{uncat} = 10^2–10^4$), presumably via a two-step mechanism involving transient acylation of the catalyst. The phosphonylation reaction itself depends on the nature of the leaving group (e.g., 4-methylsulfonylphenol), but additional stereoelectronic factors within the antibody pocket must also be important for transition-state stabilization. The covalent phosphonyl–antibody species undergoes slow hydrolysis,[63] but potent inhibition by the resulting phosphonate monoester **62** prevents turnover. As expected for a chiral binding site, antibody modification is highly stereoselective: Only 50% of racemic phosphonate **61** reacts with excess antibody, permitting stoichiometric resolution of the racemic mixture.

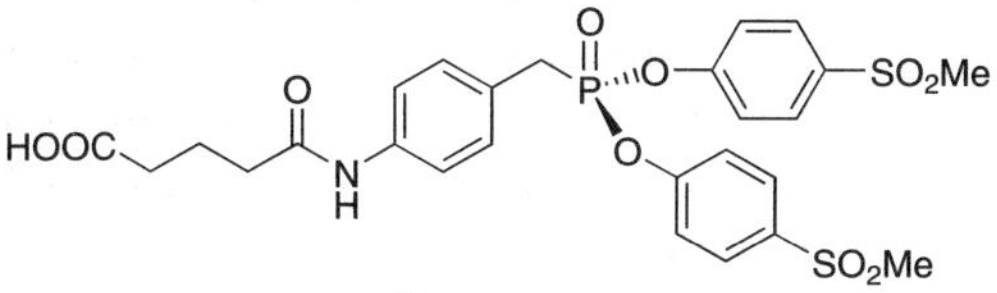

60

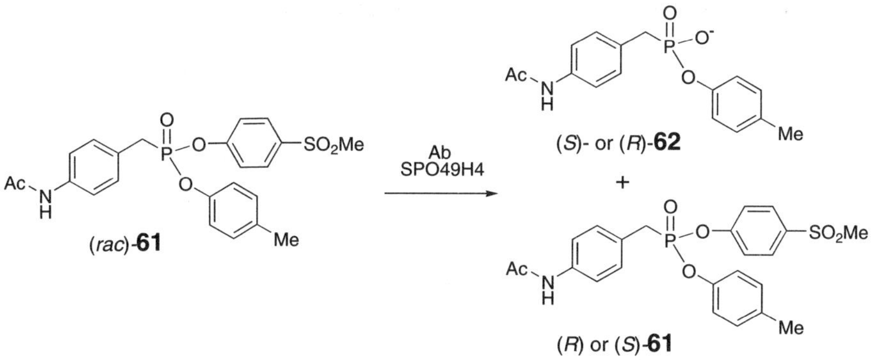

C. Additions, Eliminations, and Substitutions

The enviable efficiency of natural enzymes can be attributed to their successful orchestration of carefully timed proton transfers and their utilization of preorganized acids and bases to stabilize selectively the altered charge distribution in the transition state. The development of general methods for creating binding sites intentionally functionalized with catalytic groups has consequently been an important goal of research in the field of catalytic antibodies. One promising approach exploits charge complementarity between antibody and antigen to induce acids and bases within the immunoglobulin combining site. This strategy has been applied to the production of catalysts for a diverse set of addition, elimination, and substitution reactions.

Haptenic charge was first used by Schultz and co-workers to catalyze the elimination of HF from β-fluoroketone **63** to give the unsaturated ketone **64**.[64] The positively charged hapten **65**, in which an ammonium ion mimics a methylene group with an acidic proton, was used to elicit a complementary negatively charged carboxylate capable of serving as a base. Antibody 43D4-3D12 appears to function in accord with this design, catalyzing the elimination in a pH-dependent manner with a rate acceleration over the non-enzymatic reaction of **63** with acetate of $\approx 10^5$. Framework residue Glu46 in the heavy chain was identified as the catalytic base by affinity labeling.[65] This residue has a somewhat elevated pK_a (6.2), suggesting that it sits in a relatively apolar active site pocket. To ascertain the extent to which the inherently chiral environment surrounding the carboxylate base influences the reaction, the stereochemistry of proton abstraction and fluoride release was probed using the four possible C-3 monodeuterated diastereomers of substrate **63**.[65] Antibody 43D4-3D12 was found to exhibit surprisingly little stereofacial selectivity, abstracting either the pro-*R* or pro-*S* proton, implying considerable conformational flexibility for the substrate bound at the antibody active site. An anti elimination geometry was favored over the syn process for each of the substrate diastereomers, however. Because hapten **65** itself is a conformationally

63 **64**

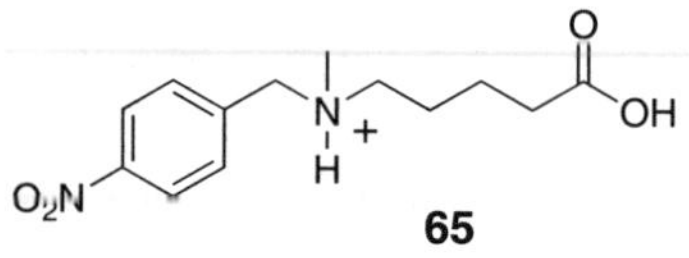

65

flexible molecule, the modest stereochemical preferences exhibited by 43D4-3D12 must reflect good fortune rather than design. Conformationally restricted haptens should, in theory, lead to better control over the reaction pathway and would be expected to benefit chemical efficiency as well.

Catalysis of an energetically demanding syn elimination by antibodies derived from the conformationally restricted hapten **66** supports this contention.[66] As in the previous example, **66** contains an ammonium group to induce a catalytic base; its bicyclic ring system was expected to elicit a binding site that would lock the flexible β-fluoroketone **67** into an energetically unfavorable eclipsed conformation. The anti-**66** antibody 1D4 converts **67** exclusively to the Z-olefin **69** via the desired syn pathway—albeit slowly ($k_{cat} = 3 \times 10^{-3}$ min^{-1}). By contrast, base-catalyzed elimination (pH 9.0, buffer) gave exclusively the *E*-olefin **68**. Further, the inability of the antibody to convert substrate **70** to **69** through an otherwise favored, staggered transition state demonstrates strong preference for the eclipsed conformation of the substrate in the binding pocket. Although this antibody-catalyzed syn elimination is not synthetically useful (the uncatalyzed elimination via the anti pathway actually produces more product under the reported assay conditions than the 1D4-catalyzed reaction), the ability of the antibody to overcome the estimated 5 kcal/mol energy difference between the syn and anti

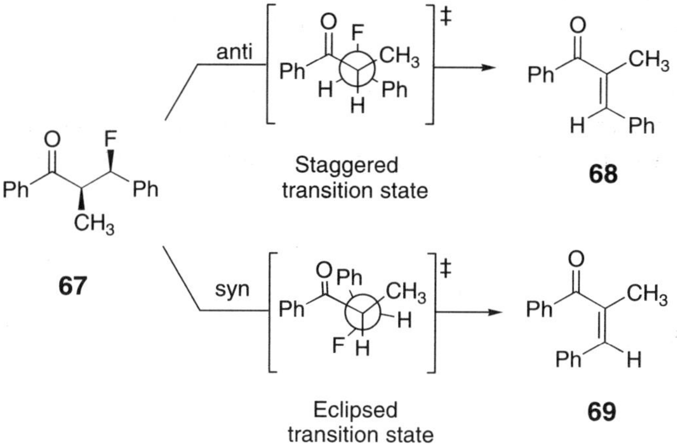

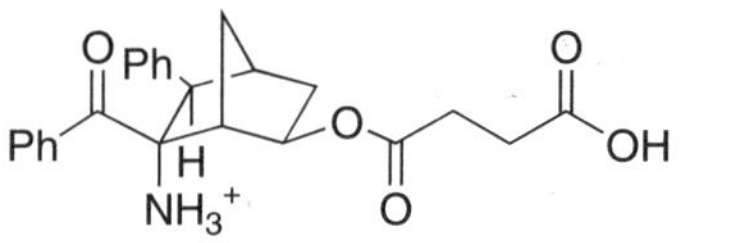

66

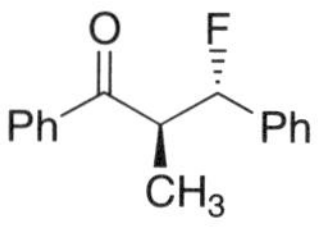

70

transition states is altogether remarkable and suggests exciting opportunities for this technology in channeling reactants down preordained reaction pathways.

Improvements in hapten design will presumably help to augment the efficiency of these catalysts. Note that the ammonium ion in **66** corresponds to the α-keto proton of **68**; so its N–H bond is not congruent with the substrate C–H. A carboxylate induced in response to this group would not be expected to be positioned optimally for catalysis, possibly explaining the low rates attained by 1D4. That highly efficient antibody catalysis of proton transfer reactions can be achieved is shown by the greater than 10^8-fold acceleration of benzisoxazole decomposition (**71** → **73**) by antibody 34E4.[67] The latter catalyst was identified by extensive screening of the immune response to the benzimidazolium hapten **74**, which effectively mimics the geometry of the reacting bonds at the transition state (**72**) and also contains a strategically placed cation for eliciting an antibody carboxylate poised to abstract the C-3 proton of substrate. Minimal resemblance to the reaction product accounts for the large number of substrate molecules (> 10^3) processed by each 34E4 active site. Although medium effects undoubtedly contribute to the unusually high efficiency of the anti-**74** antibody,[68,69] an experimental Brønsted analysis and quantum mechanical calculations indicate that preorganization of the catalytic base and the substrate within the active site, as programmed through hapten design, is crucial (K. Kikuchi, J. Na, K.N. Houk, D. Hilvert, unpublished data).

Ammonium ions have also been used as templates for creating pockets capable of stabilizing positively charged transition states. For example, antibodies raised against hapten **77** accelerate the dehydration of β-hydroxy ketone **75** to give enone **76**.[70] The ammonium group in **77** served as a surrogate for the protonated alcohol formed in the acid-catalyzed elimination reaction, and evidence obtained with

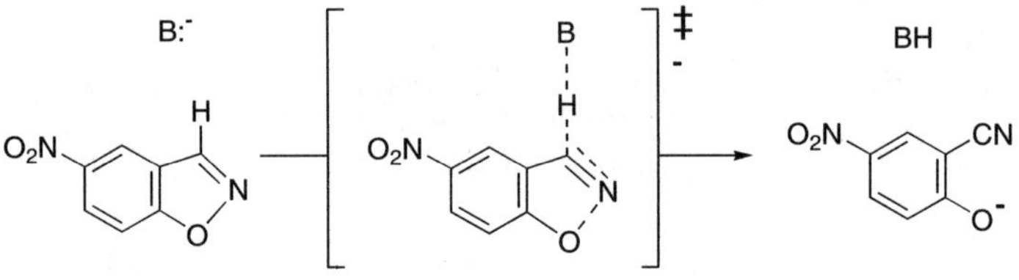

71 **72** **73**

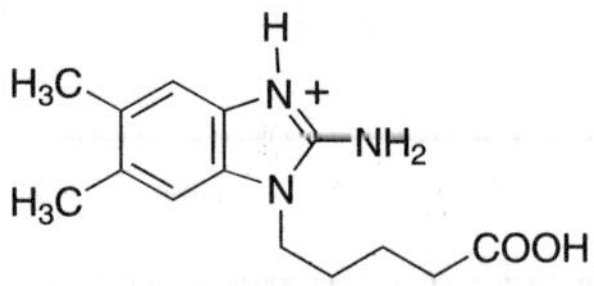

74

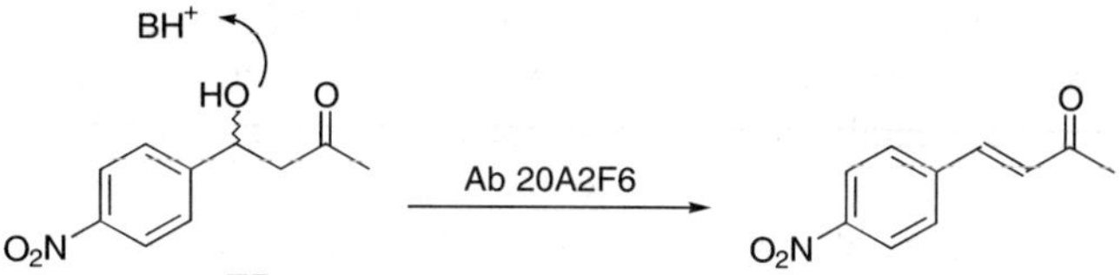

75 **76**

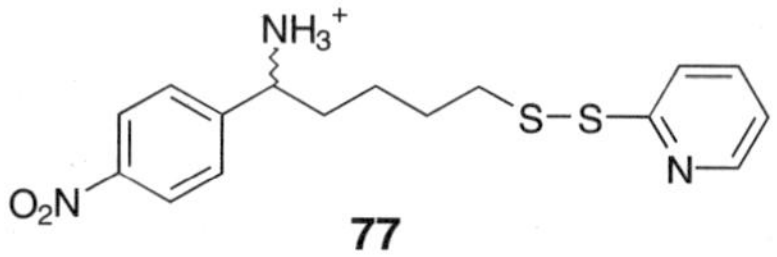

77

antibody 20A2F6 is consistent with a mechanism involving a transition state with oxonium ion character. Interestingly, 20A2F6 also catalyzes the stereoselective addition of hydroxylamine to 4-nitroacetophenone, favoring formation of the syn oxime rather than the anti isomer by 9:1.[71] Antibody 43D4-3D12, raised against **65**,[64] also catalyzes this reaction but with a reversed syn:anti ratio (1:9).[71] The pH dependence of the 20A2F6-catalyzed oxime formation indicates that an active site group with a pK_a of approximately 7.5 either stabilizes the protonated transition state or acts as a general acid. The active site of 43D4-3D12 contains an ionizable carboxylic acid, as discussed above, which may play a similar role.

Antibody 14D9, studied originally as a glycosidase mimic (see Section II.B), is also quite versatile in its ability to stabilize positively charged transition states. Recall, it was generated in response to the piperidinium hapten **52**.[52] In addition to cleaving cyclic acetals, 14D9 catalyzes the hydrolysis of ketals,[72] enol ethers,[73–75] and epoxides.[76] In each of these reactions, the antibody is believed to protonate the substrate and stabilize the resulting oxocarbenium ion with its active-site carboxylate. Although the rates of the antibody-catalyzed conversions are generally low ($k_{cat}/k_{uncat} \lesssim 10^3$), the chemo- and stereoselectivities can be quite high. The hydrolysis of enol ethers, for instance, proceeds enantioselectively with protonation at the carbon[74, 75] and has been carried out on a preparative scale.[77, 78] As shown for **78**, a proton is delivered exclusively to the *re* face of the enol ether to produce the optically active ketone (–)-(*S*)-**79**. Interestingly, the observed selectivity arises solely by discrimination of the competing enantiomeric transition states (made diastereomeric by binding at the active site), since the antibody binds the chiral products of the reaction with equal affinity ($K_{(+)\text{-}(R)\text{-}\mathbf{78}} = K_{(-)\text{-}(S)\text{-}\mathbf{78}} = 100$ μM).[75] Moreover, a broad variety of enol ethers is processed, provided they retain the aromatic group found in the original hapten as a recognition element. The best substrates also have the *Z* configuration at the double bond and an alkyl substituent at the site of protonation.

The conversion of enol ether **80** to cyclic ketal **83** in water in 12% yield exemplifies the chemoselectivity possible with 14D9.[79] Although **83** is the normal product of the acid-catalyzed hydrolysis of **80** in organic solvents, it is never observed in water because the highly reactive oxocarbonium intermediate is rapidly trapped by the solvent to give ketone **82** (via hemiacetal **81**) as the sole product. The ability of the antibody to protect the reactive oxonium ion intermediate from hydrolysis and partition it toward a product that is not typically observed under these conditions (i.e., **83**) mimics the capabilities of rather sophisticated enzymes. Extension to other reactions involving reactive, water-incompatible intermediates can be easily imagined.

Cationic cyclizations of acyclic unsaturated compounds provides another illustration of the potential of catalytic antibody technology for generating and controlling the fate of highly reactive intermediates. Lerner and colleagues used the cyclic amine oxide **86**, which loosely resembles the presumed transition state for the cationic cyclization of olefinic sulfonate ester **84** to **85**, as a hapten to elicit an immune response.[80] In addition to mimicking the conformation of the substrate in the transition state (**87**), the amine–oxide moiety was expected to induce polar functionality in the antibody that might stabilize the departing sulfonate and the developing carbocation. In the event, several anti-**86** antibodies were found to accelerate the target reaction in a biphasic medium (83:2:15 pentane:chloroform:50 mM bis-tris buffer, pH 7). Multiple products are generally observed for cationic cyclizations carried out under solvolysis conditions, but cyclohexanol **85** was formed almost exclusively (98%) in the presence of antibody

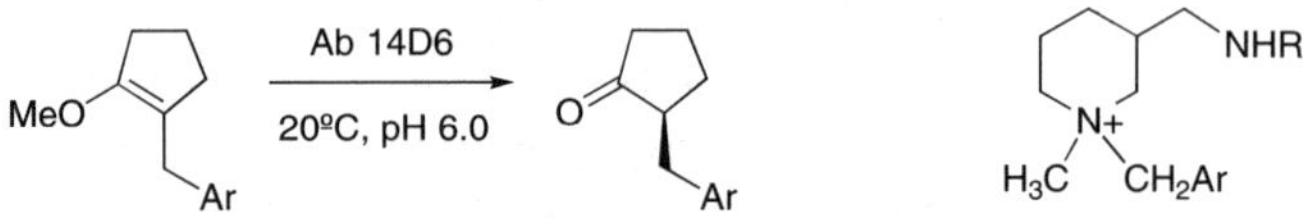

78 **79** **52**

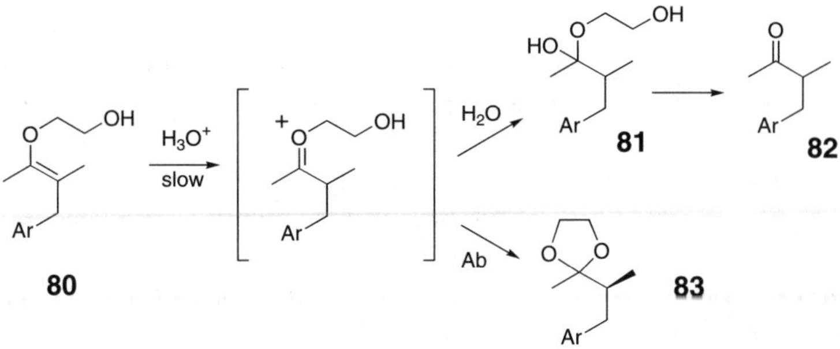

80 **81** **82** **83**

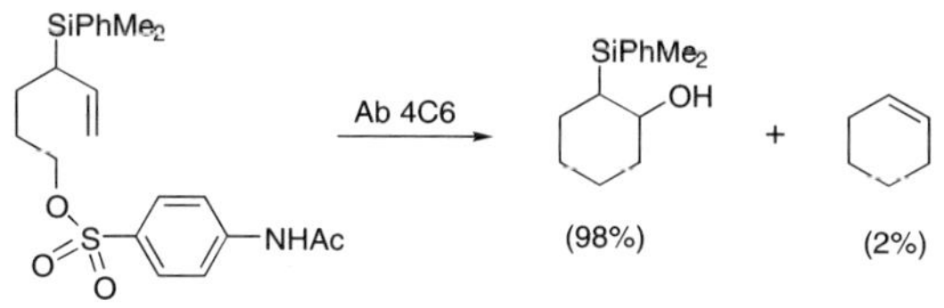

84 **85**

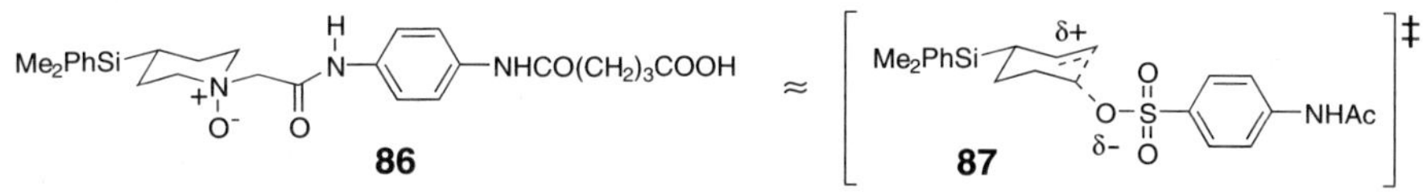

86 **87**

4C6. Moreover, although its absolute configuration is still unknown, only the trans isomer was detected. Given that many synthetic strategies require control of carbocation reactivity, these preliminary results are quite promising. Antibody-catalyzed cationic cyclizations could conceivably be useful for the formation of carbon–carbon and carbon–heteroatom bonds in a variety of multiring molecules, including steroids and heterocyclic compounds.

The related hapten **88**, which contains a quaternary ammonium ion in place of the amine oxide of **86**, also yielded an antibody, TM1-87D7, that catalyzes the cationic cyclization of **84**.[81] In this case, however, the principal product was cyclohexene (90%); trans-cyclohexanol was formed in only 10% yield. Subtle steric features at the active site of the new catalyst presumably allow the substrate to dock in a conformation that stereoelectronically facilitates the elimination of the silicon moiety from the initially formed carbocation. When the silicon atom in **84** is replaced with carbon, blocking the elimination pathway, the trans-cyclohexanol derivative **91** is formed in 60% yield.[82] Introduction of a methyl substituent at the terminal olefin influences the course of reaction in a far more dramatic way, yielding either the exocyclic alcohol **92** as a single diastereomer (80% from cis olefin **89b**) or the cyclopropane derivative **93**, again as a single isomer (63% from trans olefin **89c**).[82] The latter result is particularly notable, since formation of strained cyclopropanes is unprecedented in many studies of cationic cyclizations in simple chemical systems. Although these results could not have been predicted a priori from the structure of the hapten, they indicate the extraordinary extent to which an antibody can exert conformational and chemical control over the course of a reaction. In the present instance, the product distribution can be rationalized after the fact by invoking a protonated cyclopropane intermediate (**90**) that partitions differently depending on its pattern of substitution.[82]

Although formally a substitution reaction, the 6-*endo-tet* ring closure of epoxy-alcohol **94** to form tetrahydropyran **95** is related conceptually to the cationic

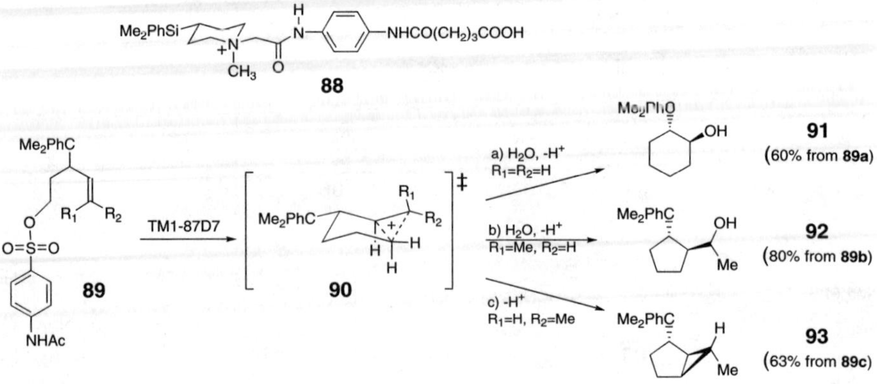

cyclization of **84**. An antibody (26D9) generated against the amine oxide **97** reverses the normal preference for formation of five-membered rings in this reaction (that is to form **96**) and gives enantiomerically pure **95**.[83] Calculations suggest that the catalyst must stabilize the disfavored 6-*endo* transition state by 3.6 kcal/mol more than the 5-*exo* transition state.[84] Antibody 26D9 also catalyzes the cyclization of the related substrate **98** via a 7-*endo-tet* process to yield the oxepane **99** in greater than 98% yield and with an enantiomeric excess of 78%.[85] Under acidic conditions, in contrast, **98** undergoes 6-*exo-tet* ring closure to give **100**. The (*S*,*S*) epoxide of substrate **94** and the (*R*,*R*) epoxide **98** were the preferred substrates for the antibody; in both cases the stereostructure of the products was consistent with direct attack of the alcohol on the epoxide with net inversion of configuration at the reacting center.

Finally, Kitazume and colleagues[86] have used antibody catalysis to promote another class of cyclization reactions: the *exo*- and *endo*-lactonization of

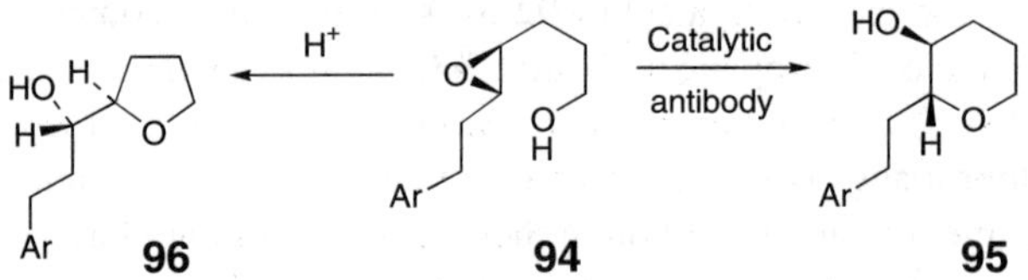

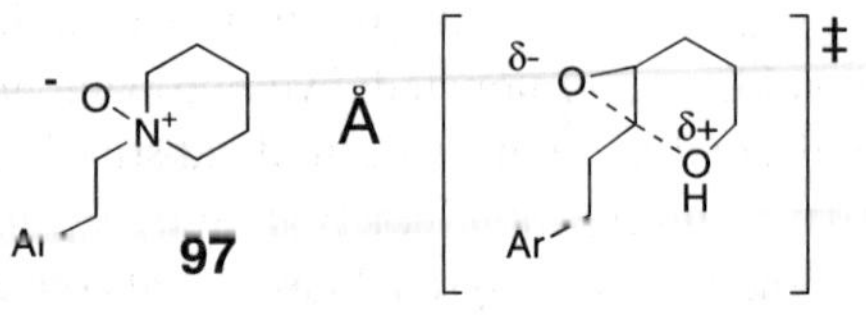

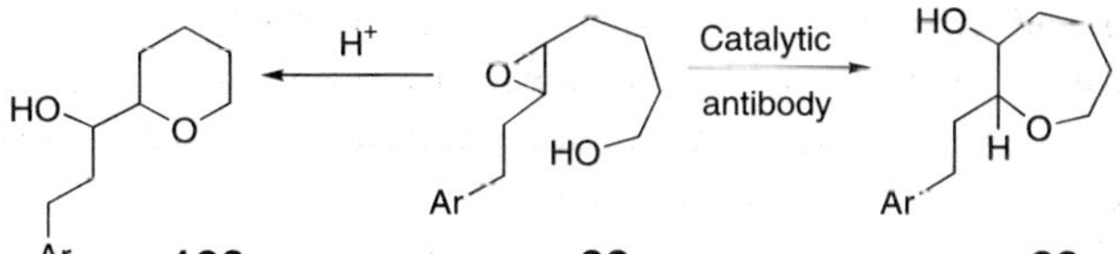

100 **98** **99**

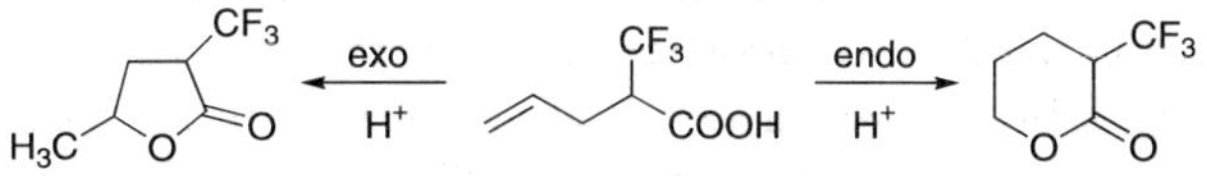

102 **101** **103**

104 **105**

α-trifluoromethyl-γ,δ-unsaturated acid (**101**). They elicited monoclonal antibodies against the sulfoxide-containing molecules **104** and **105**, which catalyzed the regiospecific formation of the γ- or δ-lactones (**102** and **103**, resp.) from the corresponding unsaturated acid. Mechanistic details were not reported; so the nature of the proton donor needed to initiate this reaction is unknown, but the antibody-dependent *exo*-lactonization reaction was found to proceed with high diastereoselectivity (>96% de). Together with the studies cited above, these antibody-catalyzed cyclizations provide striking examples of the immune system's ability to discriminate between two (or more) competing transition states and control their relative energies.

D. Aldol Reactions

Preparation of diverse glycosidase inhibitors and other highly functionalized molecules using aldolases[1] has highlighted the potential of enzyme-catalyzed carbon–carbon bond-forming reactions in organic synthesis. Antibodies capable of effecting aldol condensations (and retro aldol reactions) would be quite useful, particularly if they extended the repertoire of selectivities available from natural sources. Aldol reactions pose special difficulties for antibody catalysis, not only because they require the intervention of acids and bases, but because they proceed via a series of consecutive transition states.

Reymond and Lerner[87] have shown that the multistep nature of the aldol reaction is not an insurmountable impediment for this technology. They screened 46 antibodies elicited by the versatile piperidinium cations **52**, which previously yielded catalysts for several transformations involving acid–base chemistry (see Sections II.A and II.B), and found two that promote the intramolecular aldol condensation of keto-aldehyde **106** to give a substituted 2-benzyl-3-hydroxy-cyclohexanone **107** and subsequently **108**. As illustrated, this conversion involves four individual reaction steps. Antibody 78H6 (anti-**52a**) apparently accelerates the deprotonation reactions and the elimination of water but not the carbon–carbon bond-forming step. Thus, in the presence of the catalyst, all four possible stereoisomers of the aldol product **107** are formed in equal amounts, but the elimination sequence is catalyzed with complete anti-selectivity for a single enantiomer of substrate **107a**. Evidence for the participation of an active-site carboxylate in catalysis was presented, and even though the specificity constants are very low (k_{cat}/K_m = 0.0065 M^{-1} s^{-1})—even by antibody standards—78H6 is approximately 2×10^5 times more reactive than acetate.

Reymond and Chen[88] have investigated the same set of antibodies for their ability to catalyze bimolecular aldol condensation reactions. The antibodies were assayed individually at pH 8.0 for the formation of aldol **111** from aldehyde **109** and acetone. None catalyzed the direct reaction, but in the presence of amine **110** three anti-**52a** and three anti-**52b** antibodies showed modest activity. In analogy with natural type I aldolase enzymes, the reaction is believed to occur by formation of an enamine from acetone and the amine, followed by rate-determining condensation of the enamine with the aldehyde. As in the previous example, the catalyst, which was characterized in detail, is not very efficient in absolute terms ($k_{cat} = 3 \times 10^{-6}$ s^{-1} for the anti-**52b** antibody 72D4), but it is approximately 600 times more effective than amine alone. Moreover, the reactions with the antibody are stereoselective: The enamine adds only to the *si* face of the aldehyde to give

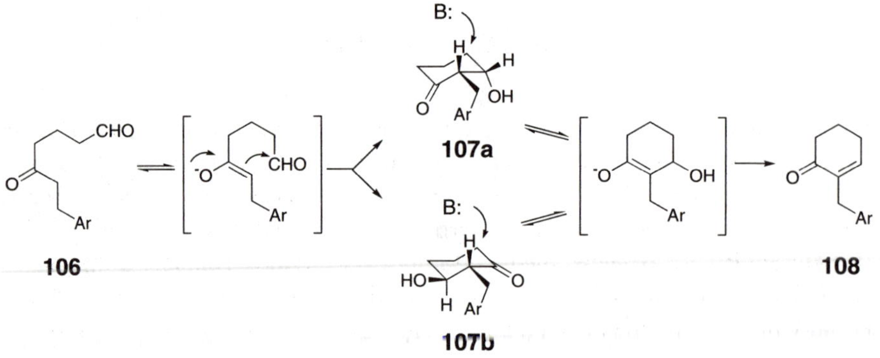

(4*S*,5*S*)-**111** (>95% de) from (*S*)-**109** and (4*S*,5*R*)-**111** from (*R*)-**109** (65% de). When racemic aldehyde is used, these products are formed in a 1:2.8 ratio.

The aldol reaction is readily reversible, with the equilibrium between starting materials and products easily controlled by adjustments in acetone concentration. Reymond[89] has shown that antibody 72D4 accelerates the retro aldol reactions of **111** in the presence of amine **110**, as expected for a true catalyst (Table 3). However, the elimination of water to form **112** competes with the retro aldol to varying extents depending on the stereoisomer examined (Table 3). With the (4*S*,5*R*)-**111** isomer, for example, this pathway is unimportant, whereas with (4*R*,5*S*)-**111** β-elimination is the only reaction observed. For the other two stereoisomers, β-elimination and aldol cleavage occur at competing rates. The two reaction pathways require different conformations of **111**, suggesting that chemoselectivity in this system is the consequence of conformational control imposed by the antibody pocket. Elucidation of the relevant factors, however, will have to await detailed structural information.

The exogenously added amine in the preceding examples serves as a cofactor for the antibodies, mimicking the role of the catalytically essential lysine in the active site of natural type I aldolases. However, the fact that the added amine is neither activated (its pK_a is unchanged within the complex[88]) nor covalently attached to the catalyst necessitates the entropically daunting task of preorganizing three reactants within the binding pocket and may explain the relatively low efficiency of catalysis. More effective antibody aldolases, which exploit a reactive lysine as well as the mechanism of the natural enzymes, have been identified by Wagner et al.[90] using the reactive diketone **113** as the hapten. This molecule reacts covalently with the ε-amino group of a lysine within the binding pocket of two different antibodies to form a stable vinylogous amide **114** (λ_{max} 316 nm, ε 15,000 M^{-1} cm^{-1}). This same amino group, which has an apparent pK_a of 7.2, reacts with ketone substrates to form an enamine that can be efficiently trapped by a variety of

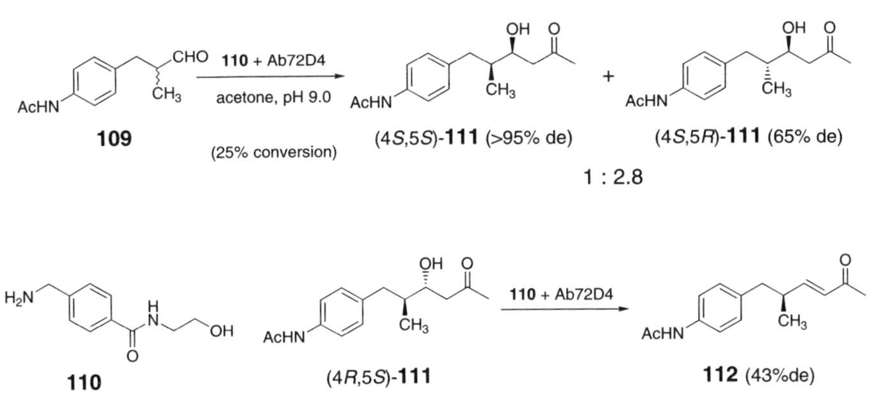

Table 3
Apparent Second-Order Rate Constants for Catalysis by Antibody 72D4 in the Presence of Amine **110**[a]

Substrate	Aldol (M^{-1} min^{-1})	Retro Aldol (M^{-1} min^{-1})	β-Elimination (M^{-1} min^{-1})
(4*S*,5*S*)-**111**	2.1	0.078	0.044
(4*S*,5*R*)-**111**	5.0	0.22	—
(4*R*,5*S*)-**111**	—	—	0.11
(4*R*,5*R*)-**111**	1.1	0.028	0.022

[a]Reference 8.

aldehydes to form a new carbon–carbon bond. The rates of reaction are typically two to three orders of magnitude faster than those reported for the same reactions by the anti-**52** antibodies,[88] presumably reflecting both the enhanced reactivity of the amine and the entropic advantage of fixing it in place within the active site. Again, even though the hapten provided no stereochemical information, the antibodies exhibit excellent stereoselectivity. Thus they catalyze the addition of acetone to the *si*-face of aldehyde **109** regardless of the configuration at C-2 of this substrate (i.e., the same selectivity as the Reymond antibodies[88]). In the reaction between racemic **109** and acetone, for example, antibody 38C2 yielded a roughly 1:1 mixture of (4*S*,5*S*)-**111** (>95% de) and (4*S*,5*R*)-**111** (83% de) after 30% conversion; the diastereofacial selectivity of the second catalyst, 33F12, was somewhat lower for the energetically more demanding formation of the anti-Cram–Felkin product (4*S*,5*R*)-**111** (65% de). Recent work from this group suggests that these aldolase antibodies exhibit unusually broad substrate specificity and can catalyze the formation of a wide range of interesting compounds.[151,152]

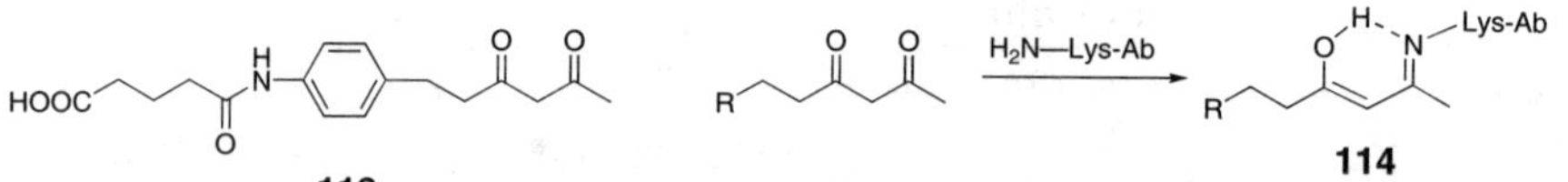

113

114

λ_{max} = 316 nm (ε 15,000 M^{-1} cm^{-1})

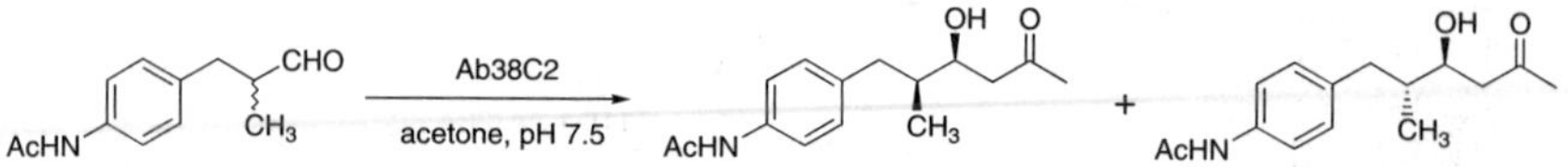

109

(4*S*,5*S*)-**111** (>95% de)

(4*S*,5*R*)-**111** (83% de)

≈1 : 1

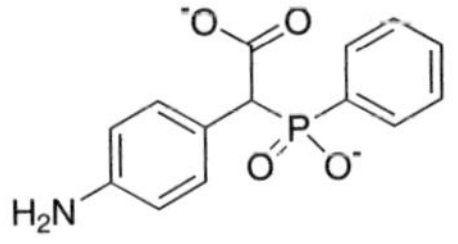

115

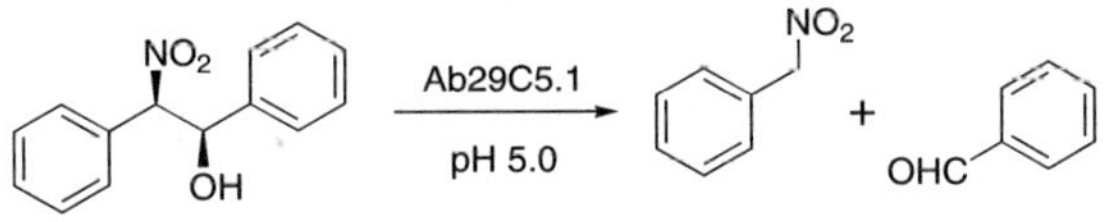

syn-**116**

Formation of the Wieland–Miescher ketone via an enantioselective Robinson annulation represents but one interesting example.[91]

Although attempts to catalyze bimolecular aldol condensations without resorting to enamine chemistry have not yet been successful, the Schultz group[92] has prepared an antibody against the phosphinate hapten **115** that catalyzes the retro aldol reaction of **116** (k_{cat}/K_m = 125 M^{-1} s^{-1}). The equilibrium in this case strongly disfavors the condensation product, and a histidine induced in response to the phosphinate may be involved in catalysis. Interestingly and in contrast to the previous examples, the stereoselectivity of the antibody is modest. The syn diastereomer of **116** was found to be the better substrate for the antibody by 2:1 over the anti diastereomer, but no evidence of enantioselectivity was observed.

E. Pericyclic Reactions

Pericyclic processes comprise a broad and important class of concerted reactions of both theoretical and practical interest. These transformations, which are especially useful in the construction of carbon–carbon bonds,[93] include electrocyclic reactions, sigmatropic rearrangements, and cycloadditions. Because they are not typically subject to general acid–general base chemistry but can be highly sensitive to strain and proximity effects, they are attractive targets for antibody catalysis.

1. Sigmatropic rearrangements

The unimolecular Claisen rearrangement of allyl enol ethers is a [3,3]-sigmatropic process in which formation of a carbon–carbon bond is accompanied by breaking of a carbon–oxygen bond. Although pericyclic reactions are rare in cellular metabolism, the conversion of chorismate (**117**) into prephenate (**119**) is an example of a biologically important Claisen rearrangement. This transformation is the committed step in the biosynthesis of the aromatic amino acids phenylalanine and tyrosine in lower organisms.[94] Analogs of the conformationally constrained transition state **118** have been developed as inhibitors of the natural enzyme, chorismate mutase.[95–97] The best of these, oxabicyclic dicarboxylic acid **120**,[96] yielded catalytic antibodies that accelerate the rearrangement 10^2- to 10^4-fold over

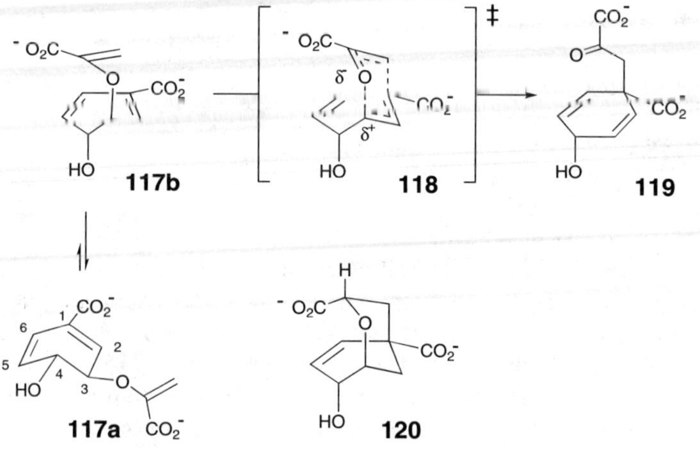

background with high enantioselectivity.[98, 99] For the two antibodies characterized in detail, only the natural (–)-isomer of chorismate is accepted as a substrate; in both cases, enantioselectivities of greater than 40:1 were observed.[100, 101] Even though the hapten was racemic, catalysts for the rearrangement of (+)-chorismate were not identified. Isolation of "unnatural" chorismate mutases will require the screening of larger libraries of antibodies raised against racemic **120**.

Spectroscopic and X-ray crystallographic investigations of the chorismate mutase antibody 1F7 have shown that the induced pocket faithfully reflects the design of the transition-state analog.[102] The active site is complementary in shape and charge to a single isomer of **120**, explaining the preferential binding of the (–)-chorismate enantiomer in the pseudodiaxial conformation (**117b**) needed for reaction. The antibody also exhibits strong mechanistic parallels to natural chorismate mutases, although it appears to be a poorer entropy trap, exploiting fewer specific hydrogen bonding interactions to fix the flexible substrate in place, and has poorer electrostatic complementarity to the transition state. Nevertheless, the antibody's selectivity for (–)-chorismate proved useful for in vitro experiments. Overproduction of 1F7 in the cytoplasm of a yeast strain lacking the natural enzyme corrected the metabolic defect and permitted cell growth in the absence of exogenously added aromatic amino acids.[103] Apparently, the antibody is able to function in vivo and participate in the normal metabolism of the cell.

Catalytic antibody technology is not limited to reactions for which natural enzymes are known. Schultz and co-workers, for example, have generated antibodies that catalyze the abiological Cope rearrangement of the disubstituted 1,5-hexadiene **121** to **122**.[104] This transformation is another [3,3]-sigmatropic process, and the cyclohexanol derivative **123** was used as a hapten to mimic the chairlike six-membered ring geometry of the corresponding pericyclic transition

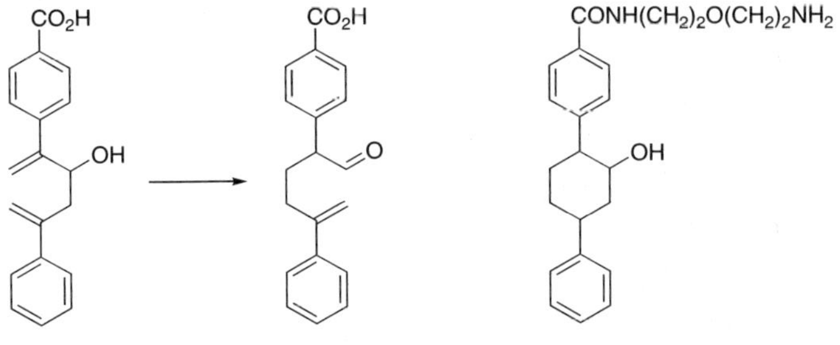

121 **122** **123**

state. As for the Claisen rearrangement of chorismate, k_{cat} values were on the order of several hundred to 5300-fold over the rate constant for the uncatalyzed reaction, and high stereoselectivities were observed.[105] Structural analysis of the antibody AZ-28 suggests that the catalytic effects are achieved by a combination of stereoelectronic effects, which are controlled by an extensive network of binding interactions in the active site.[106] Interestingly, the germ line precursor of AZ-28 is substantially more active ($k_{cat}/k_{uncat} = 3 \times 10^5$), despite its weaker affinity for the hapten.[106]

Antibody catalysts have also been developed for [2,3]-elimination reactions.[107, 108] In one example, Zhou et al. employed substituted proline derivatives **124a** and **124b** to generate a family of antibodies that accelerate the selenoxide elimination of **125** to styrenes **126**.[107] The rate enhancements over the corresponding uncatalyzed reactions ($k_{cat}/k_{uncat} \leq 10^3$) are similar in magnitude to those observed for other antibody-catalyzed sigmatropic processes, but are probably achieved principally through a medium effect in this case. To explore the

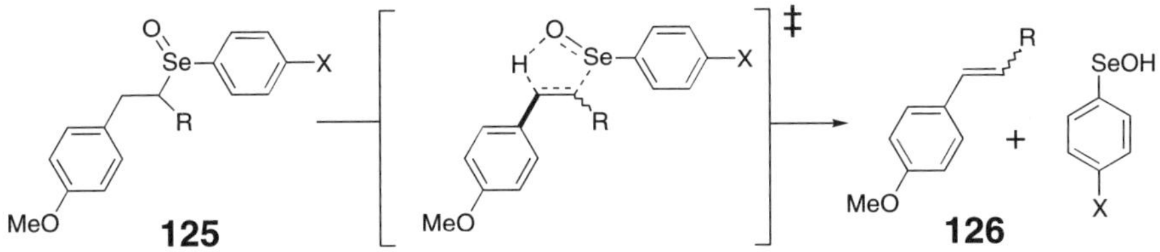

125 **126**

a $R=CO_2H$, $X=NO_2$
b $R=CH_2OH$, $X=NO_2$
c $R=CH_3$, $X=NO_2$
d R=H, $X=NO_2$
e R=H, X=H

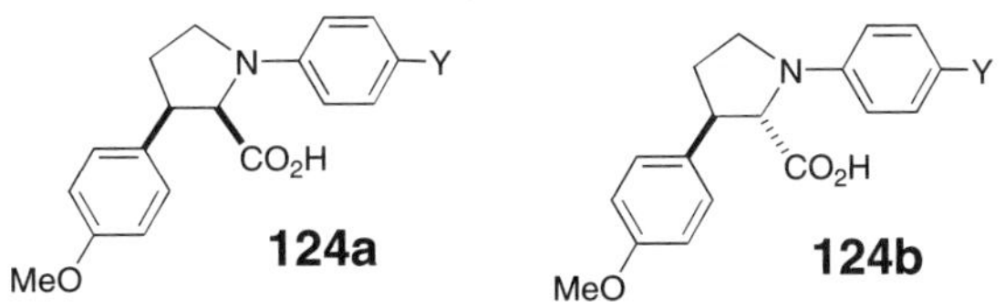

124a **124b**

extent to which conformational constraints imposed by the carboxylate and 3-aryl substituents of the hapten influence the selectivity of the induced catalysts, the secondary selenoxide substrate **125c** was examined as a substrate. The catalyst obtained with the trans-hapten **124b** (SZ-*trans*-28F8) shows no enantioselectivity with this substrate (the selenoxide center itself rapidly epimerizes under the experimental conditions), but the antibodies derived from the cis hapten (SZ-*cis*-39C11 and SZ-*cis*-42F7) convert only 50% of racemic **125c** to product. Furthermore, although formation of the trans olefin is favored in the reaction for steric reasons, SZ-*cis*-39C11 affords a 45:55 mixture of *cis*- and *trans*-anethole. The latter result follows from the hapten design and is notable in light of the severe eclipsing interactions that occur in the transition state for formation of the cis olefin.

2. *Cycloadditions*

Diels–Alder reactions are among the most versatile transformations in organic chemistry for the construction of six-membered rings. As such, they have found wide application in the chemical synthesis of diverse natural and unnatural products.[93, 109, 110] The reaction occurs via concerted addition of a diene to a dienophile, a compound containing a multiple bond, and has a highly ordered, entropically disfavored cyclic transition state with little charge separation.[111] Despite their importance in the laboratory, evidence for the existence of natural enzymes for Diels–Alder reactions has been obtained only recently.[112, 113] Antibodies raised against neutral bicyclic compounds that mimic the boatlike geometry of the Diels–Alder transition state, on the other hand, have achieved significant catalytic effects, including multiple turnovers, high effective molarities, and exacting control of reaction pathway and absolute stereochemistry.[114–116]

As for any bimolecular synthetic reaction, product inhibition is a potentially serious complication that must be addressed at the stage of hapten design. Successful approaches to minimize this problem have exploited chemical[114] or conformational[115] change to drive product release. The latter strategy is illustrated in the design of haptens for the Diels–Alder reaction between the acyclic diene **127** and olefin **128**.[116] Substituted bicyclo[2.2.2]octene derivatives **131** and **132** contain an ethano bridge that constrains the cyclohexene ring into the requisite boat conformation, allowing induction of antibody pockets capable of preorganizing the diene and olefin for reaction. Product dissociation from the active site is then facilitated by an energetically favorable conformational change of the cyclohexene, from boat to twist boat. Appropriate substitution of the bicyclooctene molecule can then be used to control the stereochemical course of the reaction. For instance, hapten **131** corresponds to the transition state for *endo* addition of acrylamide **128** to diene **127**, while hapten **132** mimics the transition state for *exo* addition. As programmed, antibodies raised against these molecules selectively catalyze the formation of either the *endo* adduct **129** (anti-**131** antibodies) or the

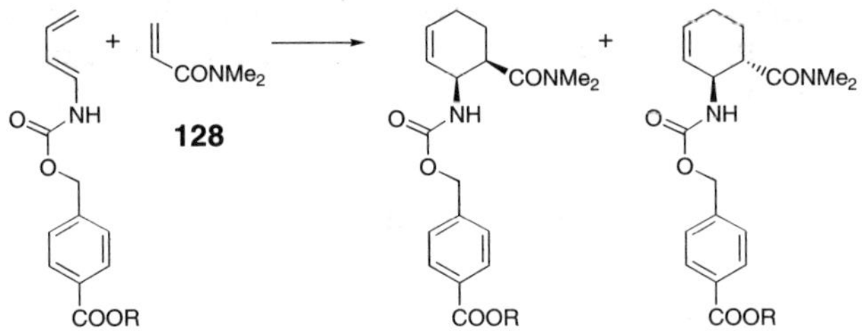

127 **129** *endo* **130** *exo*

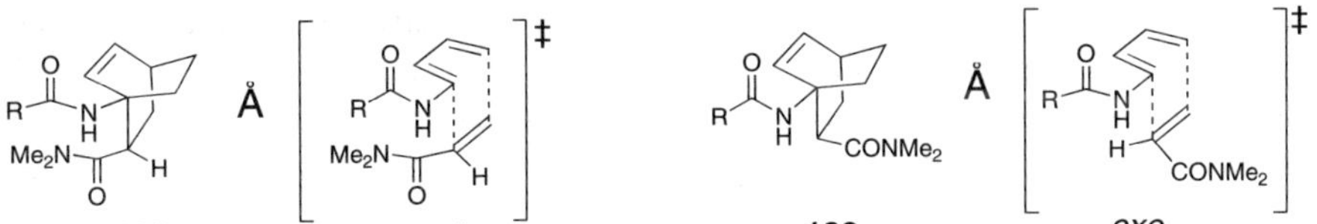

131 *endo* **132** *exo*

normally disfavored *exo* product **130** (anti-**132** antibodies).[116] In both cases, the *endo* and *exo* products are formed with greater than 98% enantioselectivity. These results show the superb control over *exo*/*endo* ratios and absolute configuration of the reaction products that can be achieved with antibodies. With appropriate modifications to the hapten it is likely that the regiochemical course of the Diels–Alder reaction can be similarly controlled.

Pandit and co-workers adopted the same strategy to catalyze a hetero-Diels–Alder reaction in which an aryl-nitroso derivative serves as the dienophile.[117, 118] Antibodies generated against the bicyclic transition-state analogs **133** and **134** accelerated the reaction between the *trans*-diene **135** and **136**, but the ratio of the two product regioisomers **137** and **138** was the same as for the uncatalyzed reaction (58:42). In experiments with stoichiometric amounts of *trans*-**135** and antibody 309-1G7, derived from hapten **134**, this ratio was altered somewhat in favor of product **138** (47:53), in accord with the structure of the

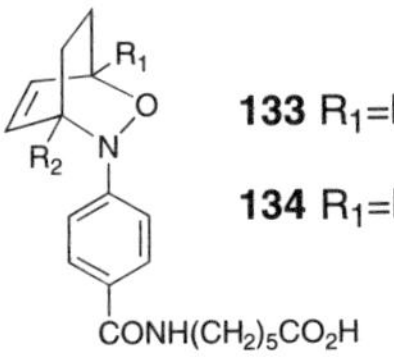

133 R_1=Me, R_2=H

134 R_1=H, R_2=Me

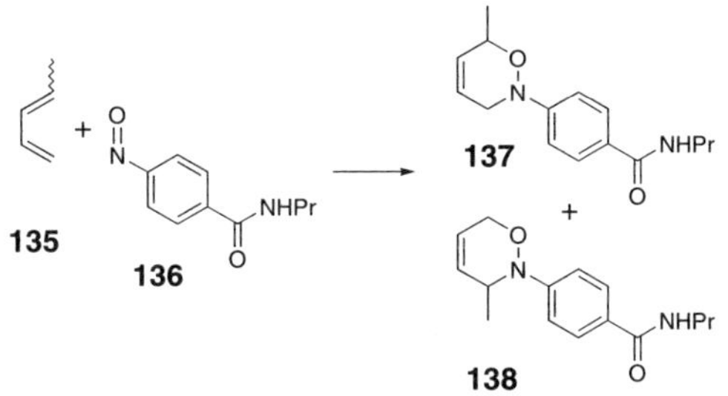

hapten. Antibody 309-1G7 also accepts *cis*-**135**, and in this case, excellent selectivity is observed: Product **138** is formed regioselectively (**137**:**138** = <5:>95) as well as enantioselectively (>82% ee).

F. Oxidation and Reduction

Naturally occurring redox enzymes have been successfully exploited for asymmetric synthesis for some years.[1] Although impressive chemo-, regio-, and enantioselectivities have been achieved in some cases, these biocatalysts have prescribed selectivity and often require expensive cofactors that must be recycled for preparative work. Catalytic antibodies offer an attractive alternative, since they are not limited a priori by Nature's choices. Thus the need for cofactor recycling can be circumvented through the use of inexpensive oxidants and reductants, and, as we have seen above, selectivity can be tailored through appropriate hapten design.

The ability of antibodies to catalyze the stereoselective reduction of ketones with hydride reagents indicates what might be possible. Arata and co-workers have shown that the highly immunogenic dansyl group can be exploited as an anchor to position a side-chain carbonyl group within the chiral environment of an anti-dansyl monoclonal antibody.[119] In the presence of stoichiometric amounts of this antibody, (*S*)-5-(dansylamino)-levulinic acid **139** is reduced by excess sodium borohydride to give 5[(dansylamino)methyl]-2-oxotetrahydrofuran **140** with 36% enantiomeric excess. The antibody thus serves as a kind of chiral auxiliary. Substoichiometric amounts of antibody were not investigated, and the rates of the reaction on and off the antibody were not compared, but this specific system is not likely to be catalytic because tight binding of the dansyl moiety (K_i = 22 nM) will preclude efficient turnover.

Schultz and co-workers have, however, successfully extended this basic approach to obtain antibodies that act as bona fide catalysts. Using phosphonate **141** as a hapten, they generated an antibody that accelerates the reduction of keto amides (**142**) by sodium cyanoborohydride at pH 5.0 with multiple turnovers.[120] In one case, the product α-hydroxy amide **143*S*** was obtained with greater than

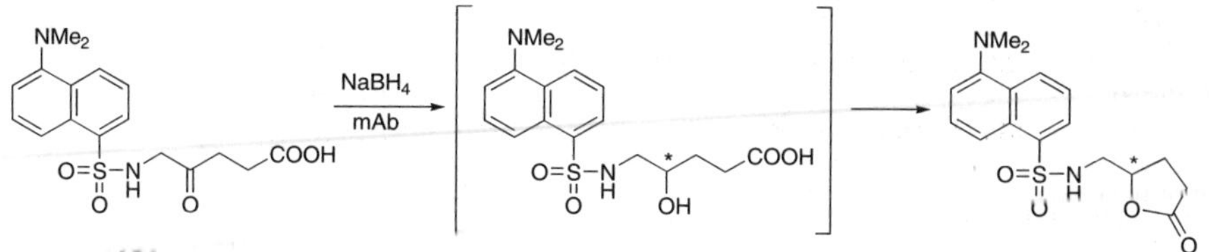

139 **140**

99% diastereoselectivity. For comparison, the corresponding uncatalyzed reaction proceeds with opposite stereoselectivity, yielding product **143*R*** with a diastereomeric excess of 56%. As in analogous experiments to generate hydrolytic antibodies, the negatively charged phosphonate moiety apparently induces an active site capable of polarizing the substrate carbonyl and activating it for attack by the external hydride reagent. Because this microenvironment is inherently chiral, the two faces of the carbonyl can be discriminated with high selectivity, presumably through simple steric shielding of one face. Although inactivation of the antibody by the borohydride reagent might have been a concern, the observation of more than 25 turnovers shows that this is not a significant problem. Further improvements in hapten design are likely to help boost catalytic efficiency (k_{cat}/k_{uncat} = 290). More extensive screening of the immune response to racemic haptens should also allow catalysts with different stereochemical preferences to be identified.

This basic strategy is attractive because it provides a simple and easily generalized method for controlling chemical reactivity. Regioselective reduction of substrates like **146** containing two nearly identical ketone moieties, for instance, is difficult to achieve using conventional chemical methods. Thus, in the absence of catalyst, sodium cyanoborohydride preferentially reduces the methoxybenzyl carbonyl group over the nitrobenzyl group by a factor of 1.35. In the presence of a properly tailored antibody, however, this preference can be easily overridden: Antibody 37B39.3, which was raised against racemic *N*-oxide **144**, accelerates the reduction of the nitrobenzyl carbonyl group of **146** with greater than 75:1 regioselectivity.[121] The antibody-catalyzed reaction is also stereoselective, affording the *S*-isomer of **147** in 96% ee. Again, the hapten serves to induce a chiral environment capable of discriminating the enantiotopic faces of the prochiral carbonyl group and of stabilizing the tetrahedral transition state resulting from

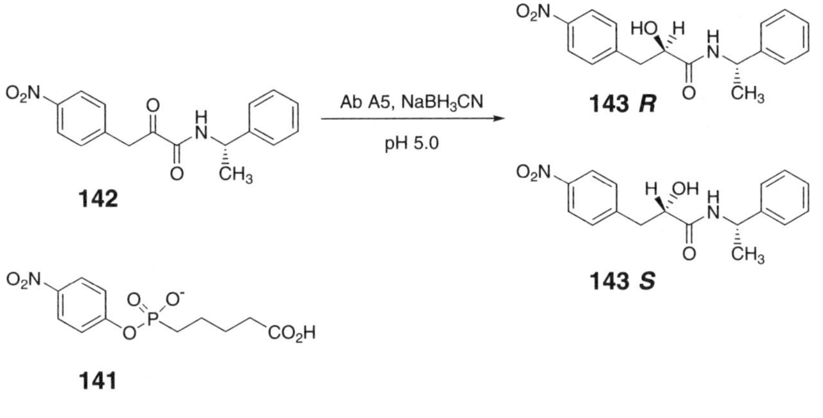

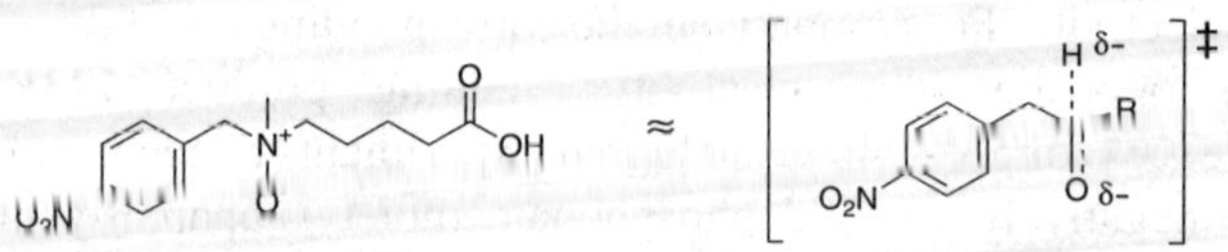

144 145

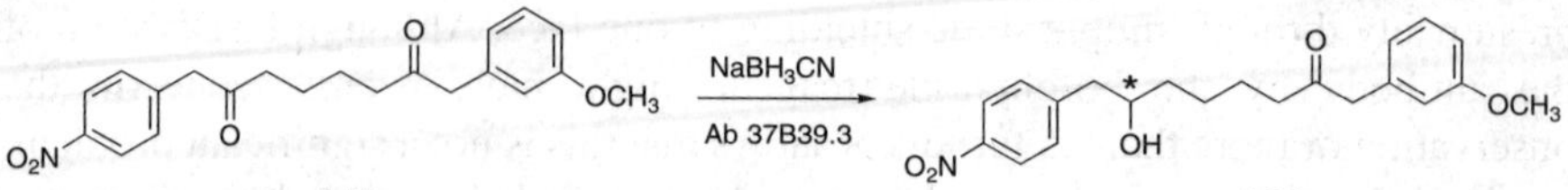

146 147

hydride attack (i.e., **145**). The amine oxide appears more effective in this capacity than the phosphonate **141** described above, as judged by the high enantioselectivities obtained with a series of nitrobenzyl ketones (86–96% ee) and the large rate accelerations observed at low substrate concentrations ($\approx 10^6$-fold over background). Of course, the comparatively large catalyst concentrations needed to drive the reaction to completion is a limitation in practice, but more extensive screening of the immune response and improvements in hapten design may again yield preparatively useful catalysts.

Antibody catalysis has also been used to promote oxidative reactions involving inexpensive oxidants as well. The asymmetric epoxidation of olefins and the oxygenation of sulfides are cases in point. Koch et al. have shown that the versatile antibodies raised against the piperidinium haptens **52** also activate alkenes toward epoxidation by peroxycarboximidic acid.[122] The oxidant, which is compatible with both the antibody and an aqueous milieu, was formed in situ from hydrogen peroxide and acetonitrile. Several of the anti-**52** antibodies modestly accelerated its reaction with a variety of substituted olefins like **148** (to form **149**), possibly as the consequence of interactions with the same active-site carboxylate/carboxylic acid required for their hydrolytic chemistry described above. Depending on the olefin substrate, the enantioselectivity of antibody 20B11, which was characterized in some detail, was estimated to be in the range 67 to >98% after correction for the large amount of racemic product formed via the uncatalyzed reaction under the

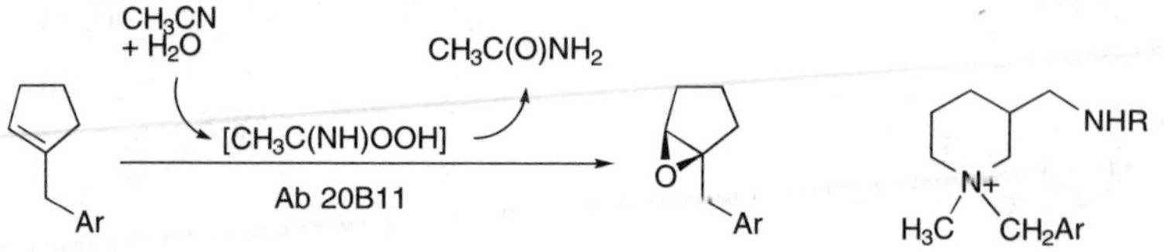

148 149 52

assay conditions. Even though the activity of 20B11 is still too low to compete with highly efficient synthetic catalysts like Jacobson's (salen)Mn system,[123] the results illustrate the ease with which a chiral antibody environment can be exploited for asymmetric reactions. Haptens that better mimic the geometry of the epoxidation reaction should yield improved catalysts, which would be particularly useful for alkenes that are poor substrates for the best synthetic catalysts.

An impressive level of catalysis has been achieved for the reaction of sulfides with sodium periodate using antibodies raised against aminophosphonic acid **150**.[124] The amino group in the hapten, which is protonated under physiological conditions, was expected to induce stabilizing interactions for the positive charge that develops on sulfur in the transition state **151**, whereas the phosphonic acid moiety was expected to elicit a binding site for the periodate. Antibody 28B4.2 accelerates the oxidation of sulfide **152** to sulfoxide **153** with a rate enhancement of 2.2×10^5-fold, indicating the success of this design. In fact, the reported turnover number k_{cat} ($8.2\ s^{-1}$) is comparable to that of some naturally occurring monooxygenase enzymes. However, selectivity is low: The enantiomeric excess for chiral sulfoxide formation by 28B4 is only 16%.[125]

Structural work on this sulfide oxidase antibody provides some insight the origins of the catalytic effects. The X-ray crystal structure of 28B4 was solved in the presence and absence of hapten **150** at 1.9 and 2.2 Å, respectively.[125] As with other antibodies raised against haptens with aromatic groups,[31] the nitroaryl moiety of **150** sits in a deep hydrophobic pocket. The aliphatic linker, in contrast, is relatively solvent exposed, and its interactions with the active site are rather nonspecific. The positively charged ammonium group is stabilized by an interaction with the aromatic π-system of a tyrosine residue, and recognition of the negatively charged phosphonate is achieved through favorable electrostatic and hydrogen-bonding interactions with the side chains of appropriately positioned

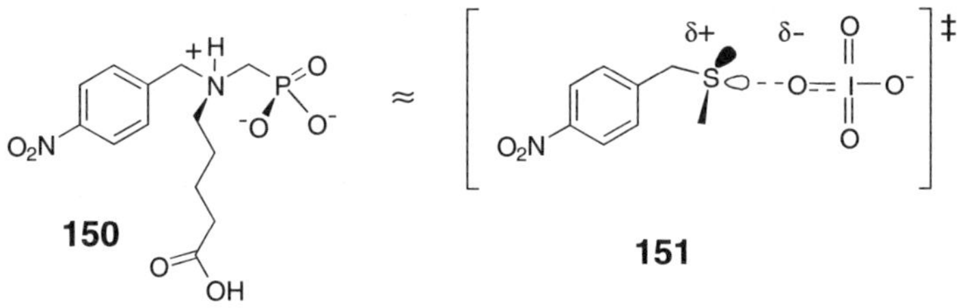

150 **151**

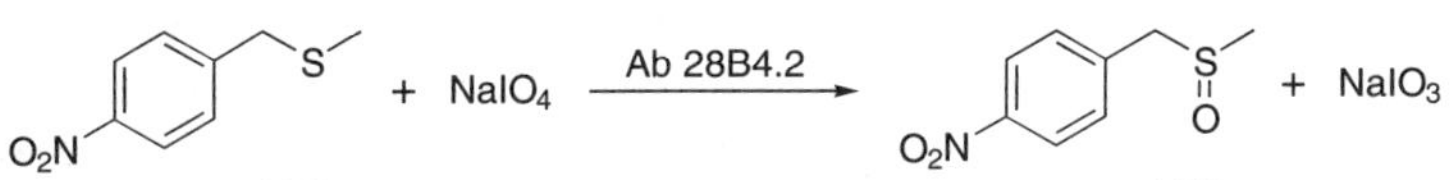

152 **153**

arginine, lysine, and tyrosine residues. The latter interactions are of direct relevance to the mechanism of action of the antibody and suggest the importance of cation–π interactions to stabilize developing charge on sulfur and complementary electrostatic effects to position the periodate properly within the active site. The structural data have been marshaled in support of the importance of precise alignment of the thioether substrate and oxidant for catalysis, but this conclusion appears at odds with the poor enantioselectivity of the catalyst. Instead, the limited complementarity between antibody and hapten in the region occupied by the linker may allow some conformational flexibility to the "short" methyl thioether moiety of the substrate. This would be consistent with the intuitive notion that high enantioselectivity requires rigid conformational constraint.[125] If so, greater enantioselectivity might be achieved in reactions of 28B4 with substrates containing longer aliphatic substituents (i.e., analogous to the linker portion of the hapten).

Many other redox reactions are potentially amenable to antibody catalysis. For example, the chemistry of the P-450 cytochromes, including the hydroxylation of alkanes and the epoxidation of alkenes, can be mimicked with synthetic porphyrins. Incorporation of such molecules into antibody active sites could conceivably yield new catalysts that combine the intrinsic reactivity of the cofactor with the tailored selectivity of the binding pocket. Work is just beginning in this area, but preliminary studies with porphyrin haptens have yielded some interesting results.[126–130] Novel redox chemistry can also be anticipated for antibodies containing metal ions, flavins, nicotinamide analogs, and other reactive moieties.

G. Cofactor-Dependent Processes

Vitamins, cofactors, and metals have the potential to broaden the scope of antibody catalysis considerably. In addition to hydrolytic and redox reactions, they facilitate many complex functional group interconversions in natural enzymes.[131] Pyridoxal, for example, plays a central role in amino acid metabolism. Among the reactions it makes possible are transaminations, decarboxylations, racemizations, and β,γ-eliminations. It is also essential for ethylene biosynthesis. Not surprisingly, then, several groups have sought to incorporate pyridoxal derivatives into antibody combining sites.

One of the earliest published attempts to create antibodies with catalytic activity had as its goal the generation of a transaminase. Raso and Stollar prepared *N*-(5-phosphopyridoxyl)-3′-amino-L-tyrosine **154** as a mimic of the Schiff's base intermediate that is formed during the pyridoxal-dependent transamination of tyrosine and showed that it was a site-directed inhibitor of the enzymes tyrosine transaminase and tyrosine decarboxylase.[132] Partially purified polyclonal antibodies, elicited against γ-globulin conjugates of the hapten, recognized both the cofactor and tyrosine portions of the hapten, and although they did not catalyze Schiff's base formation between pyridoxal and tyrosine, they enhanced the rate of tyrosine transamination by about fivefold. This modest result may reflect the use of polyclonal rather than monoclonal antibodies.

Schultz and co-workers have shown, in fact, that a monoclonal antibody prepared against the structurally analogous hapten **155** can promote aldimine formation in a stereoselective fashion.[133] Antibody 17C5-11C2 accelerates the reaction between *p*-nitro-D-phenylalanine **156** and 5′-deoxypyridoxal **157** to give the Schiff's base **158** at pH 7.0 by a factor of ≈300-fold over background as judged by comparison of apparent second-order rate constants. Surprisingly, *p*-nitro-L-phenylalanine is not a substrate for the antibody, even though the hapten was derived from this enantiomer. The fact that both hapten enantiomers inhibit catalysis ($K_{i,L}$ = 6 nM; $K_{i,D}$ = 17 nM) indicates that the antibody must differentiate the diastereomeric transition states for aminal formation or subsequent dehydration better than it differentiates enantiomers of hapten **155**. None of the antibodies catalyzed any other reactions typical of pyridoxal-dependent enzymes.

Gramatikova and Christen have similarly prepared monoclonal antibodies against N^{α}-phosphopyridoxyl-L-lysine **159**.[134, 135] To compensate for the disadvantage of using a reduced Schiff's base as a transition-state analog, they employed a two-step screening protocol in which prospective catalysts were first tested for their ability to bind the aldimine **160** and then screened directly for the easily monitored pyridoxal-dependent α,β-elimination of β-chloroalanine (Scheme 3). They identified two catalysts out of 24 that promote this reaction; both of these accelerated aldimine formation between pyridoxal and a variety of hydrophobic amino acids, and one of them (antibody 15A9) catalyzed the

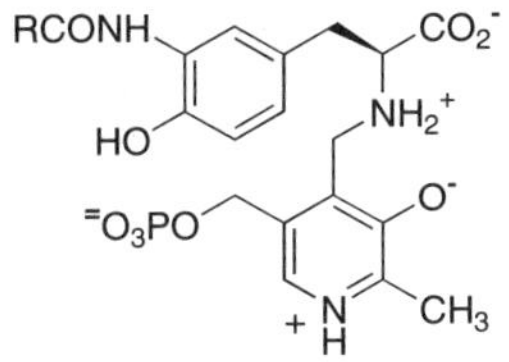

154

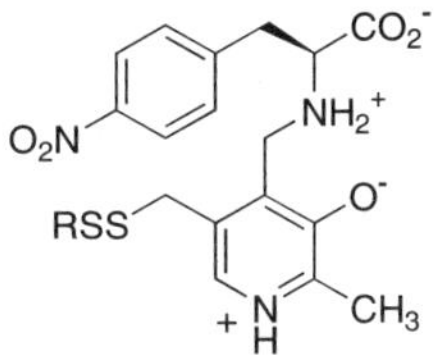

155

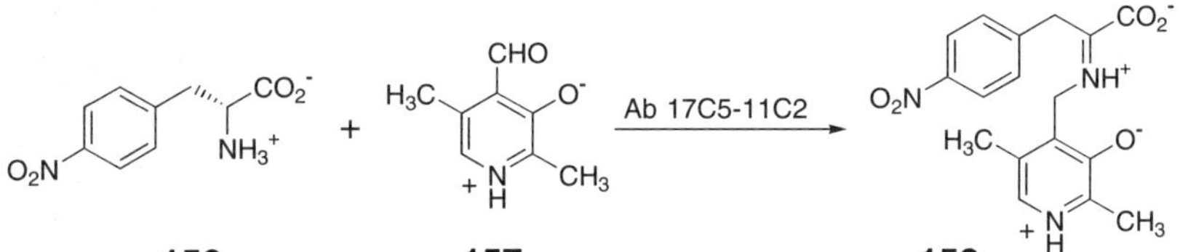

156 **157** **158**

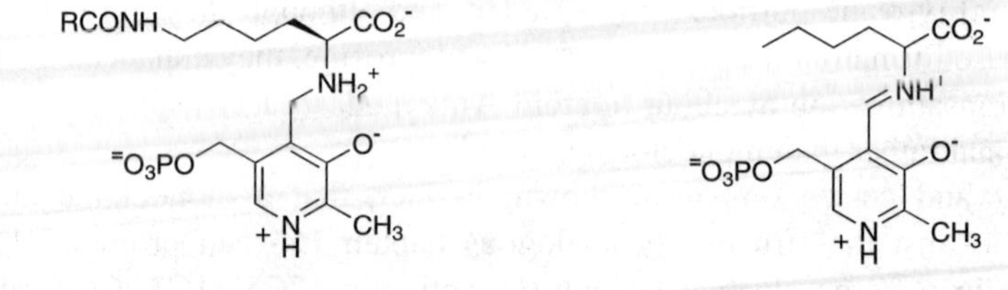

159 **160**

transamination of alanine. A third antibody only catalyzed aldimine formation. The rate accelerations achieved by 15A9 for the conversion of β-chloroalanine and alanine transamination are on the order of 10^3 over the corresponding reactions in the absence of protein. Interestingly, as seen with the Schultz antibody 17C5-11C2, only D-amino acids are accepted as substrates by this catalyst, even though both enantiomers of the hapten analog 5′-phosphopyridoxyl-alanine are recognized with comparable affinity. Since proton transfers from C(α) are essential features of both the α,β-elimination of β-chloroalanine and transamination reactions, the Cα–H bond of the L-amino acid substrate may simply be sterically shielded by the antibody surface. The importance of effective screening for the identification of catalysts with desirable attributes is perhaps the most valuable lesson to be drawn from these experiments.

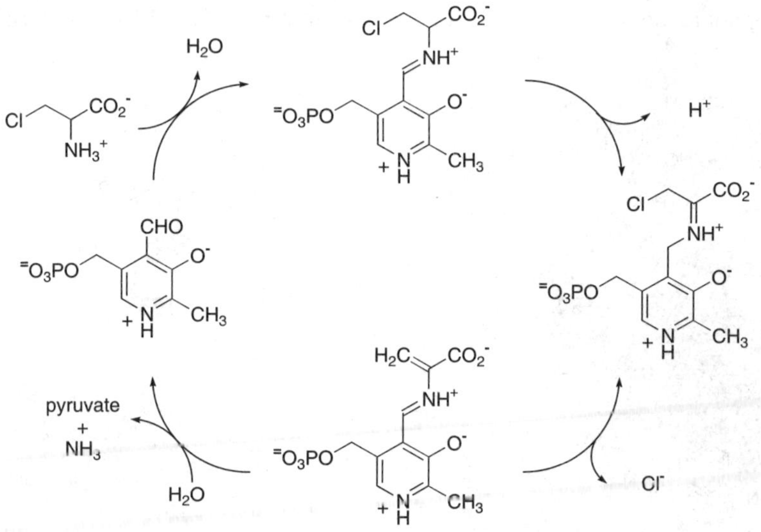

Scheme 3. Pyridoxal-dependent α,β-elimination or β-chloroalanine.

III. FUTURE CHALLENGES

The field of catalytic antibodies has made impressive gains over the past decade. The progression from simple reactions with well-studied mechanisms to processes difficult to achieve via existing chemical methods is particularly noteworthy.[11] Because both catalytic mechanism and reaction stereochemistry can be controlled at the stage of hapten design, antibody catalysis has the potential to provide practical solutions to many problems in organic synthesis. However, broad implementation of this technology will require overcoming a number of formidable obstacles.

First, the repertoire of catalyzed reactions must be greatly increased. Most of the transformations accelerated by antibodies to date have utilized small substrates of limited structural complexity, and many reactions of interest to the synthetic chemist, including aldol condensations, S_N2 displacements, and glycosyl group transfers, have been insufficiently explored. Moreover, existing strategies to generate catalysts for some reactions, like amide hydrolysis, are clearly inadequate. New approaches for tackling such transformations are needed. Narrow substrate specificity—the flip side of the high stereochemical fidelity that characterizes most catalytic antibodies—is also a liability insofar as it requires the development of a new catalyst for each new application.

Second, catalytic efficiency must be greatly enhanced. Large rate accelerations and multiple turnovers have been achieved in a few cases, but catalytic antibodies generally exhibit low activities compared with natural enzymes. This is due, in part, to their short evolutionary history and to the fact that they are selected for tight binding to imperfect transition-state analogs rather than directly for catalytic activity. As a result, even when excellent mimics of transition-state structure are available, as is the case for ester hydrolysis, the best antibodies are still substantially less efficient than their natural counterparts. Severe product inhibition further limits the efficiency of many antibodies, and strategies for facilitating product release must be considered directly in the context of hapten design. Statistically meaningful correlations between hapten structure and the probability of inducing an active antibody could be helpful in establishing useful guidelines for the design of maximally effective templating molecules.

The requirement for multiple catalytic groups—acids, bases, and nucleophiles—makes the design of effective haptens for energetically demanding reactions especially challenging. Charge complementarity has been profitably exploited to elicit catalytic groups in some cases,[67] and mechanism-based inhibitors may further refine our ability to select for specific catalytic mechanisms.[55, 63, 90] It is unlikely, however, that a single hapten will ever elicit arrays of residues as sophisticated as the catalytic triad in serine proteases. As a solution to this problem, Masamune and co-workers have developed an approach, termed *heterologous immunization*, in which two different but structurally related haptens, each containing a different functional group, are used in series to elicit an immune response.[136, 137] The subset of antibodies recognizing both haptens should ideally possess multiple catalytic groups. Preliminary results suggest that antibodies derived from a heterologous immunization experiment are superior to those generated in response to a single hapten.[136, 137] Since the individual haptens need not be structurally complex, this simple strategy can circumvent the need for elaborate hapten synthesis.

Catalytic groups can also be engineered directly into antibody active sites by site-directed mutagenesis.[138–141] This approach is likely to grow in importance with the increasing availability of structural data, although our limited understanding of structure–function relationships within the immunoglobulin scaffold may still preclude large enhancements in efficiency. Random mutagenesis coupled with genetic selection is potentially an even more powerful tool for augmenting the activity of first-generation catalytic antibodies.[103, 142, 143] Antibodies can be produced in living cells as single-chain F_v or *Fab* fragments, where they are able to participate directly in cellular metabolism, and selection schemes can be imagined for a wide range of metabolic processes as well as for transformations involving the creation or destruction of a vital nutrient or toxin.

The third challenge for catalytic antibody technology is to harness the available sequence diversity within the immune system more effectively. Specifically, more thorough screening of the immune response to individual haptens will be essential to the identification of more active catalysts. Because the most active catalysts may also be the rarest, the probability of finding them will increase with the number of candidates assayed. As many as 10^3 antibody-secreting hybridoma cell lines can be obtained using standard fusion protocols, while combinatorial libraries of immunoglobulin fragments containing $>10^6$ members can be constructed using the polymerase chain reaction (PCR). Only a tiny fraction of these libraries is assayed for catalytic activity in a typical catalytic antibody experiment, however. Sensitive immunoassays,[144, 145] tagging methods,[146] and biological selection[103, 142, 143] are likely to improve this situation greatly, enabling substantially larger populations of antibodies to be screened directly for function.

Finally, it would be helpful to make the basic techniques for generating, manipulating, and assaying antibodies more accessible to the practicing organic chemist. Production of monoclonal antibodies is time consuming, labor intensive, and expensive under the best of circumstances, and while the availability of easily manipulated combinatorial libraries of antibody fragments on phage[147, 148] may make catalytic antibodies more user-friendly, considerable expertise in molecular biology will still be necessary. In practice, the experimental protocols are relatively straightforward and highly reproducible,[149] but without special training or collaborators, they may seem insurmountably difficult. Of course, individual catalytic antibodies could be made commercially available were there sufficient demand. In principle, antibodies can be produced at very high levels in microorganisms;[150] so production costs should not be a limiting factor. However, access to catalysts described in the literature is generally limited. For this reason, establishing a central repository for catalytic antibodies with novel activities, such as the American Type Culture Collection, is desirable. Such a facility would allow interested researchers to explore the specificity and activity of novel catalysts with comparatively little investment of time and effort. It would also lend the field a degree of accountability that is currently lacking.

If these challenges can be met, catalytic antibodies could greatly extend the role enzymes play in organic synthesis. Possible applications might include regio- and enantioselective manipulations of functional groups in high-value, structurally complex molecules for which conventional methodologies are inadequate. Synthesis and/or modification of large macromolecules, such as carbohydrates, nucleic acids, and proteins, represents an especially attractive playing field for antibody catalysis given the specific recognition attainable by the large inducible antibody binding. Comparable selectivity is likely to be difficult to achieve with small synthetic catalysts. Finally, the biocompatibility of antibody molecules may make metabolic engineering of novel biosynthetic pathways in vivo a viable alternative to more conventional laboratory syntheses.

Antibodies combine some of the best features of synthetic and enzymatic catalysts. By fusing programmable design with the powerful selective forces of biology, this new technology has the potential to provide many valuable tools for stereoselective organic synthesis in the next century.

ACKNOWLEDGMENT

The author is indebted to the National Institutes of Health, the Skaggs Institute of Chemical Biology the Schweizerische Nationalfonds, and Novartis Pharma AG for generous support.

REFERENCES

1. Wong, C.-H.; Whitesides, G. M. *Enzymes in Synthetic Organic Chemistry*; Elsevier Science Inc.: Tarrytown, New York, 1994.
2. Gerhartz, W. *Enzymes in Industry. Production and Applications*; Gerhartz, W., Ed.; VCH Verlagsgesellschaft mbH: Weinheim (Germany), 1990.
3. Schultz, P. G.; Lerner, R. A. *Science* **1995**, *269*, 1835–1842.
4. Lerner, R. A.; Benkovic, S. J.; Schultz, P. G. *Science* **1991**, *252*, 659–667.
5. Alt, F. W.; Blackwell, T. K.; Yancopoulos, G. D. *Science* **1987**, *238*, 1079–1087.
6. Rajewsky, K.; Förester, I.; Cumang, A. *Science* **1987**, *238*, 1088–1094.

7. Kabat, E. A. *Structural Concepts in Immunology and Immunochemistry*; Holt, Rinehart, and Winston, Inc.: New York, 1976.
8. Landsteiner, K. *The Specificity of Serological Reactions*, 2nd ed.; Harvard University Press; Cambridge, Massachusetts, 1945.
9. Davies, D. R.; Padlan, E. A.; Sheriff, S. *Ann. Rev. Biochem.* **1990**, *59*, 439–473.
10. Wilson, I. A.; Stanfield, R. L. *Current Opinion in Struct. Biol.* **1993**, *3*, 113–318.
11. Schultz, P. G.; Lerner, R. A. *Acc. Chem. Res.* **1993**, *26*, 391–395.
12. Hilvert, D. *Chemistry & Biology* **1994**, *1*, 201–203.
13. Janda, K. D.; Benkovic, S. J.; Lerner, R. A. *Science* **1989**, *244*, 437–440.
14. Pollack, S. J.; Hsiun, P.; Schultz, P. G. *J. Am. Chem. Soc.* **1989**, *111*, 5961–5962.
15. Kitazume, T.; Lin, J. T.; Takeda, M.; Yamazaki, T. *J. Am. Chem. Soc.* **1991**, *113*, 2123–2126.
16. Kitazume, T.; Lin, J. T.; Yamamoto, T.; Yamazaki, T. *J. Am. Chem. Soc.* **1991**, *113*, 8573–8575.
17. Ikeda, S.; Weinhouse, M. I.; Janda, K. D.; Lerner, R. A.; Danishefsky, S. J. *J. Am. Chem. Soc.* **1991**, *113*, 7763–7764.
18. Ikeda, K.; Achiwa, K. *Bioorganic & Medicinal Chemistry Letters* **1997**, *7*, 225–228.
19. Landry, D. W.; Zhao, K.; Yang, G. X.-Q.; Glickman, M.; Georgiadis, T. M. *Science* **1993**, *259*, 1899–1901.
20. Yang, G.; Chun, J.; Arakawa-Uramoto, H.; Wang, X.; Gawinowicz, M. A.; Zhao, K.; Landry, D. W. *J. Am. Chem. Soc.* **1996**, *118*, 5881–5890.
21. Miyashita, H.; Karaki, Y.; Kikuchi, M.; Fujii, I. *Proc. Natl. Acad. Sci. USA* **1993**, *90*, 5337–5340.
22. Campbell, D. A.; Gong, B.; Kochersperger, L. M.; Yonkovich, S.; Gallop, M. A.; Schultz, P. G. *J. Am. Chem. Soc.* **1994**, *116*, 2165–2166.
23. Fujii, I.; Tanaka, F.; Miyashita, H.; Tanimura, R.; Kinoshita, K. *J. Am. Chem. Soc.* **1995**, *117*, 6199–6209.
24. Miyashita, H.; Hara, T.; Tanimura, R.; Tanaka, F.; Kikuchi, M.; Fujii, I. *Proc. Natl. Acad. Sci. USA* **1994**, *91*, 6045–6049.
25. Golinelli-Pimpaneau, B.; Gigant, B.; Bizebard, T.; Navaza, J.; Saludjian, P.; Zemel, R.; Tawfik, D. S.; Eshhar, Z.; Green, B. S.; Knossow, M. *Structure* **1994**, *2*, 175–183.
26. Zhou, G. W.; Guo, J.; Huang, W.; Fletterick, R. J.; Scanlan, T. S. *Science* **1994**, *265*, 1059–1064.
27. Charbonnier, J.-B.; Carpenter, E.; Gigant, B.; Golinelli-Pimpaneau, B.; Tawfik, D.; Eshhar, Z.; Green, B. S.; Knossow, M. *Proc. Natl. Acad. Sci. USA* **1995**, *92*, 11721–11725.
28. Charbonnier, J.-B.; Golinelli-Pimpaneu, B.; Gigant, B.; Tawfik, D. S.; Chap, R.; Schindler, D. G.; Kim, S.-H.; Green, B. S.; Eshhar, Z.; Knossow, M. *Science* **1997**, *275*, 1140–1142.
29. Patten, P. A.; Gray, N. S.; Yang, P. L.; Marks, C. B.; Wedemayer, G. J.; Boniface, J. J.; Stevens, R. C.; Schultz, P. G. *Science* **1996**, *271*, 1086–1090.

30. Guo, J.; Huang, W.; Scanlan, T. S. *J. Am. Chem. Soc.* **1994**, *116*, 6062–6069.

31. MacBeath, G.; Hilvert, D. *Chemistry & Biology* **1996**, *3*, 433–445.

32. Wade, H.; Scanlan, T. S. *J. Am. Chem. Soc.* **1996**, *118*, 6510–6511.

33. Tanaka, F.; Kioshita, K.; Tanimura, R.; Fujii, I. *J. Am. Chem. Soc.* **1996**, *118*, 2332–2339.

34. Iwabuchi, Y.; Miyashita, H.; Tanimura, R.; Kinoshita, K.; Kikuchi, M.; Fujii, I. *J. Am. Chem. Soc.* **1994**, *116*, 771–772.

35. Wirsching, P.; Ashley, J. A.; Benkovic, S. J.; Janda, K. D.; Lerner, R. A. *Science* **1991**, *252*, 680–685.

36. Fernholz, E.; Schloeder, D.; Liu, K. K.-C.; Bradshaw, C. W.; Huang, H.; Janda, K. D.; Lerner, R. A.; Wong, C.-H. *J. Org. Chem.* **1991**, *57*, 4756–4761.

37. Carter, P.; Wells, J. A. *Science* **1987**, *237*, 394–399.

38. Napper, A. D.; Benkovic, S. J.; Tramontano, A.; Lerner, R. A. *Science* **1987**, *237*, 1041–1043.

39. Benkovic, S. J.; Napper, A. D.; Lerner, R. A. *Proc. Natl. Acad. Sci. USA* **1988**, *85*, 5355–5358.

40. Jacobsen, J. R.; Prudent, J. R.; Kochersperger, L.; Yonkovich, S.; Schultz, P. G. *Science* **1992**, *256*, 365–367.

41. Jacobsen, J. R.; Schultz, P. G. *Proc. Natl. Acad. Sci. USA* **1994**, *91*, 5888–5892.

42. Nakatani, T.; Hiratake, J.; Shinzaki, A.; Umeshita, R.; Suzuki, T.; Nishioka, T.; Nakajima, H.; Oda, *J. Tetrahedron Lett.* **1993**, *34*, 4945–4948.

43. Pollack, S. J.; Jacobs, J. W.; Schultz, P. G. *Science* **1986**, *234*, 1570–1572.

44. Hirschmann, R.; Smith III, A. B.; Taylor, C. M.; Benkovic, P. A.; Taylor, S. D.; Yager, K. M.; Sprengeler, P. A.; Benkovic, S. J. *Science* **1994**, *265*, 234–237.

45. Smithrud, D. B.; Benkovic, P. A.; Benkovic, S. J.; Taylor, C. M.; Yager, K. M.; Witherington, J.; Philips, B. W.; Sprengler, P. A.; Smith, I., A. B.; Hirschmann, R. *J. Am. Chem. Soc.* **1997**, *119*, 278–282.

46. Stewart, J. D.; Krebs, J. F.; Siuzdak, G.; Berdis, A. J.; Smithrud, D. B.; Benkovic, S. J. *Proc. Natl. Acad. Sci. USA* **1994**, *91*, 7404–7409.

47. Janda, K. D.; Schloeder, D.; Benkovic, S. J.; Lerner, R. A. *Science* **1988**, *241*, 1188–1191.

48. Martin, M. T.; Angeles, T. S.; Sugasawara, R.; Aman, N. I.; Napper, A. D.; Darsley, M. J.; Sanchez, R. I.; Booth, P.; Titmas, R. C. *J. Am. Chem. Soc.* **1994**, *116*, 6508–6512.

49. Iverson, B. L.; Lerner, R. A. *Science* **1989**, *243*, 1184–1188.

50. Gibbs, R. A.; Taylor, S.; Benkovic, S. J. *Science* **1992**, *258*, 803–805.

51. Liotta, L. J.; Benkovic, P. A.; Miller, G. P.; Benkovic, S. J. *J. Am. Chem. Soc.* **1993**, *115*, 350–351.

52. Reymond, J.-L.; Janda, K. D.; Lerner, R. A. *Angew. Chem. Int. Ed. Engl.* **1991**, *30*, 1711–1713.

53. Yu, J.; Hsieh, L. C.; Kochersperger, L.; Yonkovich, S.; Stephans, J. C.; Gallop, M. A.; Schultz, P. G. *Angew. Chem. Int. Ed. Engl.* **1994**, *33*, 339–341.
54. Suga, H.; Tanimoto, N.; Sinskey, A. J.; Masamune, S. *J. Am. Chem. Soc.* **1994**, *116*, 11197–11198.
55. Janda, K. D.; Lo, L.-C.; Lo, C.-H. L.; Sim, M.-M.; Wang, R.; Wong, C.-H.; Lerner, R. A. *Science* **1997**, *275*, 945–948.
56. Chin, J. *Acc. Chem. Res.* **1991**, *24*, 145–152.
57. Bryant, R. A. R.; Hansen, D. E. *J. Am. Chem. Soc.* **1996**, *118*, 5498–5499.
58. Scanlan, T. S.; Prudent, J. R.; Schultz, P. G. *J. Am. Chem. Soc.* **1991**, *113*, 9397–9398.
59. Lavey, B. J.; Janda, K. D. *J. Org. Chem.* **1996**, *61*, 7633–7636.
60. Rosenblum, J. S.; Lo, S.-C.; Li, T.; Janda, K. D.; Lerner, R. A. *Angew. Chem. Int. Ed. Engl.* **1995**, *34*, 2275–2277.
61. Weiner, D. P.; Wiemann, T.; Wolfe, M. M.; Wentworth, P. J.; Janda, K. D. *J. Am. Chem. Soc.* **1997**, *119*, 4088–4089.
62. Kohen, F.; Kim, J.-B.; Barnard, G.; Lindner, H. R. *Biochem. Biophys. Acta* **1980**, *629*, 328–337.
63. Wirsching, P.; Ashley, J. A.; Lo, C.-H. L.; Janda, K. D.; Lerner, R. A. *Science* **1995**, *270*, 1775–1782.
64. Shokat, K. M.; Leumann, C. J.; Sugasawara, R.; Schultz, P. G. *Nature* **1989**, *338*, 269–271.
65. Shokat, K.; Uno, T.; Schultz, P. G. *J. Am. Chem. Soc.* **1994**, *116*, 2261–2270.
66. Cravatt, B. F.; Ashley, J. A.; Janda, K. D.; Boger, D. L.; Lerner, R. A. *J. Am. Chem. Soc.* **1994**, *116*, 6013–6014.
67. Thorn, S. N.; Daniels, R. G.; Auditor, M.-T. M.; Hilvert, D. *Nature* **1995**, *373*, 228–230.
68. Hollfelder, F.; Kirby, A. J.; Tawfik, D. S. *Nature* **1996**, *383*, 60–63.
69. Kikuchi, K.; Hilvert, D. *J. Am. Chem. Soc.* **1996**, *118*, 8184–8185.
70. Uno, T.; Schultz, P. G. *J. Am. Chem. Soc.* **1992**, *114*, 6573–6574.
71. Uno, T.; Gong, B.; Schultz, P. G. *J. Am. Chem. Soc.* **1994**, *116*, 1145–1146.
72. Sinha, S. C.; Keinan, E.; Reymond, J.-L. *Proc. Natl. Acad. Sci. USA* **1993**, *90*, 11910–11913.
73. Reymond, J.-L.; Janda, K. D.; Lerner, R. A. *J. Am. Chem. Soc.* **1992**, *114*, 2257–2258.
74. Reymond, J.-L.; Jahangiri, G. K.; Stoudt, C.; Lerner, R. A. *J. Am. Chem. Soc.* **1993**, *115*, 3909–3917.
75. Jahangiri, G. K.; Reymond, J.-L. *J. Am. Chem. Soc.* **1994**, *116*, 11264–11274.
76. Sinha, S. C.; Keinan, E.; Reymond, J.-L. *J. Am. Chem. Soc.* **1993**, *115*, 4893–4894.
77. Sinha, S. C.; Keinan, E. *J. Am. Chem. Soc.* **1995**, *117*, 3653–3654.
78. Reymond, J.-L.; Reber, J.-L.; Lerner, R. A. *Angew. Chem. Int. Ed. Engl.* **1994**, *33*, 475–477.
79. Shabat, D.; Itzhaky, H.; Reymond, J.-L.; Keinan, E. *Nature* **1995**, *374*, 143–146.

80. Li, T.; Janda, K. D.; Ashley, J. A.; Lerner, R. A. *Science* **1994**, *264*, 1289–1293.
81. Li, T.; Hilton, S.; Janda, K. D. *J. Am. Chem. Soc.* **1995**, *117*, 3308–3309.
82. Li, T.; Janda, K. D.; Lerner, R. A. *Nature* **1996**, *379*, 326–327.
83. Janda, K. D.; Shevlin, C. G.; Lerner, R. A. *Science* **1993**, *259*, 490–493.
84. Na, J.; Houk, K. N.; Shevlin, C. G.; Janda, K. D.; Lerner, R. A. *J. Am. Chem. Soc.* **1993**, *115*, 8453.
85. Janda, K. D.; Shevlin, C. G.; Lerner, R. A. *J. Am. Chem. Soc.* **1995**, *117*, 2659–2660.
86. Kitazume, T.; Takeda, M. *J. Chem. Soc., Chem. Commun.* **1995**, 39–40.
87. Koch, T.; Reymond, J.-L.; Lerner, R. *J. Am. Chem. Soc.* **1995**, *117*, 9383–9387.
88. Reymond, J.-L.; Chen, Y. *J. Org. Chem.* **1995**, *60*, 6970–6979.
89. Reymond, J.-L. *Angew. Chem. Int. Ed. Engl.* **1995**, *34*, 2285–2287.
90. Wagner, J.; Lerner, R.; Barbas, C. *Science* **1995**, *270*, 1797–1800.
91. Zhong, G.; Hoffmann, T.; Lerner, R. A.; Danishefsky, S.; Barbas III, C. F. *J. Am. Chem. Soc.* **1997**, *119*, 8131–8132.
92. Flanagan, M. E.; Jacobsen, J. R.; Sweet, E.; Schultz, P. G. *J. Am. Chem. Soc.* **1996**, *118*, 6078–6079.
93. Desimoni, G.; Tacconi, G.; Barco, A.; Pollini, G. P. *Natural Product Synthesis through Pericyclic Reactions*; American Chemical Society: Washington, D.C., 1983.
94. Weiss, U.; Edwards, J. M. *The Biosynthesis of Aromatic Amino Compounds*; Wiley: New York, 1980.
95. Andrews, P. R.; Cain, E. N.; Rizzardo, E.; Smith, G. D. *Biochemistry* **1977**, *16*, 4848–4852.
96. Bartlett, P. A.; Nakagawa, Y.; Johnson, C. R.; Reich, S. H.; Luis, A. *J. Org. Chem.* **1988**, *53*, 3195–3210.
97. Chao, H. S.-I.; Berchtold, G. A. *Biochemistry* **1982**, *21*, 2778–2781.
98. Jackson, D. Y.; Jacobs, J. W.; Sugasawara, R.; Reich, S. H.; Bartlett, P. A.; Schultz, P. G. *J. Am. Chem. Soc.* **1988**, *110*, 4841–4842.
99. Hilvert, D.; Carpenter, S. H.; Nared, K. D.; Auditor, M.-T. M. *Proc. Natl. Acad. Sci. USA* **1988**, *85*, 4953–4955.
100. Jackson, D. Y.; Liang, M. N.; Bartlett, P. A.; Schultz, P. G. *Angew. Chem. Int. Ed. Engl.* **1992**, *31*, 182–183.
101. Hilvert, D.; Nared, K. D. *J. Am. Chem. Soc.* **1988**, *110*, 5593–5594.
102. Haynes, M. R.; Stura, E. A.; Hilvert, D.; Wilson, I. A. *Science* **1994**, *263*, 646–652.
103. Tang, Y.; Hicks, J. B.; Hilvert, D. *Proc. Natl. Acad. Sci. USA* **1991**, *88*, 8784–8786.
104. Braisted, A. C.; Schultz, P. G. *J. Am. Chem. Soc.* **1994**, *116*, 2211–2212.
105. Ulrich, H. D.; Driggers, E. M.; Schultz, P. G. *Acta Chim. Scand.* **1996**, *50*, 328–332.
106. Ullrich, H. D.; Mundorff, E.; Santarsiero, B. D.; Driggers, E. M.; Stevens, R. C.; Schultz, P. G. *Nature* **1997**, *389*, 271–275.
107. Zhou, Z. S.; Jiang, N.; Hilvert, D. *J. Am. Chem. Soc.* **1997**, *119*, 3623–3624.

108. Yoon, S. S.; Oei, Y.; Sweet, E.; Schultz, P. G. *J. Am. Chem. Soc.* **1996**, *118*, 11686–11687.

109. Wasserman, A. *Diels–Alder Reactions*; Elsevier Publ. Co.: Amsterdam, 1965.

110. Boger, D. L.; Weinreb, S. M. *Hetero-Diels Alder Methodology in Organic Synthesis*; Academic Press, Inc.: San Diego, 1987.

111. Sauer, J.; Sustman, R. *Angew. Chem. Int. Ed. Engl.* **1980**, *19*, 779–807.

112. Oikawa, H.; Suzuki, Y.; Naya, A.; Katayama, K.; Ichihara, A. *J. Am. Chem. Soc.* **1994**, *116*, 3605–3606.

113. Oikawa, H.; Katayama, K.; Suzuki, Y.; Ichihara, A. *J. Chem. Soc., Chem. Commun.* **1995**, 1321.

114. Hilvert, D.; Hill, K. W.; Nared, K. D.; Auditor, M.-T. M. *J. Am. Chem. Soc.* **1989**, *111*, 9261–9262.

115. Braisted, A. C.; Schultz, P. G. *J. Am. Chem. Soc.* **1990**, *112*, 7430–7431.

116. Gouverneur, V. E.; Houk, K. N.; Pascual-Teresa, B.; Beno, B.; Janda, K. D.; Lerner, R. A. *Science* **1993**, *262*, 204–208.

117. Meekel, A. A. P.; Resmini, M.; Pandit, U. K. *J. Chem. Soc., Chem. Commun.* **1995**, 571.

118. Resmini, M.; Meekel, A. A. P.; Pandit, U. K. *Pure Appl. Chem.* **1996**, *68*, 2025–2028.

119. Kim, J. I.; Nagano, T.; Higuchi, T.; Hirobe, M.; Shimada, I.; Arata, Y. *J. Am. Chem. Soc.* **1991**, *113*, 9392–9394.

120. Nakayama, G. R.; Schultz, P. G. *J. Am. Chem. Soc.* **1992**, *114*, 780–781.

121. Hsieh, L. C.; Yonkovich, S.; Kochersperger, L.; Schultz, P. G. *Science* **1993**, *260*, 337–339.

122. Koch, A.; Reymond, J.-L.; Lerner, R. A. *J. Am. Chem. Soc.* **1994**, *116*, 803–804.

123. Jacobsen, E. N.; Finney, N. S. *Chemistry & Biology* **1994**, *1*, 85–90.

124. Hsieh, L. C.; Stephans, J. C.; Schultz, P. G. *J. Am. Chem. Soc.* **1994**, *116*, 2167–2168.

125. Hsieh-Wilson, L. C.; Schultz, P. G.; Stevens, R. C. *Proc. Natl. Acad. Sci. USA* **1996**, *93*, 5363–5367.

126. Cochran, A. G.; Schultz, P. G. *Science* **1990**, *249*, 781–783.

127. Harada, A.; Okamoto, K.; Kamachi, M.; Honda, T.; Miwantani, T. *Chemistry Letters* **1990**, 917–918.

128. Keinan, E.; Sinha-Bagchi, A.; Benory, E.; Ghozi, M. C.; Eshhar, Z.; Green, B. S. *Pure & Appl. Chem.* **1990**, *62*, 2013–2019.

129. Ohkubo, K.; Ishida, H.; Sagawa, T.; Urabe, K.; Seri, K.; Suga, M. *Chemistry Letters* **1993**, 61–64.

130. Schwabacher, A. W.; Weinhouse, M. I.; Auditor, M.-T. M.; Lerner, R. A. *J. Am. Chem. Soc.* **1989**, *111*, 2344–2346.

131. Walsh, C. *Enzymatic Reaction Mechanisms*; W. H. Freeman and Company: New York, 1979.

132. Raso, V.; Stollar, D. *Biochemistry* **1975**, *14*, 584–599.

133. Cochran, A. G.; Pham, T.; Sugasawara, R.; Schultz, P. G. *J. Am. Chem. Soc.* **1991**, *113*, 6670–6672.

134. Gramatikova, S. I.; Christen, P. *J. Biol. Chem.* **1996**, *271*, 30583–30586.

135. Gramatikova, S. I.; Christen, P. *J. Biol. Chem.* **1997**, *272*, 9779–9784.

136. Suga, H.; Ersoy, O.; Williams, S. F.; Tsumuraya, T.; Margolies, M. N.; Sinskey, A. J.; Masamune, S. *J. Am. Chem. Soc.* **1994**, *116*, 6025–6026.

137. Tsumuraya, T.; Suga, H.; Meguro, S.; Tsunakawa, A.; Masamune, S. *J. Am. Chem. Soc.* **1995**, *117*, 11390–11396.

138. Baldwin, E.; Schultz, P. G. *Science* **1989**, *24*, 1104–1107.

139. Stewart, J. D.; Roberts, V. A.; Thomas, N. R.; Getzoff, E. D.; Benkovic, S. J. *Biochemistry* **1994**, *33*, 1994–2003.

140. Roberts, V. A.; Stewart, J.; Benkovic, S. J.; Getzoff, E. D. *J. Mol. Biol.* **1994**, *235*, 1098–1116.

141. Jackson, D. Y.; Prudent, J. R.; Baldwin, E. P.; Schultz, P. G. *Proc. Natl. Acad. Sci. USA* **1991**, *88*, 58–62.

142. Smiley, J. A.; Benkovic, S. J. *Proc. Natl. Acad. Sci. USA* **1994**, *91*, 8319–8323.

143. Lesley, S. A.; Patten, P. A.; Schultz, P. G. *Proc. Natl. Acad. Sci. USA* **1993**, *90*, 1160–1165.

144. Tawfik, D. S.; Green, B. S.; Chap, R.; Sela, M.; Eshhar, Z. *Proc. Natl. Acad. Sci. USA* **1993**, *90*, 373–377.

145. MacBeath, G.; Hilvert, D. *J. Am. Chem. Soc.* **1994**, *116*, 6101–6106.

146. Lane, J. W.; Hong, X.; Schwabacher, A. W. *J. Am. Chem. Soc.* **1993**, *115*, 2078–2080.

147. Marks, J. D.; Hoogenboom, H. R.; Griffiths, A. D.; Winter, G. *J. Biol. Chem.* **1992**, *267*, 16007–16010.

148. Chiswell, D. J.; McCafferty, J. *TIBTECH* **1992**, *10*, 80–84.

149. Harlow, E.; Lane, D. *Antibodies: A Laboratory Manual*; Cold Spring Harbor Laboratory, 1988.

150. Carter, P.; Kelley, R. F.; Rodrigues, M. L.; Snedecor, B.; Covarrubias, M.; Velligan, M. D.; Wong, W. L. T.; Rowland, A. M.; Kotts, C. E.; Carver, M. E.; Yang, M.; Bourell, J. H.; Shephard, H. M.; Henner, D. *Bio/Technology* **1992**, *10*, 163–167.

151. Barbas III, C. F.; Heine, A.; Zhong, G.; Hoffmann, T.; Gramatikova, S.; Björnstedt, R.; List, B.; Anderson, J.; Stura, E. A.; Wilson, I. A.; Lerner, R. A. *Science* **1997**, *278*, 2085–2092.

152. Hoffmann, T.; Zhong, G.; List, B.; Shabat, D.; Anderson, J.; Gramatikova, S.; Lerner, R. A.; Barbas III, C. F. *J. Am. Chem. Soc.* **1998**, *120*, 2768–2779.

Stereoelectronic Effects of the Group 4 Metal Substituents in Organic Chemistry

JONATHAN M. WHITE AND CHRISTOPHER I. CLARK

School of Chemistry, The University of Melbourne, Parkville, Victoria 3052 Australia

Topics in Stereochemistry, Volume 22, Edited by Scott E. Denmark.
ISBN 0-471-25316-2

I. INTRODUCTION

The presence of donor and acceptor groups in organic molecules often results in interesting and unusual chemical and spectroscopic properties. These properties arise from interaction between filled orbitals on the donor group and electron-deficient orbitals on the acceptor group. The interactions can be broadly described as occurring directly through space, or by transmission through the organic framework (through bond), or perhaps by a combination of the two.[1, 2] The former interaction is exemplified by the structure of clivorine[3] **1** for which the close transannular contact between the amino group and the carbonyl carbon and the lengthened C=O distance is evidence for a through-space $n_N - \pi^*_{(C=O)}$ interaction.

A through-bond interaction between the nitrogen lone pair and the developing carbocation orbital is believed to be responsible for the hugely different reactivities of the bicyclic chlorides **2** and **3** towards unimolecular solvolysis.[4]

Whichever of the above types of interaction occurs, a strong dependence of the magnitude of the interaction upon the relative orientation of the donor and acceptor fragments is observed. The results of these interactions are often referred to as stereoelectronic effects, a description that reflects the electronic basis of these effects, and recognizes their dependence upon stereochemistry.

Perhaps the most well-known stereoelectronic effect is the anomeric effect,[5–7] which describes the preference for electronegative substituents at the 2-position of a tetrahydropyran ring to adopt the axial conformation (Figure 1). This conformation, which is counter to steric predictions, is believed to be favored partly because it optimizes overlap between one of the lone pairs of the ring oxygen (donor) and the low-lying C–X antibonding orbital (acceptor). More generally the anomeric effect describes the preference for the R–X–C–Y fragment to be gauche, where X is a heteroatom, Y is an electronegative substituent, and C is a saturated carbon.

The gauche effect[5, 8, 9] is a stereoelectronic effect that was initially recognized in molecules containing vicinal polar bonds. Molecules containing these fragments often show a strong bias towards conformations that are counter to steric

Structures 1-3

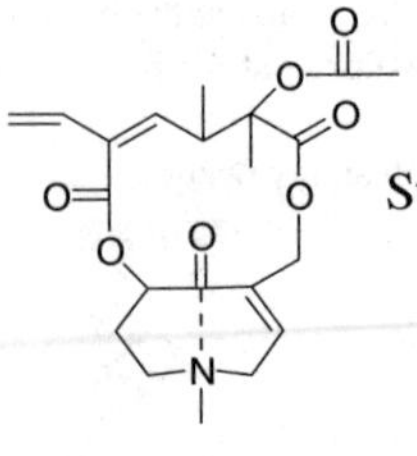

1

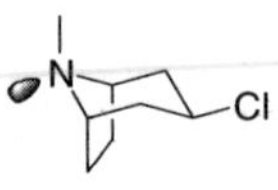

2

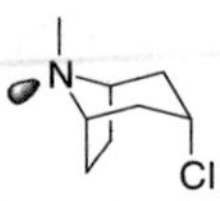

3

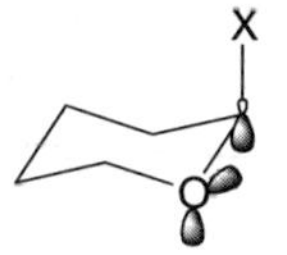

Figure 1. Axial conformation of a 2-substituted tetrahydropyran ring allowing maximum overlap between the n_O and σ^*_{C-X}.

predictions, and that result from maximization of the interaction between the best donor bond and the best acceptor bond. A simple example[9] is 1,2-difluoroethane, for which the preferred conformation is that shown in Figure 2, where there is a gauche relationship between the fluorine substituents; this conformation, which is expected to be disfavored from steric and electrostatic considerations, maximizes overlap between the best donor orbitals (C–H bonds) and the best acceptor orbitals (the C–F antibonding orbitals) across the C–C bond.

The degree of influence that stereoelectronic effects can have on the ground-state and reactivity properties of a given molecule depends not only upon the stereochemical relationship between the donor and acceptor group, but also on their intrinsic donor and acceptor abilities. A good donor is one that has weakly held electrons; that is, the electrons (or electron) occupy a high-energy orbital. Typical examples include heteroatom nonbonded electron pairs, electron-rich π systems, and high-energy sigma-bonding orbitals. A good acceptor has an unfilled low-energy orbital, and is typified by carbocation *p* orbitals, the π^* orbitals of an electron-deficient alkene or aryne, and the antibonding orbital of a C–X bond, where X is an electronegative group.

The C–M σ bonds (where M = Si, Ge, Sn, Pb) in the fragment C–MR_3 (R = alkyl, aryl) are powerful σ donor orbitals; the electrons in the C–M bond are weakly held and easily ionized, and, importantly, the C–M bonding electrons are polarized towards the carbon. As a result, the presence of group 4 metal substituents in organic compounds has long been recognized to have very profound effects on both the ground-state properties and reactivities of the compounds.[10–13] The most well known of these effects is the silicon β-effect,[10, 12] which is the

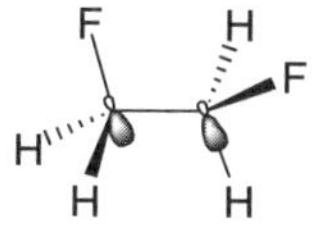

Figure 2. Gauche conformation of 1,2-difluoroethane with maximum overlap between the best donor (σ_{C-H}) and best acceptor (σ^*_{C-F}).

remarkable ability of silicon to stabilize positive charge at the β-position, and which is probably the most important electronic effect of silicon in organic chemistry. The β-effects of silicon, and of germanium and tin, in addition to other significant effects (e.g., the γ and δ-effects) that arise from the presence of these metals in organic compounds, can be rationalized on the basis of the strong donor ability of the C–M (M = Si, Ge, Sn) σ bonds and their interactions with acceptor groups and other donor groups within the molecule.

II. THE DONOR ABILITY OF THE C–MR_3 (M = Si, Ge, Sn, Pb) BOND

(Pb is included in this series for comparison purposes, although the effects of Pb in organic molecules will not be treated in any depth in this chapter.)

Information on the donor abilities of nonbonded electrons or bonded electrons can be obtained from a number of sources, one of which is photoelectron spectroscopy. According to Koopman's theorem,[14] the energy (and hence donor ability) of the C–M σ bond can be estimated from the ionization potentials of suitable model compounds containing the bond. The ionization energies of C–M bonds (M = Si, Ge, Sn) and some selected heteroatom lone-pair orbitals (for comparison) are summarized in Table 1. These values suggest that a C–Si bond has a donor ability slightly less than that of an oxygen lone pair, a C–Sn bond has a donor ability similar to a pyramidal nitrogen lone pair, and a C–Ge bond is intermediate between these two values. The strong donor abilities of the C–M bonds are also apparent from the Hammett σ^+ values for the R_3M-CH_2- substituents shown in Table 2. For example, the Hammett σ_p^+ value for the trimethylsilylmethyl substituent was determined by Eaborn et al.[15] from the measured rates of cleavage of the aryl–Si bonds in the series $(Me_3Si)_nCH_{(3-n)}$–C_6H_4–$SiMe_3$ to be –0.62. The

Table 1
Selected Vertical Ionization Energies (eV)

	IP (eV)
Et_4Si	10.5[16], 10.6[17]
Et_4Ge	9.3[17]
Et_4Sn	8.7[17]
Me_4Si	10.6[18]
Me_4Ge	10.2[18]
Me_4Sn	9.7[18]
Me_4Pb	9.1[18]
1-aza-[2.2.2]bicyclooctane	8.02[19, 20]
Me_2O	10.04[21]

Table 2
Selected Hammett Substituent Constants

Substituent	σ_p^+	Substituent	σ_p^+
Me_3SiCH_2	−0.62 [22]	Me_3PbCH_2	−1.03[24]
	−0.62[15]	Ph_3SiCH_2	−0.38[24]
	−0.76 [23]	Ph_3GeCH_2	−0.51[24]
	−0.54 [24]	Ph_3GeCH_2	−0.51[24]
Me_3GeCH_2	−0.63 [24]	Ph_3SnCH_2	−0.73[24]
Me_3SnCH_2	−0.92 [25]	Ph_3PbCH_2	−0.90[24]
	−0.82 [24]	OMe	−0.78[26]
NH_2	−1.3		

resonance component calculated from the difference $\sigma_p^+ - \sigma_p$ was −0.41. The entries in Table 2 demonstrate that the σ_p^+ values for the $R_3M{-}CH_2{-}$ substituents increase down the series M = Si < Ge < Sn < Pb and are also dependent on the nature of the substituents on the metal. The resonance electron donating power of the $R_3M{-}CH_2{-}$ substituents in the ground state is provided by the α_R^o values, which are given in Table 3, and which were determined by a combination of infrared, ^{13}C, and ^{19}F NMR. spectroscopic techniques. It is clear that the donor ability of the C–M bond increases down the group, and is dependent on the nature of the other groups attached to the metal.

Table 3
Selected σ_R^o Values

Substituent	σ_R^o
Me_3SiCH_2	−0.17 [23]
	−0.19 [27]
	−0.20 [27, 28]
Me_3GeCH_2	−0.22 [23]
	−0.20 [27]
	−0.21 [27]
Me_3SnCH_2	−0.25 [27]
	−0.26 [28]
	−0.23 [28]
Me_3PbCH_2	−0.22 [27]
Ph_3SiCH_2	−0.16 [27]
Ph_3GeCH_2	−0.17 [27]
Ph_3SnCH_2	−0.22 [27]
Ph_3PbCH_2	−0.21 [27]

III. INTERACTION OF C–MR3 BONDS WITH ACCEPTOR ORBITALS AT THE β POSITION

A. Reactivity Effects

1. Interaction of Group 4 Metal Substituents with Positive Charge at the β Position

The presence of group 4 metals β to a carbenium ion has a dramatic effect on the stability of the carbenium ion, a remarkable phenomenon known as the β-effect.[12, 29, 30] The silicon β-effect was discovered in 1937 by Ushakov and Itenberg,[31] who noted the exceptional reactivity of β-silyl-substituted halides $R_3SiCH_2CH_2X$ towards elimination (Scheme 1). The effect has since been the subject of many mechanistic[32–36] and theoretical studies[37–39] and several reviews.[10–12, 29]

The majority of these studies support an E1-like mechanism involving rate-limiting C–X bond heterolysis forming a (β-silyl–stabilized) carbocation intermediate followed by the rapid loss of the metal substituent. The β-effects of the lower group 4 metals (Ge and Sn) are even larger than that of silicon, and are among the strongest neighboring group effects known. The magnitude of the group 4 metal β-effects is nicely demonstrated by the relative rates of unimolecular solvolysis of the β–Me_3M–substituted cyclohexyl esters **4**,[40] **5** and **6**,[41] which react at rates 10^{12}, 10^{13}, $>> 10^{14}$ relative to the corresponding unsubstituted esters.

In addition to a relatively small inductive effect, two modes of stabilization of positive charge by the β-R_3M substituent have been suggested. The first involves the classical cation **7**, where the positive charge is stabilized by hyperconjugation (σ–*p* conjugation) between the C–M σ bonding orbital and the vacant carbocation

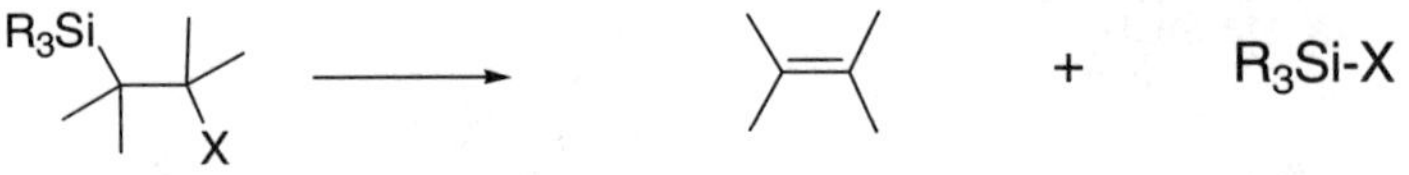

Scheme 1.

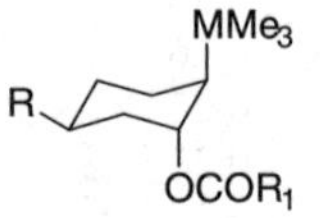

4: R = *t*-Bu, M = Si, R_1 = CF_3
5: R = H, M = Ge, R_1 = CF_0
6: R = H, M = Sn, R_1 = CH_3

p orbital; this is often referred to as vertical stabilization because it does not require significant nuclear movement. The second mode of stabilization involves the nonclassical ion **8** and is often referred to as nonvertical stabilization. Both modes of stabilization involve the donation of electron density from the C–M σ orbital to the *p* orbital on the electron-deficient carbon, and simply differ in the degree of interaction between these two orbitals. Experimental[22, 40] and theoretical studies[37, 38] have been carried out (particularly for M = Si) to differentiate between the two ions **7** and **8**. For example, in the gas phase the nonclassical ion **8** is calculated by high-level ab initio methods to be 2.4 kcal/mol more stable than the classical ion **7** for the β-silyl–substituted primary carbenium ion **9**.[37] Both cations are significantly more stable (ca. 38 kcal/mol) than the corresponding silicon-free carbenium ions. For the secondary and tertiary carbenium ions **10** and **11** the order is reversed, with the open cation of type **7** calculated to be slightly more stable than the nonclassical ion of type **8**.[38] Again, both cations are significantly more stable (28 and 18 kcal/mol, resp.) than the corresponding silicon-free cations. The structures of the β-silyl cations suggested from these theoretical studies are qualitatively supported by solution studies. For example, the unimolecular solvolysis of trimethylsilylethyl bromide **12**[33] exhibits a secondary deuterium isotope effect considered to be consistent with the formation or partial formation of a bridged cation for this primary system.

The bridged cation was proposed to be involved on the path to the rearranged products **12** and **13** when the alcohol **14** was treated with PBr_3 (Scheme 2).[34] It was

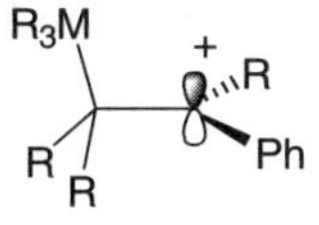

7

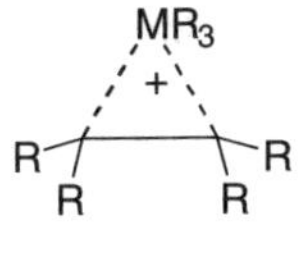

8

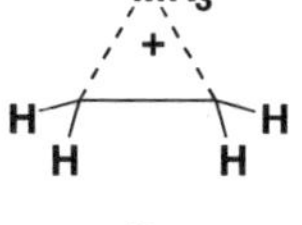

9

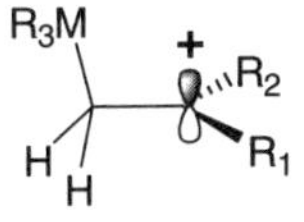

10: R_1 = H, R_2 = CH_3
11: R_1, R_2 = CH_3

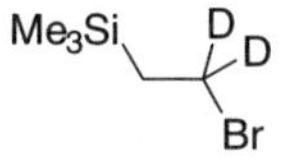

12

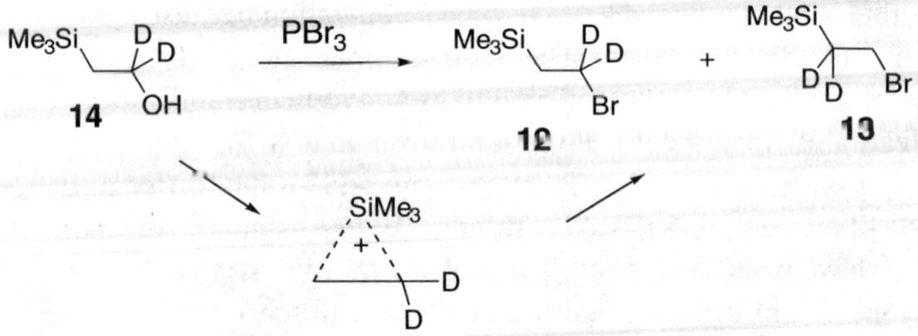

Scheme 2.

recognized, however, that a rapid 1,2 migration of the trimethylsilyl substituent occurring in the open cation would give the same result. Yukawa–Tsuno analysis of the rates of solvolysis of a series of 2-(aryldimethylsilyl)ethyl chlorides **15**[42] in aqueous acetone gave an r value of -1.42 with a high resonance demand ($r = 1.48$). These values suggested a significant buildup of positive charge on the silicon at the transition state, which was proposed as providing support for a bridged cation intermediate **16**. However, a significant charge buildup on the silicon is also expected for the open cation **17** due to the σ–p interaction. Evidence for the open cation structure **7** in a secondary β-silyl cation was provided by Davis and Janocks.[43] They found that the second silicon in the acid-catalyzed deoxysilylation of 1,3-bis(trimethylsilyl)-2-propanol had about the same β effect as the single silicon substituent in 1-trimethylsilyl-2-propanol. Whereas both silyl substituents can concurrently stabilize a carbenium ion by hyperconjugation (Figure 3), it is

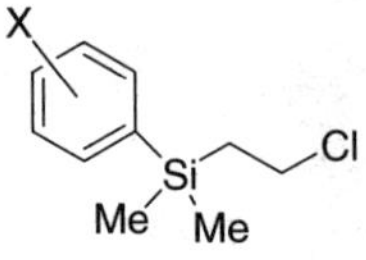

15

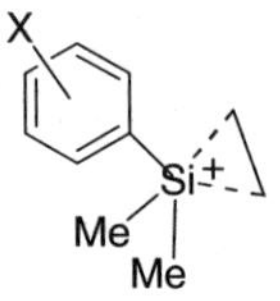

16

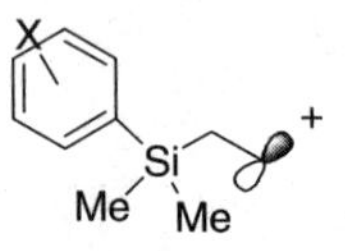

17

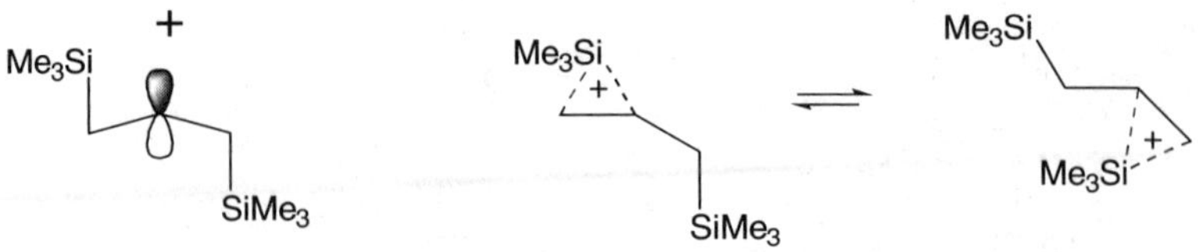

Figure 3. Vertical and nonvertical stabilization in the 1,3-bis(trimethyl- silyl)-2-propyl cation.

likely only one can realistically stabilize it by bridging. Traylor came to similar conclusions based on the measured rates of hydride abstraction from 1,3-bis-Me_3M–substituted propanes.[44]

Lambert and co-workers measured a secondary α-deuterium isotope effect k_H/k_D of 1.17 in 97% trifluoroethanol for the β-silyl ester **18**,[45] which is consistent with a transition state leading to a classical carbenium ion such as **19**, which is stabilized by hyperconjugation. A displacement like transition state leading to the nonclassical carbenium ion **20** was expected to show small or negative isotope effects, typically in the range 0.95–1.05. Shimizu and co-workers[46] observed a similar secondary α-deuterium isotope effect k_H/k_D of 1.18–1.19 in the unimolecular solvolysis of the β-silyl benzyl trifluoroacetates **21**; notably this was almost identical to the isotope effect observed in the solvolysis of the nonsilylated substrate **22**, giving further support for the open cation structure. Although secondary α-deuterium isotope effects in the range 1.17–1.20 are consistent with the formation of classical ions in the solvolysis of **18** and **21**, there has been some discussion as to whether this indeed rules out the involvement of the nonclassical ion in these solvolyses.[47] The bridged cation is a three-center two-electron species, which might formally be described as an R_3Si^+ cation coordinated to an alkene ligand. Those neighboring group effects for which kinetic isotope effects in the range 0.96–1.05 are observed involve the formation of a more conventional three-membered ring structure (e.g., the bromonium ion) with a total of four electrons in the two σ bonds.

Attempts to observe β-silyl carbenium ions of the general structure **23** in solution directly have generally been thwarted by the ready desilylation of these species.[48] Desilylation is believed to result from attack of the silaphilic counterions

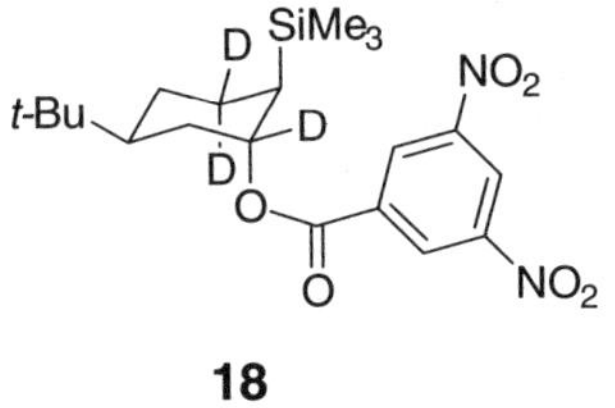

18

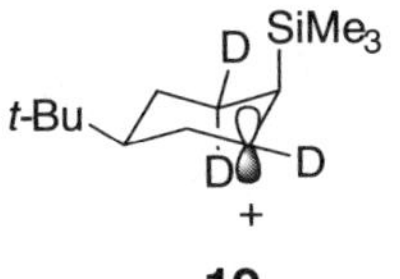

19

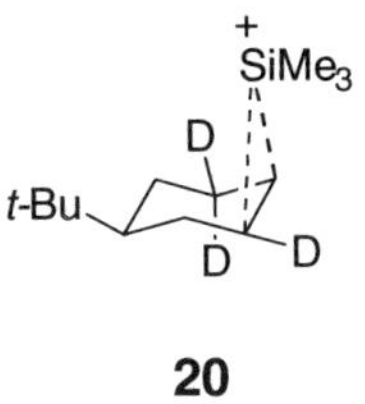

20

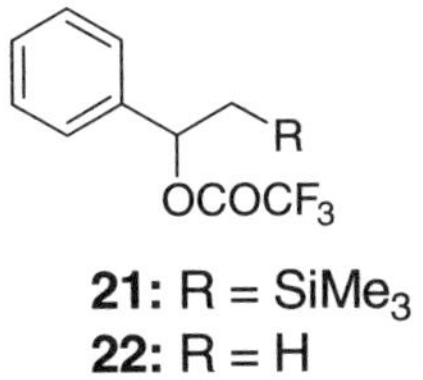

21: R = $SiMe_3$
22: R = H

at the silicon.[48, 49] For example, subjection of the alcohol **24** to super acid conditions gave the methyl-substituted cation **25** after desilylation and protonation. Lambert and Zhao[50] avoided the problem of desilylation by generating the β-silyl carbenium ion **26** by addition of Et_3Si^+ to the neutral precursor **27** in the presence of the non-nucleophilic counterion tetrakis(pentafluorophenyl)borate, $(C_6F_5)_4B^-$. The ^{29}Si and ^{13}C NMR data for the ion **26** were argued to be consistent with an open structure, with some transfer of positive charge to the silicon by hyperconjugation. In particular, the ^{29}Si signal at δ 46.2 ppm, which is significantly deshielded compared to the neutral precursor **27**, and the methylene carbon ^{13}C resonance, which occurs at δ 56.2 ppm, is 26 ppm higher in frequency than the methyl carbon in cation **25**. Both indicate significant contributions of the resonance structure **28**. The observation of the classical ion structure for **26** is not surprising for this system, where the open cation is also doubly benzylically stabilized. The silicon analog of the nonclassical 2-norbornyl cation **29** has recently been generated in solution by hydride abstraction from the silane precursor **30**; on the basis of its ^{13}C and ^{1}H NMR data, the cation **29** was described as an intramolecularly π-complexed silicon cation.[51]

In a series of experiments utilizing geometrically constrained five- and six-membered rings, Lambert et al. examined the dependence of the unimolecular solvolysis rates of β-trimethylsilyl–substituted esters upon the Si–C–C–OR dihedral angle. The substrates **31**,[52] **32**,[12, 40] **33**,[53] **34**,[52] and **4**,[12, 40] which had

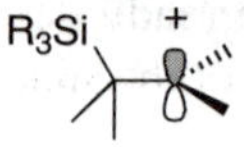

23

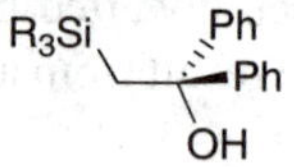

24

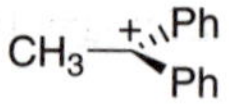

25

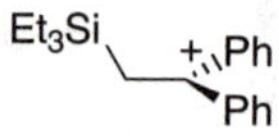

26

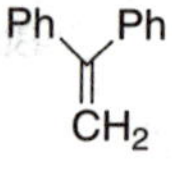

27

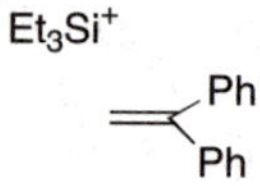

28

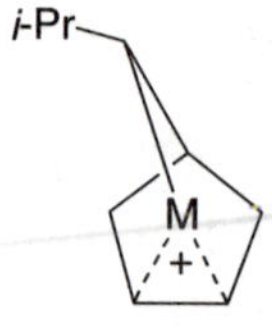

29: M = $SiMe_2$

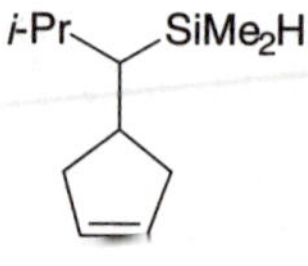

30

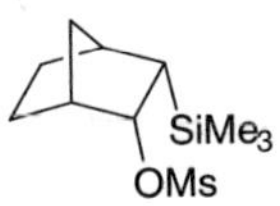

31

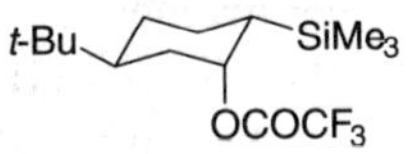

32

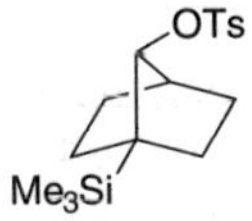

33

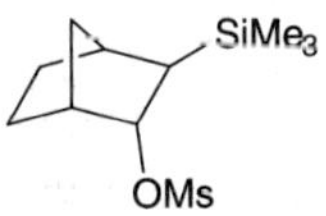

34

dihedral angles constrained to be ca. 0°, 60°, 90°, 120°, and 180°, respectively, were used in these studies. The β-effect was determined for each geometry by comparing the reaction rates with the corresponding silicon-free esters. A plot of the β-effect versus dihedral angle revealed a cosinelike dependence (Figure 4)[53] that is flattened on the *syn*-periplanar side. The cosinelike dependence of reaction rate upon dihedral angle is qualitatively consistent with hyperconjugative stabilization of the β-cation.

The β-effect is much weaker for *syn*-periplanar[52] than for *anti*-periplanar geometries. It has been suggested that the diminished β-effect in the *syn*-periplanar geometry could be due to steric repulsion between the leaving and trimethylsilyl groups hindering optimal σ–σ^* hyperconjugation on the way to the solvolysis transition state. While this suggestion has not been disproven, the relatively weaker *syn*-periplanar β-effect is consistent with the σ–σ^* hyperconjugation model. Extended Hückel calculations[54] predict that orbital interactions across a C–C bond are stronger in the *anti*-periplanar geometry than in the *syn*-periplanar geometry,

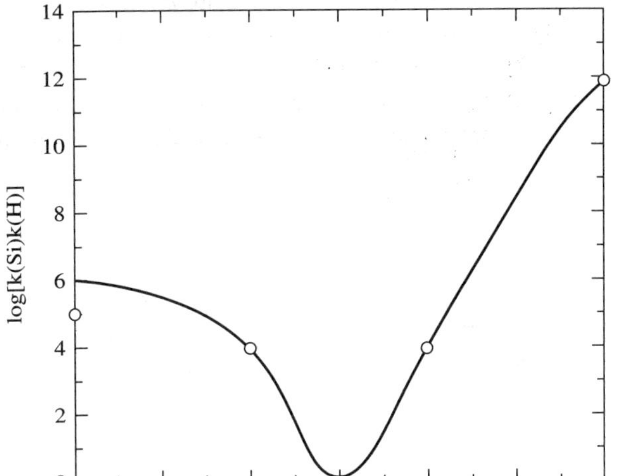

Figure 4. The dihedral angle dependence of the β-effect of silicon. Reprinted from J. B. Lambert, X. Zhao, *J. Organomet. Chem.* **1996**, *521*, 203, with permission from Elsevier Science, Kidlington, UK.

leading to the prediction that vertical stabilization should be more effective in the *anti*-periplanar geometry than the *syn*-periplanar geometry.

Further examination of Figure 4 shows that the silicon β-effect in the orthogonal geometry is negligible, suggesting that the contribution of the inductive effect of the silicon substituent to the stability of the carbenium ion intermediate is very small in these secondary systems; this is in agreement with conclusions from theoretical calculations.[38]

A similar dependence of reactivity towards unimolecular solvolysis upon the R_3M–C–C–OR dihedral angle has also been observed for β-germyl and β-stannyl esters,[41] demonstrating the importance of hyperconjugation in stabilizing the β-cation by Ge and Sn. For example, esters derived from *cis*-2-trimethylgermylcyclohexanol **35** , where the Ge–C–C–OR dihedral angle is constrained to be ca. 60°, solvolyze 10^5 times faster than the corresponding unsubstituted derivatives, but very much slower than the *trans* isomer **36**, for which the germyl substituent provides an acceleration of 10^{13}. (This includes a correction factor of 400 to allow for the small population of the reactive diaxial conformation.) The β-stannyl substituent provides an acceleration of 10^{11} in esters derived from *cis*-2-trimethylstannylcyclohexanol **37**, which has an enforced synclinal geometry, whereas the esters derived from the *trans* β-stannyl alcohol **38** are too reactive to measure accurate reaction rates, but a lower limit for the accelerating effect of the β-stannyl of 10^{14} in the *anti*-periplanar geometry was estimated. The rates of the thermally induced elimination of the neat β-tributylstannyl esters **39**, **40**, and **41** were determined by Jousseaume et al.[55] The ester derivatives of *trans*-2-(tributylstannyl)cyclohexanol (**40**), which have an *anti*-periplanar Sn–C–C–OR dihedral angle, were too reactive to be isolated, whereas derivatives of (**39**) (Sn–C–C–OR 60°) and (**41**) (Sn–C–C–OR ca. 0°) were much less reactive.

The unimolecular solvolysis of α-(pentamethyldisilanyl)benzyl halides provides evidence for stabilization of positive charge by hyperconjugation with a

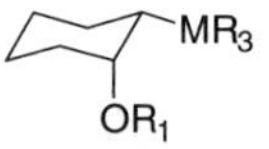

35: M = Ge, R = Me, R_1 = H
37: M = Sn, R = Me, R_1 = H
39: M = Sn, R = Bu, R_1 = C(O)*t*-Bu

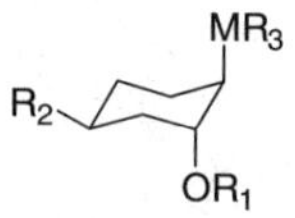

36: M = Ge, R = Me, R_1, R_2 = H
38: M = Sn, R = Me, R_1 = H, R_2 = *t*-Bu
40: M = Sn, R = Bu, R_1 = C(O)*t*-Bu, R_2 = H

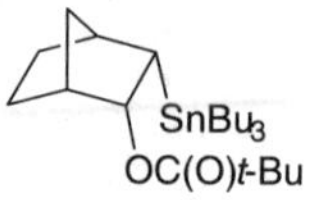

41

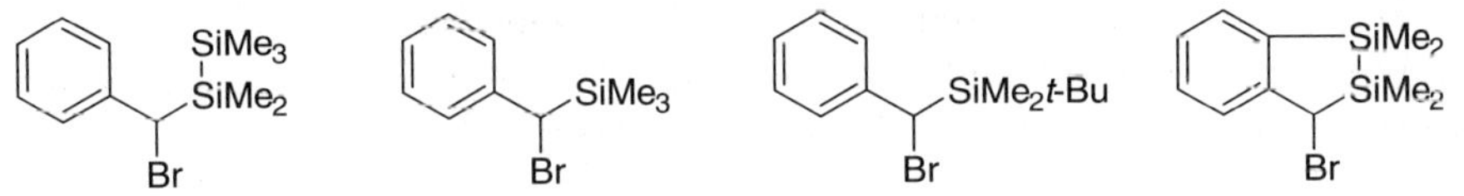

42 **43** **44** **45**

Si–Si bond.[56] For example, the relative rates of unimolecular solvolysis of α-(pentamethyldisilanyl)benzyl bromide **42**, α-(trimethylsilyl)benzylbromide **43**, and α-(*t*-butyldimethylsilyl)benzyl bromide **44** are 1.07×10^5:0.54:1.0 in trifluoroethanol at 25°C. When the Si–Si bond is constrained to be close to orthogonal to the developing carbocation orbital, as is the case in the solvolysis of the disilaindane derivative **45**, significant rate enhancements were not observed.

2. *Reactions Involving C–H Bonds β to Group 4 Metal Substituents*

The C–H bonds at the β position relative to R_3M– (M = Si, Ge, Sn, Pb) substituents are activated towards attack by electrophilic reagents. Two types of electrophilic attack at a β C–H bond are considered in Scheme 3: Path 1 involves hydride abstraction by the electrophile, resulting in the formation of a carbenium ion intermediate, a process that is assisted by the metal β-effect. Such a pathway might be expected to be followed by strongly Lewis acidic reagents, such as carbenium ion reagents.

Path 2 involves the insertion of an electrophilic carbene reagent into the β C–H bond. Although this pathway does not involve the development of a formal positive charge on the β carbon, it is well known that there is a significant buildup of positive charge on the carbon at the transition state for the insertion reaction.[57] Both pathways should, then, display significant β-metal effects. In a hypothetical situation in which there is more than one C–H bond β to the metal substituent, the

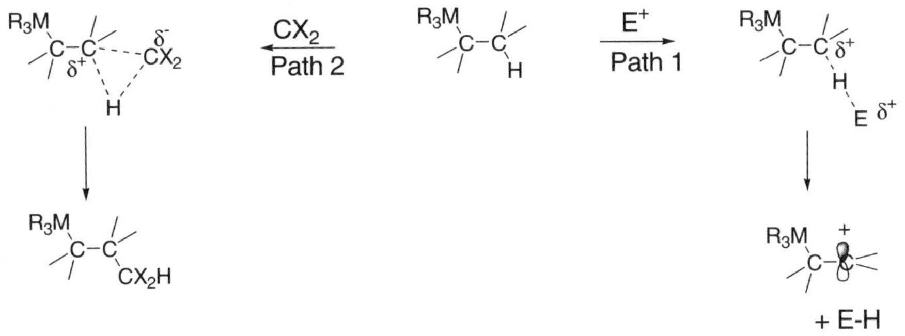

Scheme 3.

C–H bond that is *anti*-periplanar to the metal substituent should to be more activated towards electrophilic attack because the developing carbocation orbital receives maximum stabilization by hyperconjugation to the β C–M bond.

a. Hydride Abstraction Reactions of Group 4 Metal-Substituted Hydrocarbons. Traylor[58, 59] demonstrated that the organometallic compounds Et_4M (M = Si, Ge, Sn, Pb, Hg) react readily with the trityl cation to give ethylene by a mechanism involving abstraction of the β-hydrogen followed by demetallation of the resulting cation (Scheme 4). The reaction rates (M = Si < Ge < Sn < Pb) correlated with the charge-transfer frequencies for the complexes R_3MCH_2Ph–tetracyanoethylene, consistent with the driving force for hydride abstractions being the vertical stabilization of the intermediate cation by the metal substituent. Simultaneous vertical stabilization of the developing carbenium ion during hydride abstraction by two metal substituents was suggested from the rates of hydride abstraction from a range of 1,3-dimetallo-substituted propanes $Me_3MCH_2CH_2CH_2MMe_3$ (M = Si, Ge, Sn). The rates of hydride abstraction were found to correlate with the sum of the $\sigma^+_{CH_2\text{-}MMe_3}$ constants for both metal substituents.[44] The preference for an *anti*-periplanar geometry between the metal substituent and the hydride that is abstracted is suggested by the relative rates of hydride abstraction from **46** and **47** of ca. 3.45:1; the latter must adopt the twist boat conformation **48** to fulfill this requirement.[58] In later studies the preference for hydride abstraction *anti*-periplanar to the trimethylstannyl substituent was more clearly demonstrated by the reaction of *threo*-3-deuterio-2-(trimethylstannyl) butane **49** with trityl cation,[59, 60] which gave *trans*-2-deuterio-2-butene **50** and *cis*-2-butene **51** as the only products arising from hydride abstraction at C3 (Scheme 5). The *trans* alkene **50** arises from hydride abstraction *anti*-periplanar to the tin substituent from conformation **49a**, while the *cis* alkene arises from

46

47

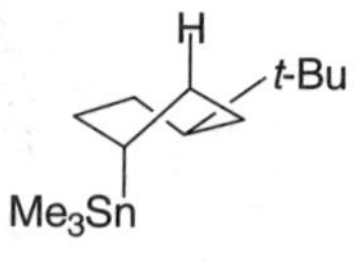

48

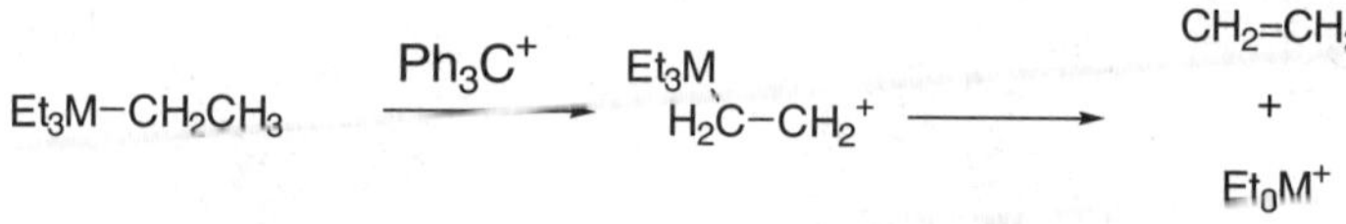

Scheme 4.

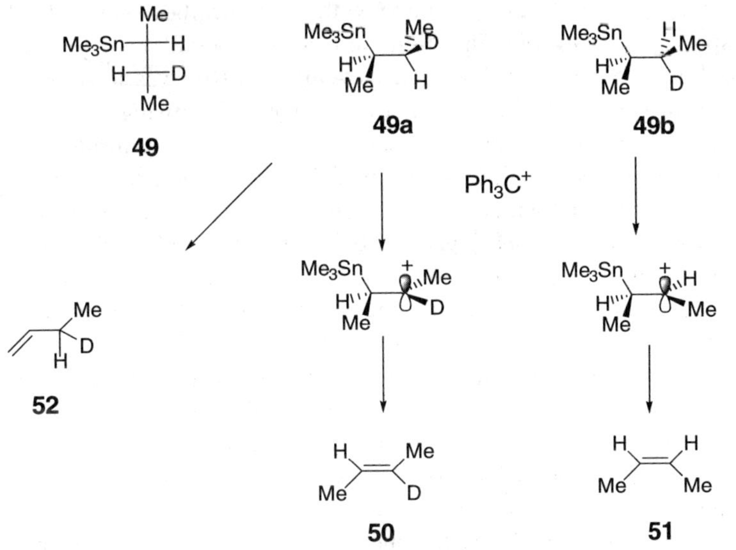

Scheme 5.

conformation **49b**. Small amounts of 3-deuterio-1-butene (**52**) arose from hydride abstraction from C1.

b. Directing Effects of Group 4 Metals on Carbene C–H Insertion Reactions. Seyferth et al. investigated the reaction of dichlorocarbene derived from phenyl(bromodichloromethyl)mercury with a wide range of tetraalkylsilanes and stannanes.[61, 62] It was found that both silicon and tin substituents strongly directed dichlorocarbene insertion into the C–H bonds at the β position. For

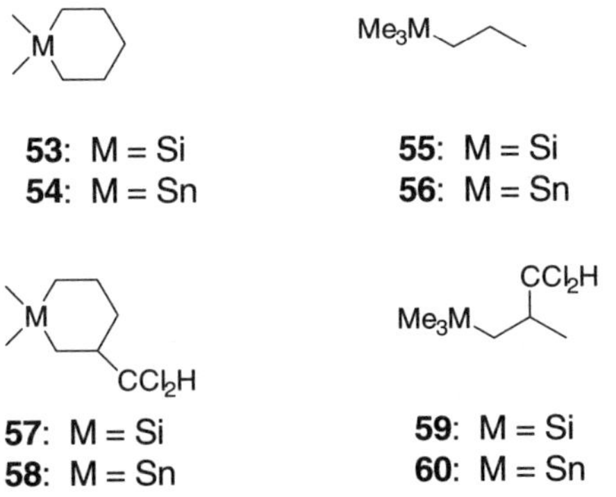

example, reaction of **53**, **54**, **55**, and **56** with dichlorocarbene gave **57**, **58**, **59**, and **60**, respectively, as the only insertion products in generally good yields. Insertion into the β C–H bond of 1,1-dimethylsilacyclobutane (**61**) was accompanied by the formation of a large amount of the silacyclopentane **62**[63] arising from insertion into the strained C–Si bond of the four-membered ring, a similar result was observed more recently in the reaction of **61** with $^1CH_2$.[64] Reaction of the corresponding di-*tert*-butyl analog **63** with dichlorocarbene and dibromocarbene gave only the β C–H insertion products **64**,[65] presumably due to the bulky substituents on the silicon disfavoring insertion into the strained C–Si bond. Stannyl substituents were found to be more effective at directing dichlorocarbene insertion to the β-C–H bond than silicon.[61] For example, in **65** where the β-C–H bond is competing with a tertiary C–H bond further displaced from the tin substituent, only insertion into the β-C–H bond is observed, whereas in the analogous silicon compound **66** a mixture of products arising from insertion into the β-C–H and the tertiary C–H bond are obtained. On the basis of their studies, Seyferth et al. proposed that the directing effects of these substituents can be best explained by hyperconjugation between the C–M bond (M = Si, Sn) and the partial positive charge on the β-carbon at the transition state for the insertion reaction.

Tentative evidence that the insertion of dichlorocarbene occurs preferentially into the C–H bond which is *anti*-periplanar to a β–silicon or tin is suggested from an examination of the stereochemistry of its reaction with *sec*-butyltrimethylsilane **67**: M = Si and *sec*-butyltrimethylstannane **67**: M = Sn (Scheme 6).[66] Both substrates gave an uneven mixture of the *erythro*-**68** and *threo*-**69** diastereoisomers, but the diastereomeric product ratio derived from the silicon compound was opposite to that obtained from the tin compound. The diastereoselectivity was rationalized as reflecting the preferential insertion of dichlorocarbene into the *anti*-periplanar C–H bond (with retention of configuration) in the most stable conformation. By the predictions of molecular

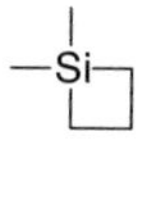

61

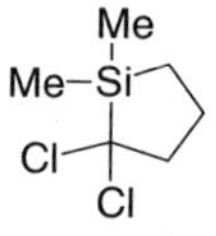

62

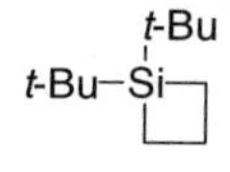

63

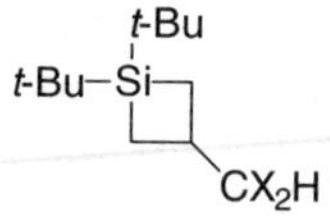

64: X = Cl, Br

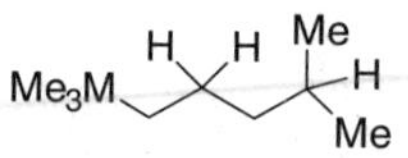

65: M = Sn
66: M = Si

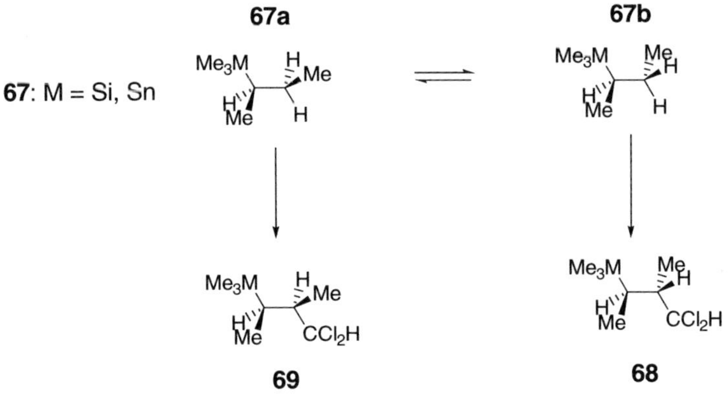

Scheme 6.

mechanics,[66] this is conformation **67a** for *sec*- butyltrimethylsilane and conformation **67b** for *sec*-butyl-trimethylstannane. Thus the major insertion product for **67**: M = Si was tentatively assigned as the *threo* diastereomer **69**, while the major isomer obtained from the stannylated substrate **67**: M = Sn was assigned as the *erythro* diastereomer **68**.

Reaction of the 6-*endo* deuterium-labeled silabicyclo[2.2.1]heptane derivative **70** with dichlorocarbene gave the product **71** by insertion exclusively into the C–H bond *anti*-periplanar to the C–Si bond (Scheme 7).[67] The stereoselectivity in this case was rationalized in terms of stabilization of the partial positive charge in the *exo* direction in the transition state, due to σ–p conjugation with the strained C–Si σ bond, in addition to a steric effect favoring approach to the *exo* C–H bond.

Activation by silicon of a β-C–H bond to an intramolecular carbene insertion reaction is exemplified by the silicon-directed Bamford–Stevens reaction.[68] For example, thermal decomposition of β-trimethylsilyl *N*-aziridinyl imines **72** in toluene (Scheme 8) [with or without $Rh_2(OAc)_4$ catalyst] results in the formation of allylic silanes **73** as major or exclusive products by the preferential insertion of the carbene intermediate into the C–H bond β to the silicon substituent.

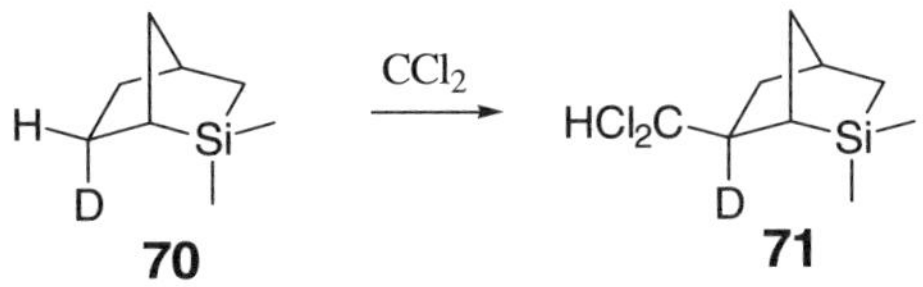

Scheme 7.

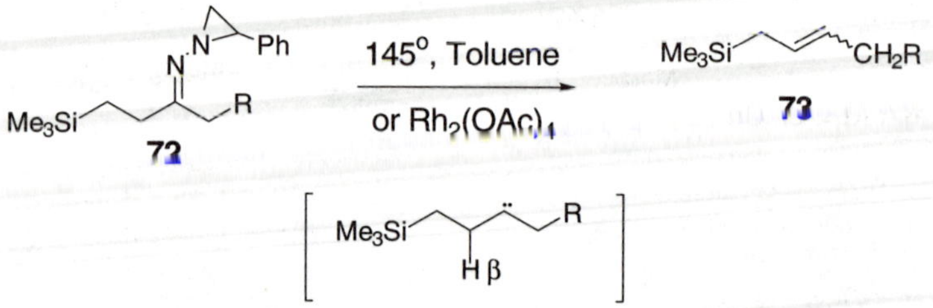

Scheme 8.

3. *Group 4 Metal-Directed Reactions*

Of the group 4 metal β-effects, it is the silicon β-effect that has been most exploited in synthetic organic chemistry. Silicon has been shown to be very useful for controlling the course of reactions that involve carbenium ion (or developing carbenium ion) intermediates. These reactions generally follow a pathway that (when possible) places the positive charge β to the silicon substituent.[69–82] For example, the electrophilic addition to vinylsilanes and allylsilanes[81, 82] occurs at the site that generates a cationic center β to the silicon atom; subsequent loss of the silyl group results in the formation of a substituted alkene (Scheme 9). Vinylsilanes and allylsilanes have proven to be very useful synthetic intermediates, particularly as vinyl and allyl transfer reagents. The silyl substituent not only controls the regiochemical course of addition of electrophiles to allylsilanes but also has the potential to control the stereochemical outcome.[83] This is demonstrated by the stereochemical course of the electrophilic substitution of the optically active (*R*)-(*E*) allylsilanes **74** and (*R*)-(*Z*) allylsilanes **75**, which gave the products (*S*)-**76** and (*R*)-**77**, respectively, in high enantiomeric enrichments (Scheme 10). The stereochemical course of the reaction was consistent with electrophilic attack on the double bond selectively *anti* to the trimethylsilyl substituent in an S_E' reaction.

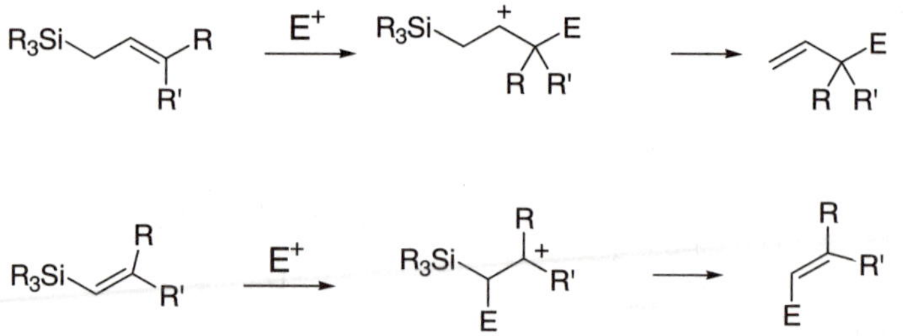

Scheme 9.

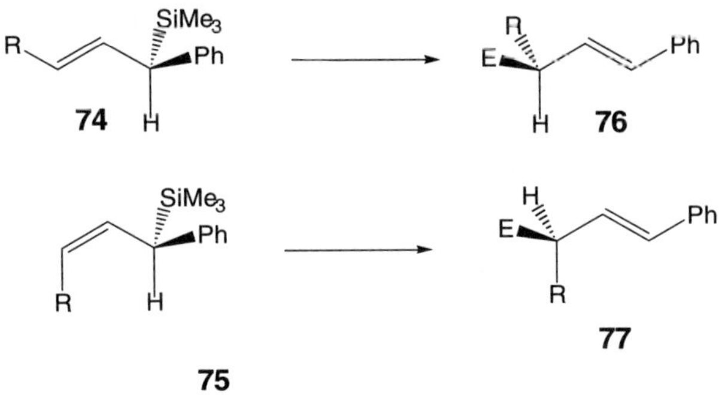

Scheme 10.

This outcome is rational since the (*R*)-allylsilanes **74** and **75** are expected to exist in conformation **78** (Scheme 11), in which the C–Si bond overlaps with the π bond. This conformation maximizes the σ–π interaction between the C–Si bond and the π system[84] (see also later section on allylsilanes). Attack of the electrophile from the opposite face to the trimethylsilyl group in **78** gives the carbenium ion intermediate **79**, which is stabilized by a σ–π interaction with the C–Si bond substituent[12,37,38]; loss of silicon from this intermediate gives the *E* olefin products. *Anti* selectivity has also been observed in the trifluoroacetolysis of 2-cyclohexenylsilanes, germanes, and stannanes.[85]

Trialklysilyl groups control the course of the Nazarov cyclization.[78–80] For example, the presence of the trimethylsilyl substituent in **80a** results in the exclusive formation of the less stable cyclopentenone isomer **81**[78] (Scheme 12) arising from desilylation of the β-trimethylsilyl carbenium ion intermediate **82**, whereas in the absence of the silyl substituent in **80b** the thermodynamically more stable cyclopentenone **83** is obtained.

A series of angular-substituted octalins **84** and **85** were studied using electron-impact mass spectroscopy.[86] It was found that the major fragmentation pathway of the derivatives **84** having a trans relationship between the silyl substituent and the angular substituent involved loss of the angular substituent

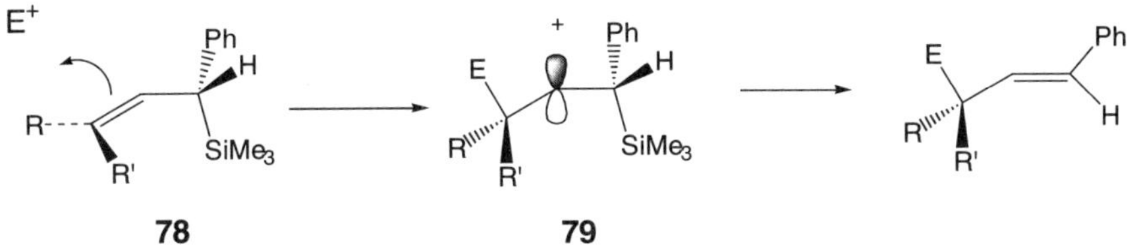

Scheme 11.

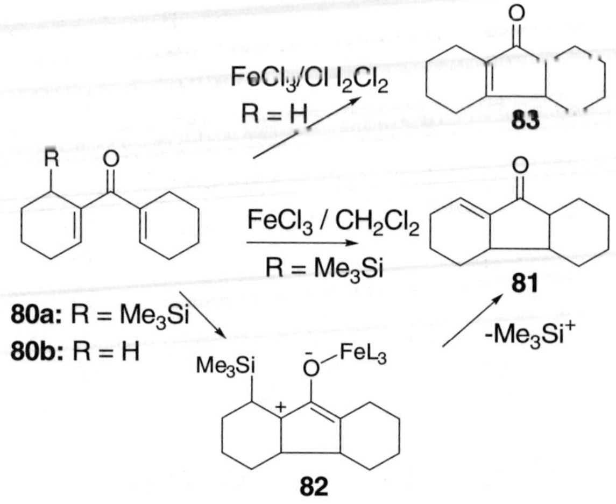

Scheme 12.

radical (Scheme 13). In contrast, the major fragmentation pathway of derivatives **85** having a cis relationship between the silyl and angular substituents involved loss of a methyl radical from the silicon substituent. This differing behavior was explained in terms of stabilization by hyperconjugation between the C–Si bond and

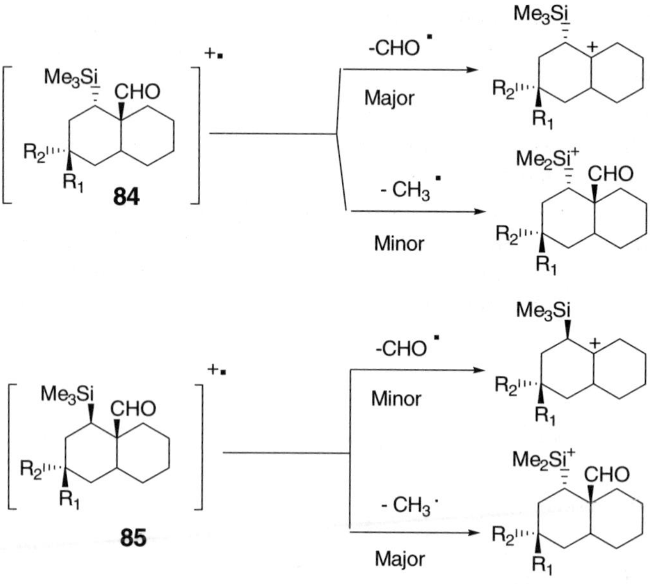

Scheme 13.

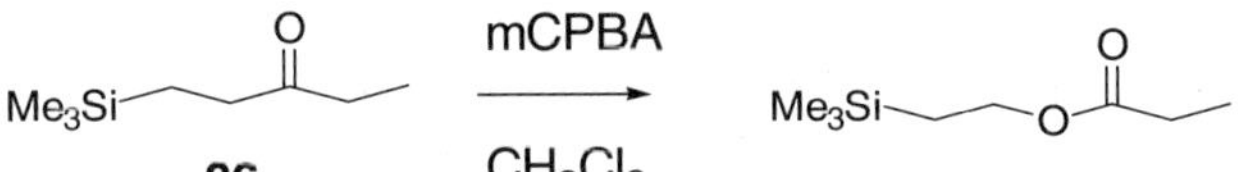

86 **87**

Scheme 14.

the developing carbenium ion at the bridgehead carbon: The trans derivatives **84** have favorable geometry for this interaction, whereas the less than optimal alignment of orbitals in the cis isomers **85** allows other fragmentation pathways to compete.

A β-trialkylsilyl moiety enhances the migratory aptitude of groups in the Baeyer–Villiger reaction,[87, 88] as is demonstrated by the reaction of 1-trimethylsilylpentan-3-one (**86**) (Scheme 14) with *m*-chloroperoxybenzoic acid, which results in the exclusive formation of 2-trimethylsilylethyl propanoate (**87**). The migratory aptitude of groups in the Baeyer–Villiger reaction is related to the ability of the migrating group to bear a positive charge. The presence of β-silicon is expected to enhance this ability, and in fact the migratory aptitude of the primary carbon in $Me_3SiCH_2CH_2$ was shown to be between that of secondary and tertiary alkyl groups.[89]

Oximes derived from β-silyl and β-stannyl ketones undergo unusual Beckmann rearrangement, giving rise to nitriles as predominant products. For example, Beckmann rearrangement of 3-trimethylsilylcyclohexanone oxime **88** (Scheme 15) proceeds to the formation of the acyclic alkenyl nitrile **89** as the major product.[90] The nitrile is believed to be formed by fragmentation of the protonated oxime, giving the β-silyl–stabilized cation **90**, followed by desilylation. Our

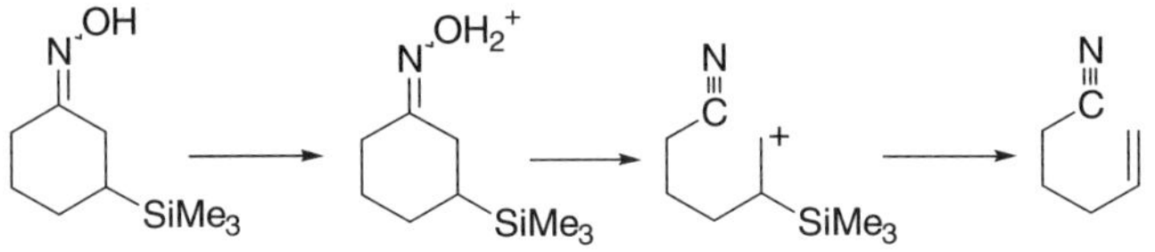

88 **90** **89**

Scheme 15.

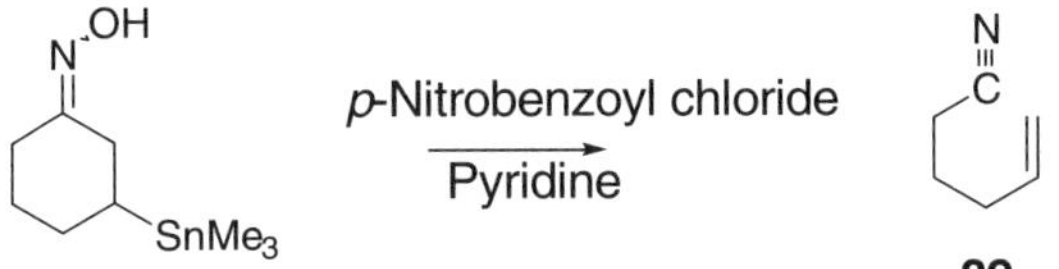

91 **89**

Scheme 16.

attempts[91] to prepare the p-nitrobenzoate ester derived from 3-trimethylstannylcyclohexanone oxime **91** for X-ray analysis resulted in the exclusive formation of **89**, presumably by a similar mechanism (Scheme 16).

B. Ground-State Effects

1. *β-Trimethylmetal-Substituted Esters, Alcohols, and Ethers*

The presence of an R_3M (M = Si, Ge, Sn) β to an oxygen substituent in organic compounds has a profound influence on the structure and preferred conformation of the molecule. These effects can be attributed to the interaction between the C–M σ bond and the C–O σ* orbital. Manifestations of these effects are described below.

a. Structural Studies of β-Trimethylsilyl Esters and Ethers. The X-ray crystal structures of the 2-trimethylsilylethyl esters **92**, **93**,[92] and **94**[93] reveal that all exist in the solid state in the expected *anti* conformation. The C_1–O bond distances for **92–93**, which were 1.464(2), 1.466(2), and 1.476(2) Å, respectively, were all significantly lengthened (by ca. 0.015 Å) relative to the corresponding distances for the unsubstituted derivatives **95–97**. The origin of the bond lengthening is believed to be the σ–σ* interaction between the C–Si bonding orbital and the C–O antibonding orbital. The *anti*-periplanar relationship between the C–Si bond and the C–O of **92–94** ensures maximum interaction between the two orbitals.

Similar studies have been carried out using the cyclohexane framework to control the stereochemical relationship between the silicon and oxygen substituents. Low-temperature X-ray structural studies[94–97] on the esters and ethers **98–102**, in which the trimethylsilyl and oxygen substituents are constrained to be *anti*-periplanar, so maximizing the overlap between the C–Si bond and the C–O antibonding orbital, again reveal significant lengthening of the C(alkyl)–O bond compared to the C–O bond distance (observed[98] or predicted[93]) for corresponding silicon-free derivatives **103–107**. If, however, the C–Si bond is constrained to be gauche to the C–O bond, as is the case for structures **108–111**,[94] then no systematic lengthening of C–O bond distance is observed. This result reflects the poor overlap

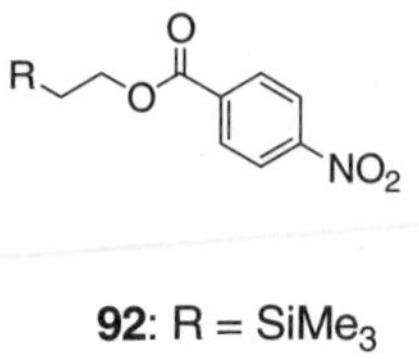

92: R = SiMe$_3$
95: R = H

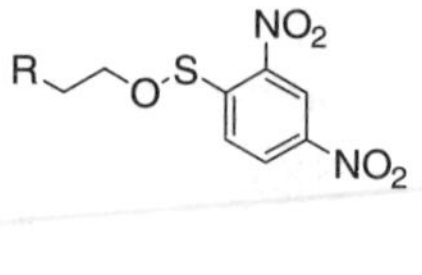

93: R = SiMe$_3$
96: R = H

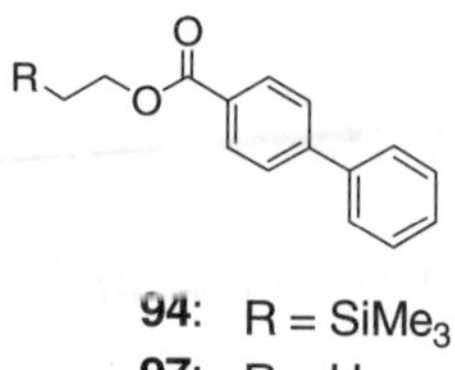

94: R = SiMe$_3$
97: R = H

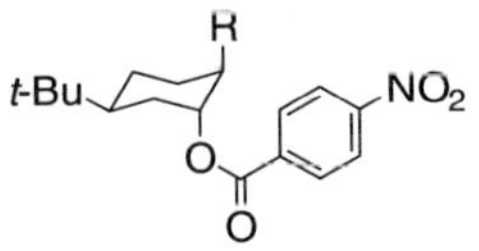

98: R = $SiMe_3$
103: R = H

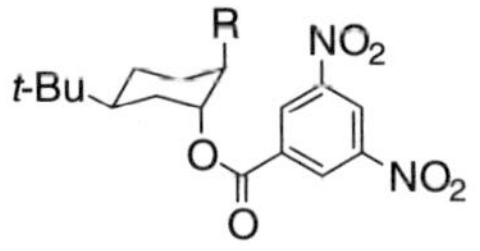

99: R = $SiMe_3$
104: R = H

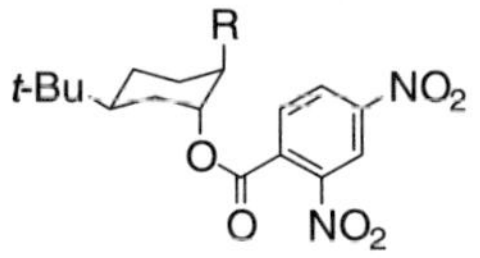

100: R = $SiMe_3$
105: R = H

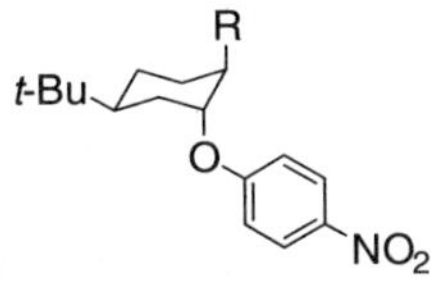

101: R = $SiMe_3$
106: R = H

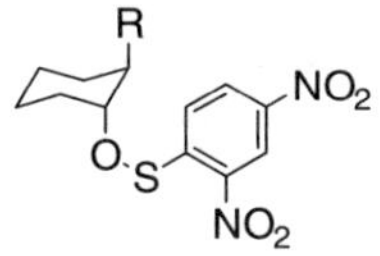

102: R = $SiMe_3$
107: R = H

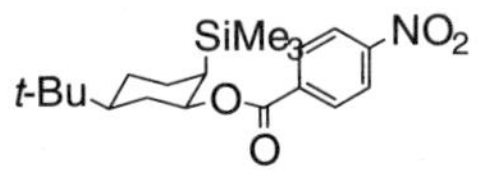

108

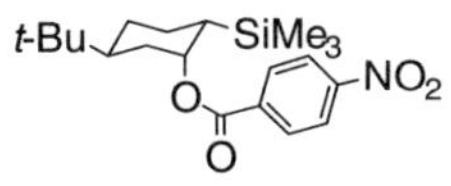

109

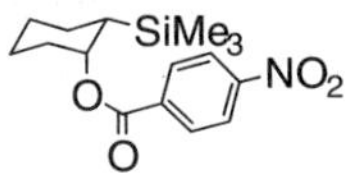

110

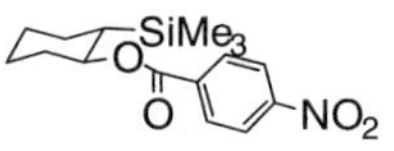

111

between the C–Si bonding orbital and the C–O antibonding orbital in this geometry.

Constraining the C–Si bond to be *syn*-periplanar to the C–O bond has been achieved within the norborneol framework.[95, 99] The X-ray structures of the esters

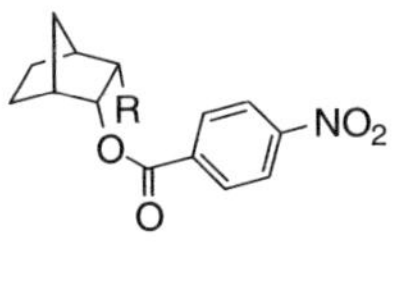

112: R = $SiMe_3$
114: R = H

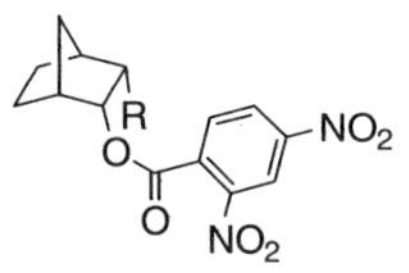

113: R = $SiMe_3$
115: R = H

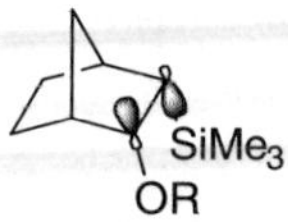

Figure 5. The $\sigma_{C\text{–}S}$–$\sigma^*_{C\text{–}O}$ interaction in the *syn*-periplanar geometry.

112 and **113** derived from *endo*-trimethylsilyl-*endo*-2-norborneol, for which the Si–C–C–O dihedral angles were –17.5(2) and –9.4(2) respectively, showed no significant lengthening relative to the silicon-free analogs **114** and **115**.[100] This result suggests that the interaction between the C–Si σ orbital and the C–O σ^* orbital in the *syn*-periplanar geometry is very small when compared to the *anti*-periplanar geometry, and is very likely the result of the reduced overlap between these two orbitals in the *syn*-periplanar geometry (Figure 5).

The interaction between the C–Si σ orbital and the C–O σ^* orbital can be qualitatively represented by the molecular-orbital interaction diagram in Figure 6. The strength of the interaction is inversely proportional to the energy difference between the C–Si σ orbital (donor) and the C–O σ^* orbital (acceptor).[101] The energy of the $\sigma^*_{C\text{–}OR}$ orbital decreases with increasing electron demand of the substituent OR. An increase in the electronegativity of –OR therefore should increase the strength of the $\sigma_{C\text{–}Si}$–$\sigma^*_{C\text{–}OR}$ interaction. A measure of the relative electronegativities of the RO– substituents is provided by the pK_a values for the parent acids ROH. A plot of the C–OR bond distances versus the pK_a value for ROH (Figure 7, for the series **98–102**) is linear over the pK_a range studied (ca. 1–7.5) and shows that the C–OR bond-length increases as the electronegativity of the OR– substituent increases.[97] The C–OR distance is related to the pK_a value for ROH by the equation:

$$r(\text{C–OR}) = 1.502 - (5.338 \times 10^{-3})\, \text{p}K_a(\text{ROH}) \quad (1)$$

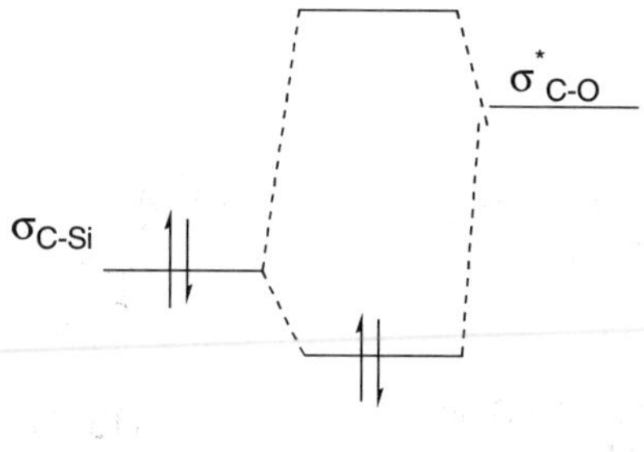

Figure 6. Interaction between a C–Si bonding orbital and a C–O antibonding orbital.

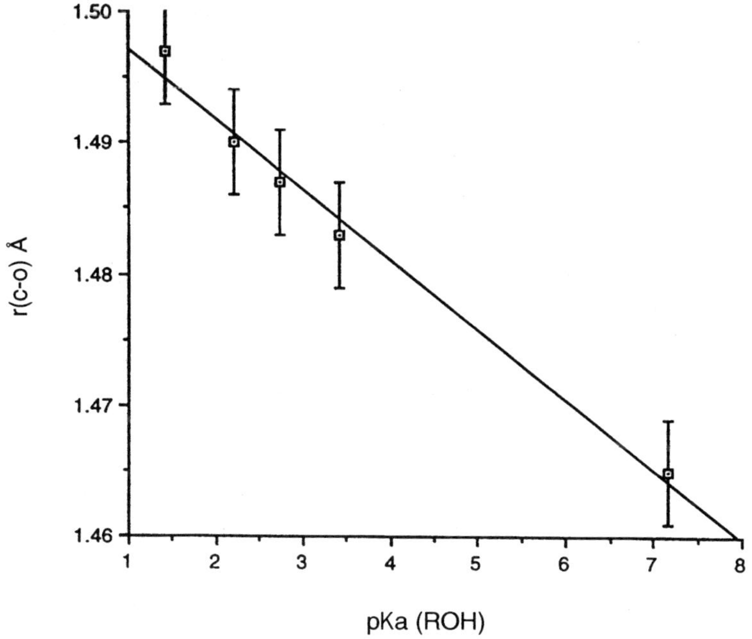

Figure 7. Plot of C–O bond distance versus pK_a for the compounds **98–102**.

Associated with the increase in the C–OR bond distance with increasing electronegativity of the –OR substituent is a smaller but nonetheless significant increase in the C–Si distance. These structural effects are consistent with the σ_{C-Si}–σ^*_{C-O} interaction; as the electronegativity of the –OR substituent increases, the energy of the σ^*_{C-OR} orbital decreases; the degree of σ–σ^* interaction increases; consequently there is greater electron donation from the C–Si σ orbital, and so the C–Si bond also lengthens. These systematic structural effects represent the early stages of the heterolytic bond breaking of the C–OR bond, which occurs in the unimolecular solvolysis of β-trimethylsilyl esters (Scheme 17). Linear relationships between C–OR bond distances and pK_as for ROH have been demonstrated in other systems.[102, 103] For example, esters and ethers derived from cyclohexanol

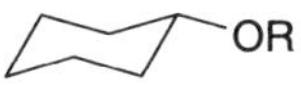

116

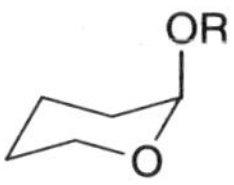

117

118

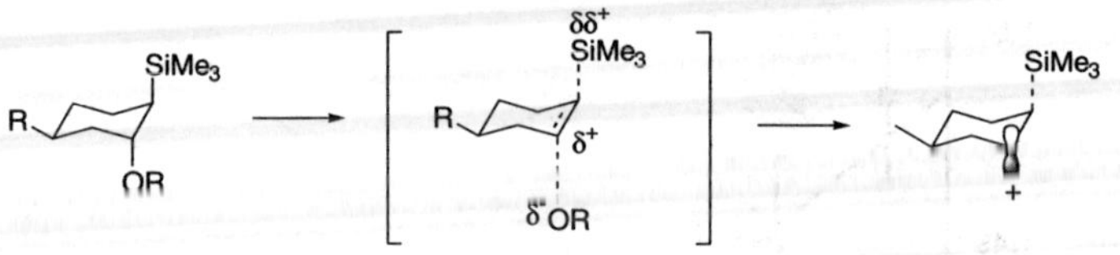

Scheme 17.

116[93, 102] and 2-hydroxytetrahydropyran **117**[103] show linear responses of the C–OR bond distances with pK_a (ROH).

$$\textbf{116};\quad r(\text{C–OR, Å}) = 1.475 - (2.90 \times 10^{-3})\, pK_a(\text{ROH}) \tag{2}$$

$$\textbf{117};\quad r(\text{C–OR, Å}) = 1.493 - (6.495 \times 10^{-3})\, pK_a(\text{ROH}) \tag{3}$$

The differences in the constants of these relationships reveal useful information. The slopes indicate the sensitivity of the C–OR bond distance to the electron demand of the substituents –OR, and the intercepts are the predicted bond distance for C–OR when pK_a (ROH) is zero.

The strongest response (largest slope) of the C–OR bond distance to the electron demand of the OR substituent occurs for **117**, which has an oxygen lone pair *anti*-periplanar to the C–OR bond. The cyclohexyl esters **116** in which the C–C bond is *anti*-periplanar to the C–OR bond show a much weaker response of the C-OR distance to the electron demand of the OR substituent. Thus the slopes of the lines [*r*(C–OR) vs. pK_a (ROH)] appear to provide qualitative measures of the donor abilities of orbitals that are *anti*-periplanar to the C–OR bond.

Comparison of Eq. (1) derived from the *r*(C–OR) versus pK_a (ROH) plot for the *anti*-periplanar β-trimethylsilyl esters **98–102** with Eq. (3) derived for the 2-substituted tetrahydropyran derivatives **117** reveals a similar response of the C–OR bond distance to the electron demand of the OR substituent. This is consistent with the similar energies (and hence donor abilities) of a C–Si bond and an oxygen lone-pair orbital (Tables 1 and 2). Thus the above structural data suggest that the σ_{C-Si}–σ^*_{C-O} interaction in **98–102** is similar in magnitude to the n_O–σ^*_{C-O} present in **117**. Also consistent with the present interpretation is the plot of C–OR distance versus pK_a (ROH) for structures derived from the gauche β-trimethylsilyl alcohol **118**,[97]

$$r(\text{C–OR}) = 1.48 - (2.571 \times 10^{-3})\, pK_a(\text{ROH})$$

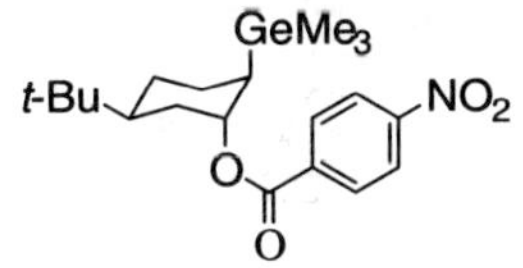

119

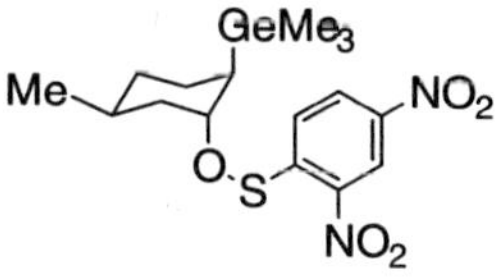

120

which indicates a much weaker response of C–OR distance to the electron demand of the substituent OR than was observed for the *anti*-periplanar β-trimethylsilyl esters **98–102** but is similar to that observed for the derivatives of cyclohexanol **116**.

b. Structural Studies of β-Trimethylgermyl Esters. Structural parameters obtained from accurate low-temperature structural studies on the β-trimethylgermyl esters **119** and **120**[96] reveal that the presence of the β-trimethylgermyl substituent results in significant lengthening of the C–O bond distance when compared with the corresponding unsubstituted derivatives **103** and **107**. This lengthening is consistent with the expected structural effects of the σ_{C-Ge}–σ^*_{C-O} interaction. Comparison of the C–$GeMe_3$ bond distance in the two structures shows that the C–Ge bond distance is slightly longer in the more electron-demanding sulfenate ester **120** than in **119**. The C–O bond-lengthening effects observed for **119** and **120** are similar in magnitude to the effects observed in the corresponding β-trimethylsilyl esters **98** and **102**, suggesting similar ground-state effects of trimethylsilyl and trimethylgermyl substituents. Despite the fact that the C–Ge bond is a better donor orbital than a C–Si bond, the σ_{C-Ge}–σ^*_{C-O} interaction does not lead to any extra effects on the C–OR bond distance compared to the σ_{C-Si}–σ^*_{C-O} interaction. Given that the relative reactivities of β-trimethylgermyl esters, β-trimethylsilyl esters, and corresponding nonsubstituted esters towards unimolecular solvolysis are ca. 10^{13}:10^{12}:1, and given that a trimethylsilyl substituent lengthens C–O bonds at the β position by ca. 0.017 Å relative to the corresponding nonsubstituted esters, any extra effects of a trimethylgermyl substituent are not surprisingly too small to be detected by X-ray crystallography.

c. Conformational Studies of β-Trimethylsilyl Alcohols and Esters. The stabilizing σ_{C-Si}–σ^*_{C-O} interaction described above is maximized when the C–Si and C–OR bond are *anti*-periplanar. Therefore, incorporation of the Si–C–C–OR moicty into a flexible organic framework should result in a bias towards the conformation having these two bonds *anti*-periplanar. This effect is demonstrated by the conformational analysis of the mobile β-trimethylsilyl alcohol **121** and its derived esters **122a–m**, which incorporate the C5-methyl as a counterpoise substituent.[104] The position of the equilibrium (Figure 8) is determined by steric

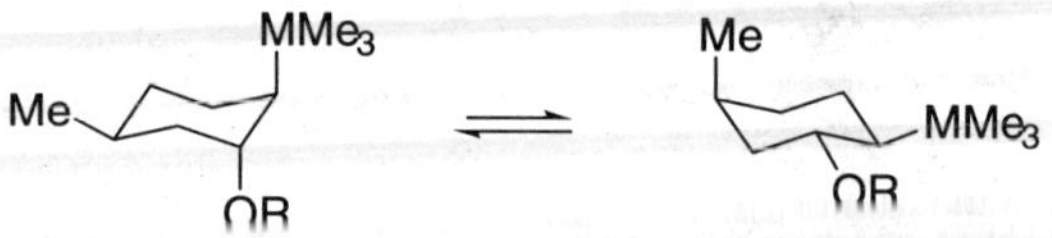

M = Si, Ge

Figure 8. Conformation equilibrium of β-trimethylmetalcyclohexanol and derivatives.

and dipole effects, which favor the diequatorial conformation, and the $\sigma_{C-Si}-\sigma^*_{C-O}$ interaction, which is maximized in the diaxial conformation. The parent alcohol **121** and the derived esters **122a–m** were predicted by molecular mechanics calculations (which take into account steric and dipole effects only) to favor the diequatorial conformation. The conformational equilibria of **121** and **122a–m** were measured by the coupling constant method,[105] and the resulting conformational equilibrium data are summarized in Table 4. The ester derivatives in Table 4 all showed a clear preference for the diaxial conformation consistent with the presence of $\sigma_{C-Si}-\sigma^*_{C-O}$ interactions, and, furthermore, there was a distinct relationship between the electron demand of the ester function and the proportion of the diaxial conformation. For example, the sequential replacement of the methyl hydrogens of the acetate derivative **122b** with chlorines (Table 4, entries 2–5) was

Table 4
Conformational Data for **121** and esters **122a–122m**

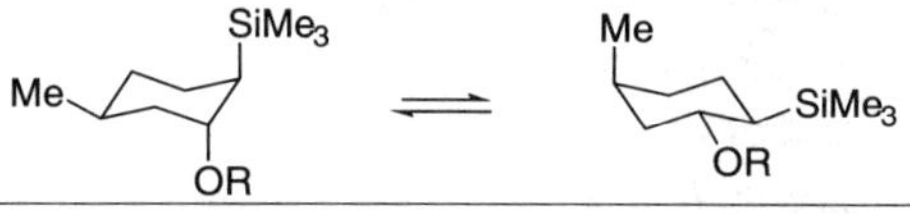

Compound	% Diaxial	% Diequatorial	$-\log K_{eq}$
121	41.2	58.8	−0.154
122a	58.0	42.0	0.140
122b	70.0	30.0	0.367
122c	80.8	19.2	0.624
122d	87.0	13.0	0.825
122e	68.5	31.5	0.337
122f	70.0	30.0	0.367
122g	67.0	33.0	0.307
122h	70.0	30.0	0.367
122i	72.9	27.1	0.429
122j	81.8	18.2	0.647
122k	74.5	25.5	0.466
122l	73.0	27.0	0.431
122m	72.9	27.1	0.429

accompanied by an increasing proportion of the diaxial conformation. This type of relationship was also apparent with the substituted benzoate esters (entries 6–14); a plot of log K_a for the parent acids against log K_{eq} for the corresponding esters showed a reasonably linear relationship with a slope of 0.1. A similar relationship was observed for the β-trimethylsilyl esters **124a–d** derived from 2-trimethylsilylethanol **123**. The solution conformational equilibria were determined by analysis of the AA′BB′ splitting pattern in the ^{1}H NMR spectra and determination of the J_{AX} and $J_{AX'}$ coupling constants.[106] It was found that in all cases studied the *anti* conformation was preferred, and that the relative proportions of the gauche and *anti* conformers depended on the electron demand of the ester. Thus as the electronegativity of the ester increased, so did the proportion of the *anti* conformation.

The relationship between the proportion of the *anti*-periplanar conformation and the electron demand of the ester substituent in flexible systems containing the Si–C–C–OR moiety is consistent with the σ_{C-Si}–σ^*_{C-OR} interaction: Increasing the electron demand of the ester substituent decreases the energy of the σ^*_{C-O} orbital, resulting in a better energy match with the σ_{C-Si} orbital and hence a stronger σ_{C-Si}–σ^*_{C-OR} interaction. Furthermore, the coefficient of the σ^*_{C-OR} orbital on the carbon increases with increasing electron demand of OR, leading to more effective overlap with the adjacent σ_{C-Si} orbital.

d. Conformational Studies of β-Trimethylgermyl Alcohols and Esters. Conformational analysis of the β-trimethylgermyl alcohol **125**[107] and its derived esters **126a–g** revealed a bias towards the diaxial conformation (Figure 8), in

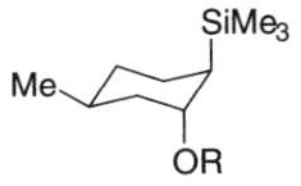

121: R = H
122a: R = CH_3CO
122b: R = CH_2ClCO
122c: R = $CHCl_2CO$
122d: R = CCl_3CO
122e: R = PhCO
122f: R = 3-$MeOC_6H_4CO$
122g: R = 4-$MeOC_6H_4CO$
122h: R = 3-$NO_2C_6H_4CO$
122i: R = 4-$NO_2C_6H_4CO$
122j: R = 2,4-$(NO_2)_2C_6H_3CO$
122k: R = 3,4-$(NO_2)_2C_6H_3CO$
122l: R = 3,5-$(NO_2)_2C_6H_3CO$
122m: R = 4-CNC_6H_4CO

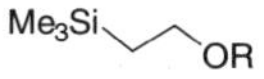

123: R = H
124a: R = 4-$NO_2C_6H_4CO$
124b: R = 3,5-$(NO_2)_2C_6H_3CO$
124c: R = 2,4-$(NO_2)_2C_6H_3CO$

Table 5
Conformational Data for Germanium Alcohol **125** and Derived Esters **126a–h**

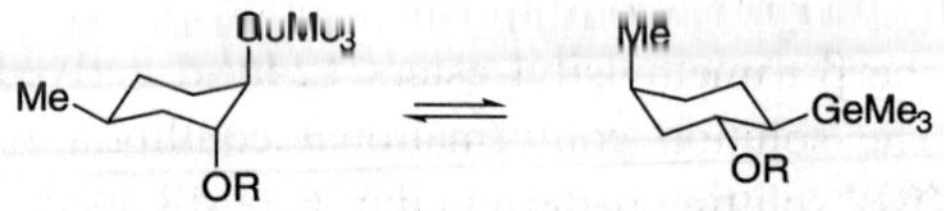

Compound	% Diaxial	% Diequatorial	$-\log K_{eq}$
125	52.1	47.9	0.04
126a	67.3	32.7	0.31
126b	78.5	21.5	0.56
126c	78.5	21.5	0.56
126d	81.3	18.7	0.64
126e	81.3	18.7	0.64
126f	84.2	15.8	0.73
126g	87.5	12.5	0.85

contrast with that predicted by molecular mechanics (Table 5). This is consistent with the presence of stabilizing $\sigma_{C-Ge}-\sigma^*_{C-OR}$ interactions, which are optimum in the diaxial conformation. Comparison with the corresponding β-trimethylsilyl esters **126a–g** shows a slightly greater bias towards the diaxial conformation for the β-germyl esters, due in large part to the smaller conformational *A* value for the trimethylgermyl substituent (2.1 kcal/mol)[108] compared with the trimethylsilyl substituent (2.5 kcal/mol).

2. *α-Stannylated Methyleneboranes*

Studies on the distannylmethyleneborane **127**[109] using ^{11}B and ^{13}C NMR spectroscopy provides evidence for strong C–Sn hyperconjugation with the boron

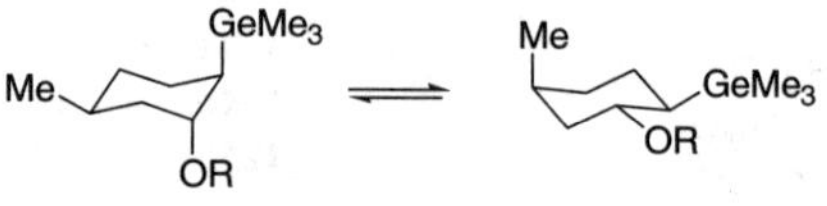

125: R = H
126a: R = CH_3CO
126b: R = 4-$MeOC_6H_4CO$
126c: R = 3-$NO_2C_6H_4CO$
126d: R = 3-CNC_6H_4CO
126e: R = 4-$NO_2C_6H_4CO$
126f: R = 2,4-$(NO_2)_2C_6H_3CO$
126g: R = 3,5-$(NO_2)_2C_6H_3CO$

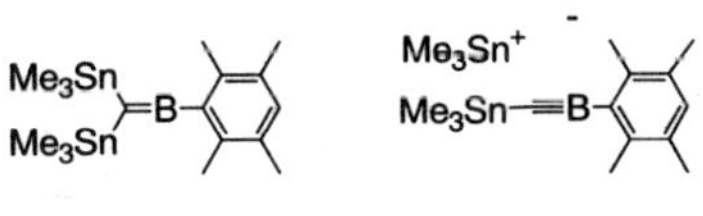

127 127a

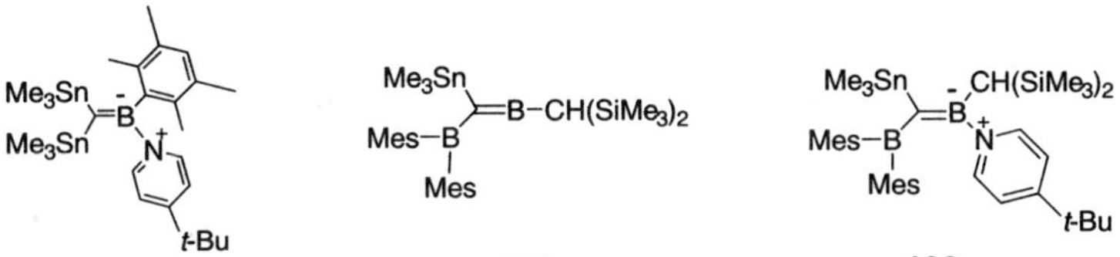

128 129 130

p orbital. Systematic effects on the ^{11}B and ^{13}C δ values suggested significant contributions of the resonance form **127a** to the ground-state structure of **127**. Also consistent with this interpretation was the small coupling constant ^{1}J (^{119}Sn, ^{13}C) of **127**, which is only 198 Hz. In the 4-*tert*-butylpyridine adduct **128**, where the boron *p*-orbital is no longer available for hyperconjugation with the C–Sn bond, the corresponding coupling constant of 325 Hz lies in the range typical for C–Sn bonds in geminal trimethylstannyl groups (ca. 298 Hz). Structural evidence for strong C–Sn hyperconjugation with a *p* orbital on boron is provided by the crystal structure of the stannylmethyleneborane **129**.[110] Hyperconjugation between the C–Sn bond and the *p* orbital on the boron of the B=C double bond is reflected in (1) a small B–C(sp^2)–Sn bond angle of 105.6°, (2) a long C(sp^2)–Sn bond distance of 2.16 Å, and (3) a short B=C double bond distance of 1.31 Å. In the 4-*tert*-butylpyridine adduct **130**, the C–Sn distance is shortened (2.12 Å), the C=B distance is lengthened (1.42 Å), and the B–C(sp^2)–Sn bond angle opens up to 119°.

IV. INTERACTION OF C–MR3 BONDS WITH ACCEPTOR ORBITALS AT THE γ POSITION

A. Reactivity Effects

1. *Interaction of Group 4 Metal Substituents with Positive Charge at the γ Position*

The possibility that a γ-silicon substituent could stabilize positive charge was recognized as early as 1946 by Sommer and Whitmore[111] when they found that the γ-chlorosilane **131** was more reactive than the α-isomer **132**. A quantitative

131

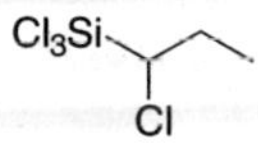

132

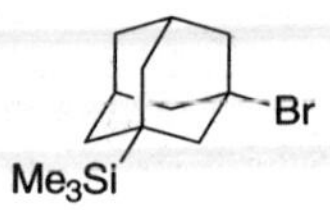

133

134

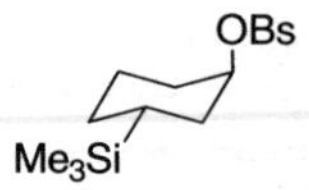

136

135: R = H
137: R = *t*-Bu

understanding of the mechanism by which silicon and tin substituents interact with γ-positive charge has been the subject of more recent studies.[112–119] Grob concluded from substituent effects on the unimolecular solvolysis rates of 1-adamantyl bromide[118] that rate enhancements by the γ-trimethylsilyl substituent in **133** were greater than predicted from its σ_I value. This suggested some form of conjugative electron release from the trimethylsilyl substituent to stabilize the developing carbocation. Studies of the unimolecular solvolyses of stereochemically well-defined γ-Me_3M (M = Si, Sn) cyclohexyl esters demonstrated that the γ-effect of silicon and tin is strongly dependent upon geometry.[112, 113, 117] For example, *cis*-3-trimethylsilylcyclohexyl-4-bromobenzenesulfonate **134** (in 97% trifluoroethanol/water) reacts at a rate 452 times faster than the corresponding carbon analog **135**[112, 113]; this contrasts with the trans isomer **136**, which shows no significant rate enhancement. Important differences in the solvolyses of **134** and **136** also manifest both in the secondary deuterium isotope effects exhibited by these substrates and in the nature of their solvolysis products. Whereas the solvolyses of **134** and **136** show similar secondary α-deuterium isotope effects, they have vastly different secondary β-deuterium isotope effects. The trans isomer **136** exhibits a large secondary β-deuterium isotope effect ($k_H/k_D = 2.86$), indicating significant involvement of the β C–H/D bonds at the transition state, leading to the carbocation intermediate consistent with reaction occurring from the conformation having the OBs leaving group axial. Similar results were also obtained for both the carbon analogs **135** and **137**.

The major reaction product formed from the solvolysis of the trans γ-silyl ester was cyclohexene formed from the carbocation **138** (Scheme 18) by 1,2-hydrogen migration to give the β-silyl cation **139** followed by loss of the silicon substituent. In contrast to the behavior shown by **136**, the cis γ-silyl ester **134** exhibited a small inverse β-d_4 isotope effect ($k_H/k_D = 0.97$) attributed to the inductive effect of the β deuteriums and implies that there is very little hyperconjugative stabilization of

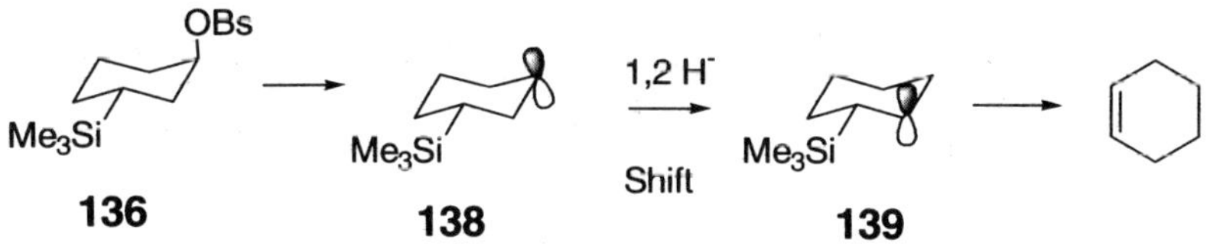

Scheme 18.

the developing carbocation by the β C–D/H bonds. It was concluded that **134** reacts from the diequatorial conformation having a W relationship between the C–Si bond and the C–OBs bond. Based on these results, it was suggested that during the solvolysis of the cis isomer **134** the γ-silyl substituent stabilizes the developing carbocation by a through-space interaction between the back lobe of the C–Si bond and the developing carbocation *p* orbital. This type of participation is often referred to as percaudal (through the tail) homohyperconjugation,[121] and the carbenium ion can be represented by the valence-bond forms in Scheme 19. The tail of a C–M bond (M = Si, Ge, Sn) has significant electron density due to the polarization of the C–M bond, which is in the direction of the carbon. This mode of participation is supported by retention of configuration in the major product **141**, and by the consistent formation of small quantities of bicyclo[3.1.0]hexane **140** as products of the solvolysis of **134** (Scheme 20).[112, 113]

Inductively enhanced C–C hyperconjugation (Scheme 21) may also contribute to the higher reactivity of the cis isomer **134** compared to the trans isomer **136**.[118] Information that allows the relative contributions of inductively enhanced C–C hyperconjugation (Scheme 21) and homohyperconjugation (Scheme 20) to the γ effects of silicon and tin has been provided by a number of experimental and theoretical studies:

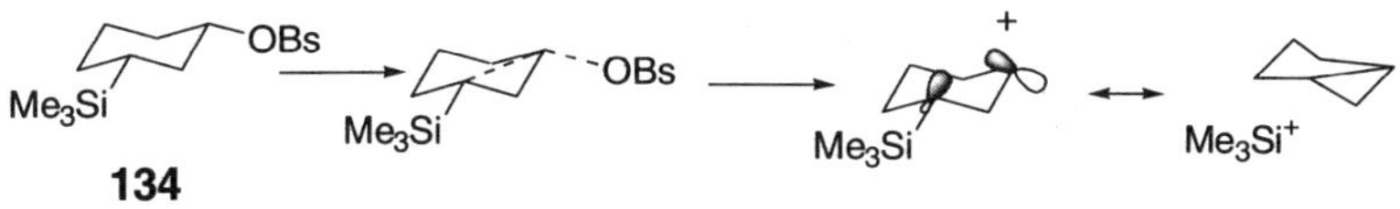

Scheme 19.

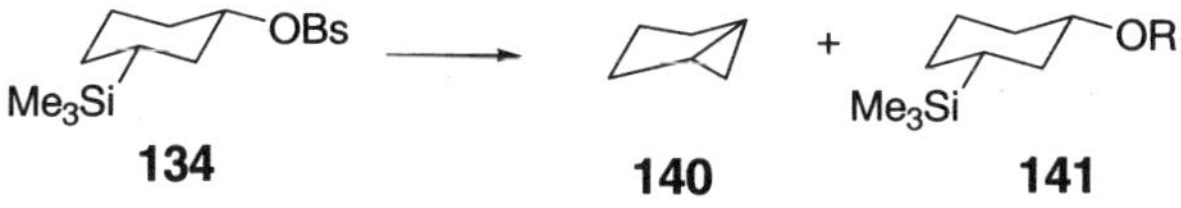

Scheme 20.

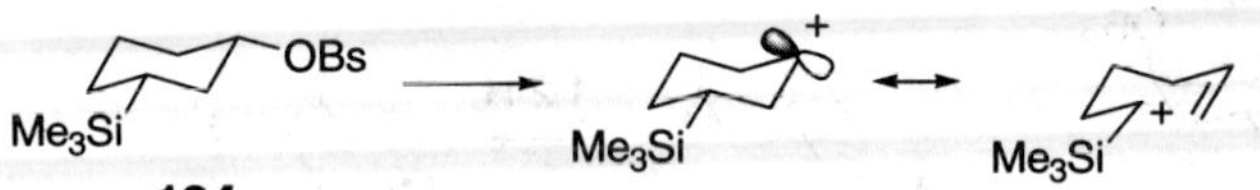

134

Scheme 21.

1. The γ-effect of silicon in the adamantane framework is small by comparison with the γ-effect in the relatively flexible cyclohexane framework **134**.[112, 113, 117] For example, the relative rates of unimolecular solvolysis of 1-bromo-3-trimethylsilyladamantane (**133**) and the nonsilicon analog are only 8.6:1.[118, 120] The reduced γ-effect of silicon in the rigid adamantane skeleton is consistent with there being reduced overlap between the backlobe of the C–Si bond and the developing carbenium ion. If the mode of stabilization of γ-positive charge was substantially due to inductively enhanced hyperconjugation, the γ-effects in the adamantane and cyclohexane frameworks would be expected to be similar.

2. A theoretical study of the 3-silaprop-1-yl cation **142**[116] suggested partial bridging with a distance of 1.82 Å between the γ-carbon and the carbenium ion, together with a high degree of positive charge on the silicon, consistent with the percaudal interaction. The carbenium ion **142** was calculated to be 10 kcal/mol more stable than the 1-propyl carbenium ion **143**, and the C2–C3 bond distance was no longer than the corresponding distance for the 1-propyl carbenium ion **143**, indicating that the stabilizing effect of the silicon in **142** is not due to increased C–C hyperconjugation in this case.

3. The relative rates of unimolecular solvolysis of the cis- and trans-γ-trimethylstannylcyclohexyl nosylates **144** and **145** versus the nonstannylated derivative **146** were determined to be > 53,000:3.1:1, respectively.[122] Whereas the developing carbenium ions formed during the solvolyses of **144** and **145**

142

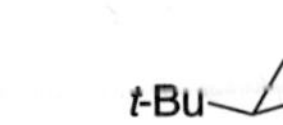

143

144

145

146

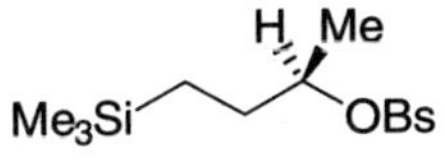

147a

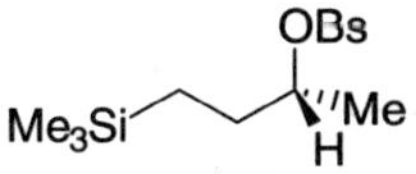

147b

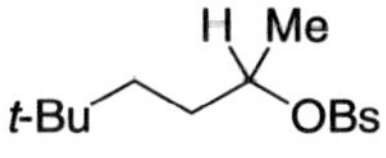

148

potentially are both stabilized by C–C hyperconjugation, the cis isomer **144** is also stabilized by Sn–C percaudal homohyperconjugation; this suggests that inductively enhanced C–C hyperconjugation is of minor importance in this system also.

Shiner et al. studied the γ-effect of silicon using the optically active acyclic substrate (*R*)-(4-trimethylsilyl)-2-butyl 4-bromobenzenesulfonate (**147**).[114, 115] The γ-trimethylsilyl–substituted substrate **147** showed significant rate enhancements compared with the carbon analog **148** (130:1). The racemic nature of the solvolysis products was interpreted as arising from participation from the back lobe of the γ-carbon–silicon bond from both the "W" conformation **147a** and the *endo* sickle[123] conformation **147b** in equal proportions. This result contrasts with the cyclic systems discussed above for which only the W geometry is attainable.

The presence of bridgehead Me_3M (M = Si, Sn) substituents has a huge effect on the stability of the 1-norbornyl cation **149**.[124] The relative rates of unimolecular solvolysis for **152**, **151**, and **150** are 20,486:1,319:1, respectively, in 97% trifluoroethanol/water, which indicates that γ-tin has a substantially larger effect than γ-silicon. The stabilization of the 1-norbornyl cation by the bridgehead Me_3M (M = Si, Sn) substituents was attributed predominantly to the through-space back-lobe interaction (Scheme 22), with perhaps some stabilization due to through-bond, double hyperconjugation involving the ethano bridges (Scheme 23).

149: R = H
153: R = Me_3Si
154: R = Me_3Sn

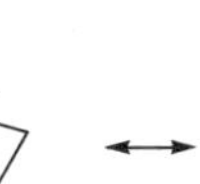

150: R = H
151: R = Me_3Si
152: R = Me_3Sn

155

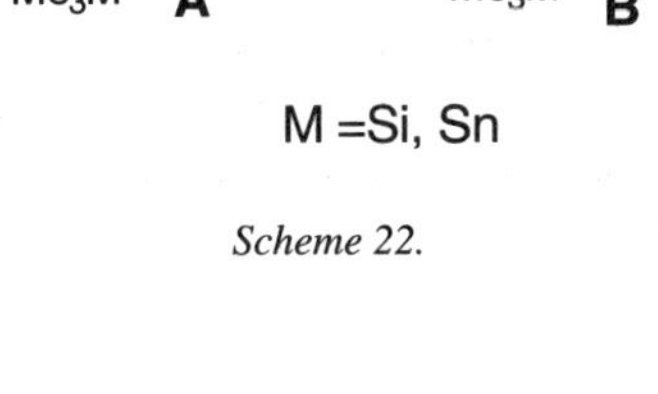

Scheme 22.

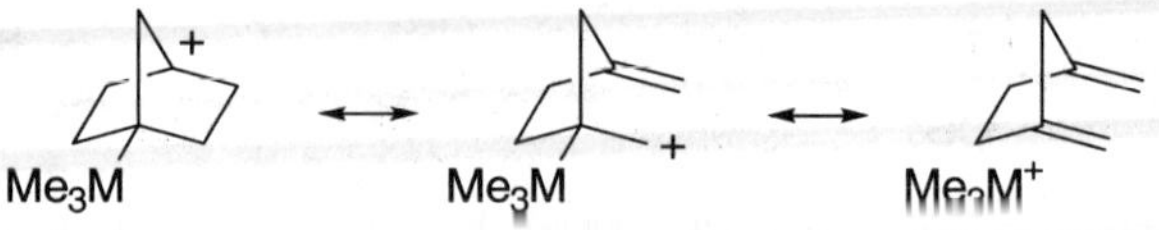

Scheme 23.

The through-bond and through-space interactions, however, have differing symmetry requirements.[2]

In the parent norbornyl system the C1· · ·C4 distance is ca. 2.28 Å, and the internuclear angle is favorable for the through-space interaction, while the geometry for the through-bond interaction is less than optimum.[2] High-level ab initio calculations carried out on the cations derived from **151** and **152** were presented as part of the same study. These provided evidence for substantial involvement of the resonance contributor **B** (Scheme 22), a result of the interaction of the back lobe of the C–M bond with the cation *p* orbital. The greater involvement of the resonance contributor **B** (Scheme 22) for M = Sn is suggested by the greater contraction of the C1· · ·C4 distance in the formation of the stannylated cation **154** (1.802 Å) compared with the silylated cation **153**, for which the corresponding distance is 1.866 Å.

The greater stabilizing effect of tin on γ positive charge compared with silicon suggests that the C–SnR_3 bond donates electron density through the back lobe more effectively than the C–SiR_3 bond and would imply a greater contribution of cyclopropanelike resonance forms (e.g., Schemes 19 and 22 **B**) to the overall structure of the cation. It is interesting to note therefore that cyclopropane-containing products feature much more prominently in the reaction products of γ-tin systems compared with their γ-silicon analogs. For example, bicyclo[3.1.0]hexane (**140**) makes up only 8% of the products from the solvolysis of the γ-silyl ester **134** in 97% trifluoroethanol/water,[112, 113] whereas under the same conditions the corresponding γ-tin ester **155** gives the bicycle **140** as the sole product.[117]

Fleming noted that γ-stannyl tertiary and benzylic alcohols form cyclopropanes stereospecifically on treatment with acid, with inversion of configuration at both carbon atoms,[125, 126] whereas γ-silyl tertiary alcohols undergo a silicon-controlled carbenium ion rearrangement.[127] Cylopropanes feature in many reactions of organic compounds bearing a stannyl substituent γ-disposed to a leaving group, provided the W geometry can be attained.[128–132]

The γ-effect of silicon is implicated in the silicon-promoted Nef reaction.[133] The Nef reaction of 5-nitrobicyclo[2.2.1]hept-2-ene **156** yields none of the expected ketone **157** (Scheme 24), whereas Nef reaction of the corresponding trimethylsilyl-substituted nitro compound **158** results in the formation of high yields of the ketone **159**. The silicon is believed to promote the breakdown of the hydrated intermediate **160** by stabilizing the γ-carbenium ion **161** (Scheme 25).

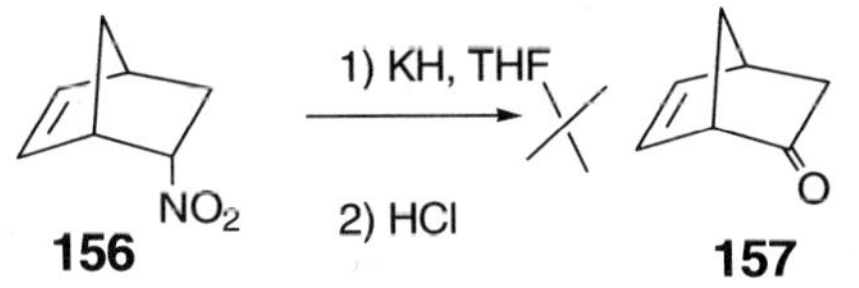

Scheme 24.

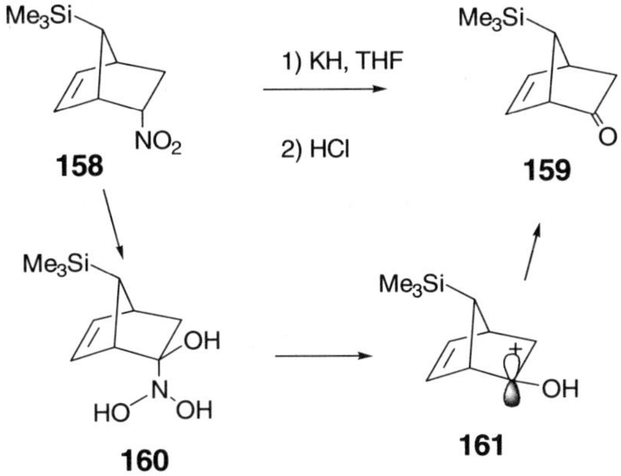

Scheme 25.

B. Ground-State γ-Effects

Analysis of the ^{19}F NMR substituent chemical shifts[134, 135] (SCS) for a range of 3-substituted 1-fluoroadamantane derivatives **162** demonstrate the presence of significant residual contributions (^{19}F SCS-$r_F s_F$), when R = $SiMe_3$ and $SnMe_3$.

These effects were interpreted in terms of either (1) homohyperconjugation, where electron density is donated from the back lobe of the C–M (M = Si, Ge, Sn) σ-bonding orbital into the C–F σ* orbital (as represented by the resonance form **162a**), or (2) C–C hyperconjugation involving the resonance form **162b**, where the Group 4 metal substituent stabilizes the resonance form **162b** by inductive electron donation. Low-temperature X-ray structural studies on the γ-silyl **163** and γ-stannylcyclohexyl esters **144** and **164**, which have a W relationship between the γ C–M (M = Si, Sn) bond and the C–OR bond,[122] reveal small but significant lengthening of the C–OR bond distance relative to nonmetal-substituted esters **165**[98] and **166**.[136]

In contrast, the esters **145** and **167** show no significant effects on the C–OR bond distance. These structural effects are consistent with a significant percaudal interaction, as represented by the resonance form **168**.

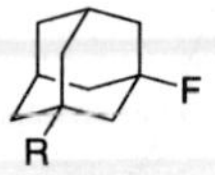

162

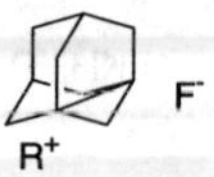

162a

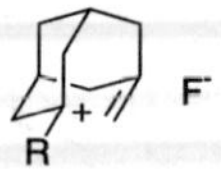

162b

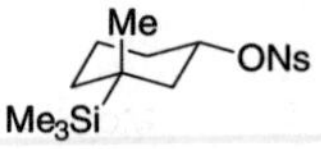

163

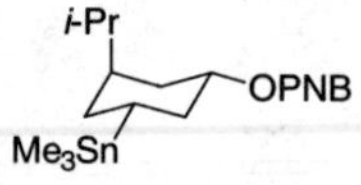

164

165

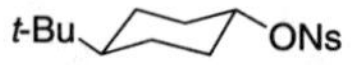

166

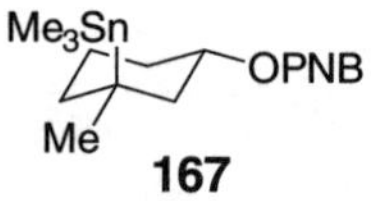

167

169 **168**

The lack of any effects in **145** and **167** implies that any inductive effects that the trimethylstannyl substituent has on the σ–σ^* interaction between the C2–C3 σ bond and the C–OR σ^* orbital represented by the resonance form **169** is relatively small.

V. INTERACTION OF C–M BONDS WITH ACCEPTOR ORBITALS AT THE δ POSITION

A. Reactivity Effects

1. Interaction of Group 4 Metal Substituents with Positive Charge at the δ Position

Solvolysis of 4-metalloidal [$M(CH_3)_3$, M = Si, Ge, Sn] substituted bicyclo [2.2.2]oct-1-yl mesylates (**170**)–(**172**)[137] provides evidence for the stabilization of

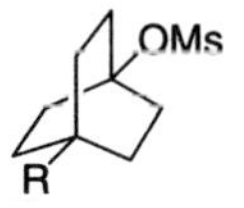

170: R = Me_3Si
171: R = Me_3Ge
172: R = Me_3Sn
173: R = H

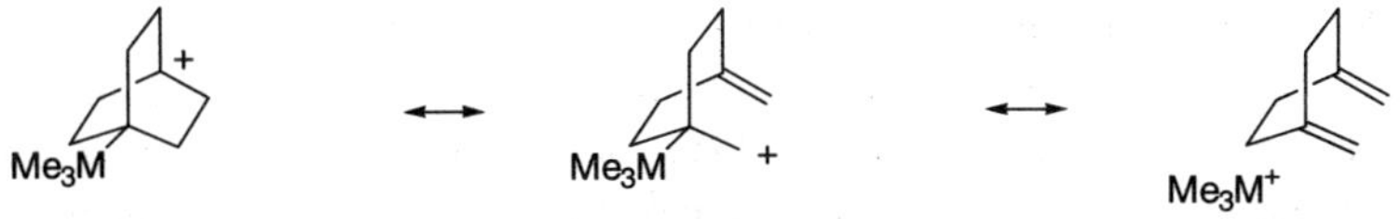

174 **174a** **174b**

carbenium ions at the δ position by silicon and to a larger extent by germanium and tin. Thus the relative rate of solvolysis (in 97% trifluoroethanol/water) of the 4-trimethylsilyl mesylate **170** and the nonsubstituted parent compound **173** is 49:1; the trimethylgermyl derivative **171** is slightly more reactive at 71:1 (relative to **173**, and the trimethylstannyl derivative **172** is by far the most reactive (2841:1). The stabilization by δ-silicon, germanium, and tin substituents is believed to be through double hyperconjugation (through-bond stabilization); such an interaction can be represented by the resonance structures **174**, **174a**, and **174b**, and was proposed to account for the very large δ-deuterium isotope effect (k_H/k_D = 1.050 in trifluoroethanol/water) observed for the solvolysis of 4-D-bicyclo[2.2.2]oct-1-yl mesylate.[137] Support for this interpretation came from calculations on 4-substituted bicyclo[2.2.2]oct-1-yl cations.[138]

Further evidence for the through-bond stabilization of positive charge by δ-Me_3M substituents was obtained from solvolysis of *E* and *Z* 5-(trimethylsilyl)- and 5-(trimethylstannyl)-2-adamantyl sulfonates.[139] Significant rate accelerations were observed for the *E* isomers **175** and **176** (with $Me_3Sn > Me_3Si$) (for which the

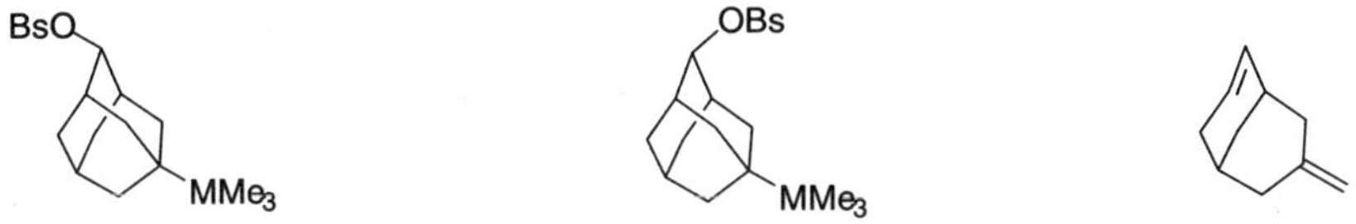

175: M = Si
176: M = Sn

177: M = Si
178: M = Sn

179

180

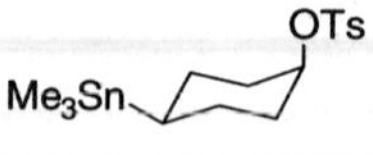

181

stereoelectronic requirements of the through-bond interaction are fulfilled)[2, 140] compared with the *Z* isomers **177** and **178** (for which through-bond participation in the developing carbenium ion are not possible). The major product obtained from the (*E*)-stannyl-substituted isomer **176** was the fragmentation product 7-methylenebicyclo[3.3.1]non-2-ene **179**, providing further support for the through-bond mode of participation of the δ-stannyl substituent in this geometry.

Lambert et al.[117] demonstrated the presence of a small rate-enhancing effect of the trimethylstannyl substituent on the unimolecular solvolysis of *trans*-4-trimethylstannylcyclohexyl tosylate **180** compared to the *cis*-isomer **181**.

B. Ground-State δ-Effects

The transmission of polar electronic effects over the three bonds in the adamantane ring system has been monitored by ^{19}F NMR studies on a range of 5-substituted 2-fluoroadamantanes **182** and **183**.[141] The ^{19}F substituent chemical shifts for the isomer **182** were factored into polar field contributions $\rho_F\sigma_F$ and residual contributions (^{19}F SCS-$\rho_F\sigma_F$) referred to as $\rho_x\sigma_x$; the latter contribution predominated in the series **182** but not for the series **183**. In the trimethylsilyl (R = Me_3Si) and trimethylstannyl (R = Me_3Sn) substituted derivatives the $\rho_x\sigma_x$ contributions to the ^{19}F SCS suggested contributions from the resonance structures **182a** and **182b** to the ground-state structure; here the C–Si (C–Sn) σ-bonding electrons are donated by a "double hyperconjugative" mechanism through the intervening C–C framework bonds to the C–F σ* acceptor orbital. The σ_{C-F}–σ_{C-C}–σ_{C-Si} and σ_{C-F}–σ_{C-C}–σ_{C-Sn} orbitals, whose interactions are responsible for these effects, are optimally aligned in the isomer **182** but not the isomer **183**; hence the residual effects for the isomer **183** are insignificant. The $\rho_x\sigma_x$ contributions for the Me_3Si and Me_3Sn substituents to the ^{19}F SCS of 4-substituted bicyclo[2.2.2]oct-1-yl fluorides (**184**)[142] similarly suggest contributions from the resonance structures **184a** and **184b** (shown for one of the ethano bridges).

Accurate low-temperature X-ray structural studies on the *trans*-4-trimethylmetalcyclohexyl nosylates (**185**: M = Si, and **186**: M = Sn)[143] gave no evidence of any significant structural effects associated with a through-bond interaction between the C–M σ orbital and the C–ONs σ^* orbital. Thus the

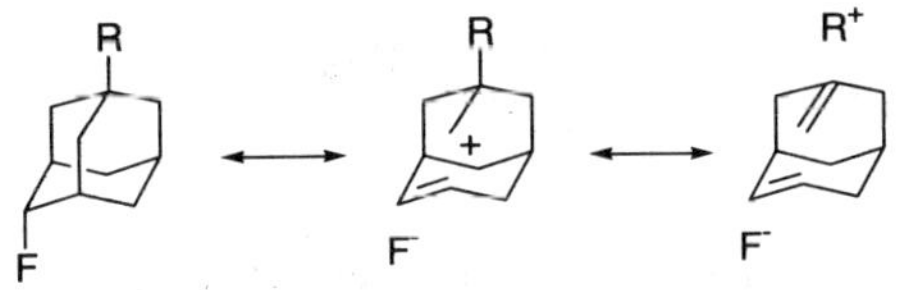

182 **182a** **182b**

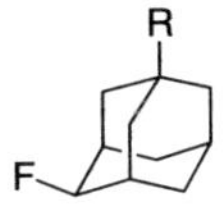

183

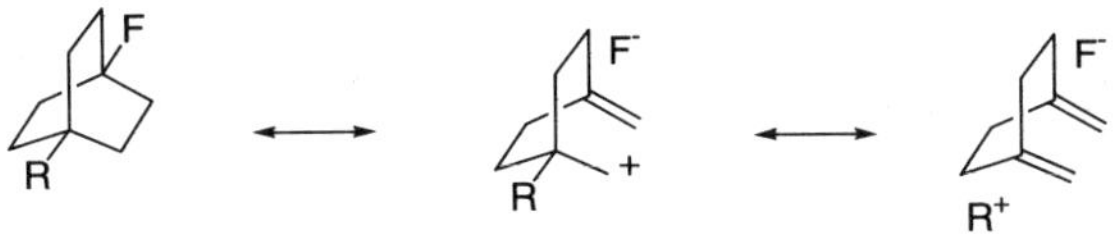

184 **184a** **184b**

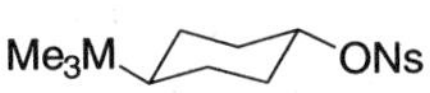

185: M = Si
186: M = Sn

187

C–ONs distance in **185** and **186** was not significantly different from the *cis*-stannyl derivative **187** or the 4-*tert*-butyl derivative **166**.

VI. INTERACTION OF $C-MR_3$ BONDS WITH REMOTE ELECTRON-DEFICIENT ORBITALS

Interaction of donor bonds with remote acceptor orbitals conceivably can occur by through-bond or through-space mechanisms. In an effort to find evidence for a through-6-bond interaction (or triple hyperconjugation), with a trimethylstannyl substituent as the donor and a carbenium ion *p* orbital at the z (zeta) position as the acceptor, Lambert et al. investigated the solvolyses of the *cis,trans,cis*-6-(trimethylstannyl)-2-decalyl tosylate **188**.[144] **188** showed no rate enhancement compared to the *cis,trans,trans*-stereoisomer **189** and the corresponding nonstannylated analogs **190** and **191**, and its rate was found to be sensitive to

188: R = Me_3Sn
190: R = H

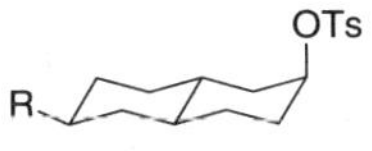

189: R = Me_3Sn
191: R = H

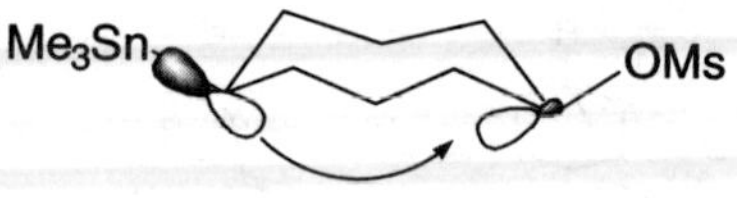

Figure 9. The transannular percaudal interaction in *cis*-5-trimethylstannylcyclooctyl mesylate **192**.

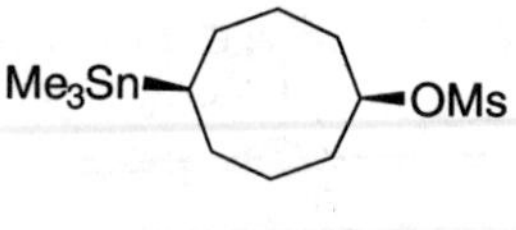

192

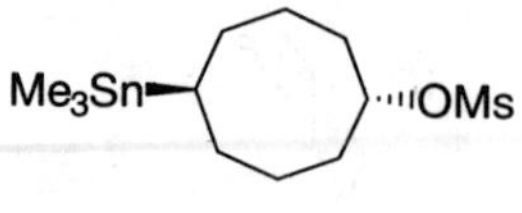

193

solvent nucleophilicity, suggesting a bimolecular process. Thus in this system the effect of the stannyl substituent was not great enough to raise the rate of the unimolecular solvolysis reaction for **188** above that of the competing bimolecular reaction.

Kitching et al. have provided evidence for significant carbon–tin σ participation in the solvolysis of *cis*-5-mesyloxycyclooctyltrimethylstannane **192**.[145] The mesylate **192** was found to react 800 times faster than the trans isomer 193 in 80% ethanol/water and gives rise to bicyclo[3.3.0]octane as the only product. The rate enhancement observed for the cis isomer **192** was accounted for by a stabilizing transannular percaudal interaction (Figure 9) (through space), which is formally an ε-effect; this interpretation was supported by the formation of bicyclo [3.3.0]octane as the major solvolysis product. The trans isomer **193** cannot achieve a conformation that allows the percaudal interaction to assist the departure of the leaving group.

VII. INTERACTION OF C–MR$_3$ BONDS WITH π SYSTEMS

A. Allyl Group 4 Metal Derivatives

The structures and properties of allyl-MR$_3$ (M = Si, Ge, Sn) are dominated by the σ–π interaction between the high-lying C–M σ orbital and the π* orbital of the double bond. This interaction is maximized in the conformation where the C–M bond is orthogonal to the π system (Figure 10). The observed conformation of allyl and allyl-like metal systems tends to represent a compromise between the requirements of the σ–π interaction and steric repulsion. For example, the observed conformation of allyltrimethylsilane **194**[146, 147] is characterized by the Si–C–C=C dihedral angle, which lies in the range 102–106°; this conformation represents a

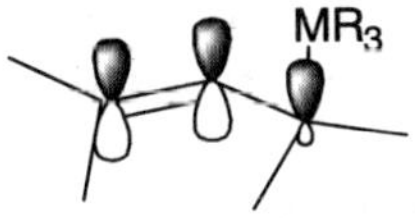

Figure 10. Conformation of allyl group 4 metal derivatives that maximizes σ–π conjugation.

compromise between the skew conformation (Si–C–C=C 120°) and the conformation required to maximize σ–π overlap (Si–C–C=C 90°). Evidence that allyltrimethylgermane **195** and allyltrimethylstannane **196** exist in a similar conformation to **194** has been provided by photoelectron spectroscopy studies.[17] Crystal structures of the allylic stannane derivatives **197**, **198**, **199**, and **200**[148–151] illustrate the dominating influence that the σ_{C-Sn}–π^* interaction has on the preferred ground state conformations of these structures.

In all these structures the Sn–C–C=C angle is close to optimum for σ–π interaction between the C(allyl)–Sn σ bond and the π system. In the two crystallographically independent molecules of cyclopenten-2-yl-triphenylstannane **197** and cyclohepten-2-yltriphenylstannane **199**, the stannyl substituent takes up a pseudoaxial orientation. Although cyclohex-2-enyltriphenylstannane **198** exists in the solid state with the stannyl substituent in a pseudoequatorial orientation (Sn–C–C=C 120°), it exists in solution predominantly in a pseudoaxial conformation. Analysis of the structural parameters in **198** and **199** provides some tentative structural evidence for the presence of the σ–π interaction in these compounds; the Sn–C(allyl) bond distances, which are 2.189(5) and 2.182(5) Å, respectively, are slightly longer than the normal Sn–C(aliphatic) distance, which is typically of the order of 2.13 Å, consistent with the expected structural effects of electron donation from the σ_{C-Sn} orbital into the π^* orbital.

The σ–π interaction manifests in a number of other interesting ground-state properties of allyl group 4 metal compounds. For example, the energy of the

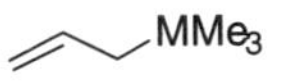

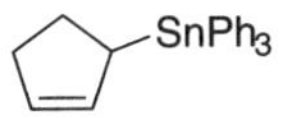

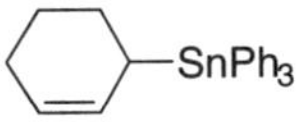

194: M = Si
195: M = Ge
196: M = Sn

197

198

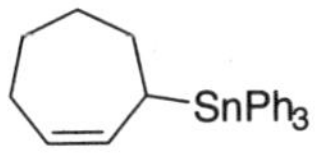

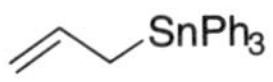

199

200

HOMO of the π system is raised, as is clear from photoelectron spectra of the allyltrimethylmetal compounds **194**, **195**, and **196**.[17, 152–156] The first ionization energy of allyltrimethylsilane appears at 9.0 eV, that of allyltriethylgermane at 8.8 eV and that of allyltributylstannane at 8.4 eV, and these values can be compared to propene (10.0 eV) and ethylene (10.5 eV).[156] The energies of the isolated C–M σ orbitals involved in the σ–π interaction as obtained from the photoelectron spectra of the Et_4M compounds[16, 17] are 10.5 eV for C–Si, 9.3 eV for C–Ge, and 8.7 eV for C–Sn. Thus the closest energy match between the interacting orbitals occurs in the σ_{C-Si}–π interaction of allyltrialkylsilanes, where the HOMO is raised by 1.5 eV. The σ_{C-Ge}–π interaction raises the HOMO by 0.5 eV in allyltriethylgermane (relative to the σ_{C-Ge} orbital energy) and σ_{C-Sn}–π interaction raises the HOMO by 0.3 eV in allyltributylstannane (relative to the σ_{C-Sn} orbital energy). The strength of the σ–π interaction is also dependent on the nature of the other substituents attached to the metal inasmuch as these influence the energy of the M–C(allyl) σ bond and hence the energy match with the π^* orbital.[155] The σ_{C-M}–π interaction in allylsilanes, -germanes, and -stannanes manifests also in increased the basicity of the C=C double bond relative to propene; this has been demonstrated from the O–H stretching frequencies in the infrared spectra of the phenol complexes of these unsaturated systems.[157]

The effects of the σ–π interaction on the ground-state properties of allyltrimethylmetal compounds are paralleled by the effect on reactivity towards electrophilic reagents. Mayr demonstrated that allyltrialkylsilanes, allyltrialkylgermanes, and trialkylstannanes react with diphenylcarbenium ions at rates 10^5, 5.6×10^5, and 10^9, respectively, relative to propene.[158] The reaction rates were also found to be sensitive to the inductive effects of the other substituents attached to the metal. A theoretical evaluation of the factors determining the regiochemistry and stereochemistry of electrophilic addition to allylsilanes and other allyl systems is reported by Hehre et al.[159] They predict a preference for electrophilic attack *anti* with respect to the silane substituent, a prediction that is supported by many experimental studies.[82, 160]

B. Benzyl Group 4 Metal Derivatives

The C–M bonds (M = Si, Ge, Sn, Pb) interact by strong hyperconjugation with the adjacent π system of an aromatic ring in the neutral ground state.[11,161,162] This has been referred to as sacrificial or isovalent hyperconjugation[163] and can be represented by the valance structures **201a–201d**. The aromatic ring of benzyl group 4 metal derivatives is therefore very electron rich. Evidence for this is provided by ^{19}F,[15, 27, 164–166] ^{119}Sn,[167] ^{119}Hg,[167] ^{13}C,[27, 168, 169] ^{1}H NMR,[11] infrared spectroscopy,[170–174] electron spectroscopic studies of the charge-transfer complexes of benzyl group 4 derivatives with π-deficient species,[11, 22, 23, 25, 175, 176] and the influence of these substituents on the magnetic circular dichroism of the benzene L_b band.[177] Studies of ^{13}C chemical shifts in benzyl-, α-naphthyl-, and β-naphthyl-substituted group 4 metals, and of ^{19}F chemical shifts in the corresponding fluorinated derivatives, established the rather surprising order for electron release in the neutral ground state for the R_3MCH_2 substituent as Pb ~ Sn > Ge ~ Si.[27] The dependence of the σ–π interaction on ΔE, which is the difference $E(\pi)-E(\sigma_{C-M})$ would suggest that the order of electron release should be Pb >> Sn >> Ge >> Si, since DE decreases in distinct steps from M = Si to Pb (cf Table 1), however, the σ–π interaction depends also on other factors, including $a^2(\pi)$ and $a^2(\sigma)$, which are the electron densities in the *p* orbitals of the connecting atoms of the interacting π and σ systems.[27] It was proposed that the order of electron release for C–M bonds in the neutral ground state results from variation of the $a^2(\sigma)$ parameter from M = Si to Pb in a manner that offsets the trends dictated by the ΔE term. In contrast with the neutral ground state, hyperconjugation of C–M bonds with positively charged π systems does follow the order Pb >> Sn >> Ge >> Si,[15, 22–26] suggesting that the ΔE term is more important when the electron demand is high. This is evident from the σ^+ values for the R_3MCH_2 substituents determined from electronic spectroscopy of charge-transfer complexes of group 4 benzyls with various acceptors, and from the rates of electrophilic substitution of R_3M-CH_2-substituted aromatics. ^{19}F substituent chemical shift studies of 4-(trimethylsilylmethyl)fluorobenzene **202** and the bicyclic structures **203** and **204** demonstrate the stereoelectronic requirements of σ–π conjugation.[166] The ^{19}F chemical shift observed for **202** (where the favored conformation has the C–Si bond aligned with the π system) indicates significant electron release by the trimethylsilyl substituent, whereas for **203** and **204**, where the five- and

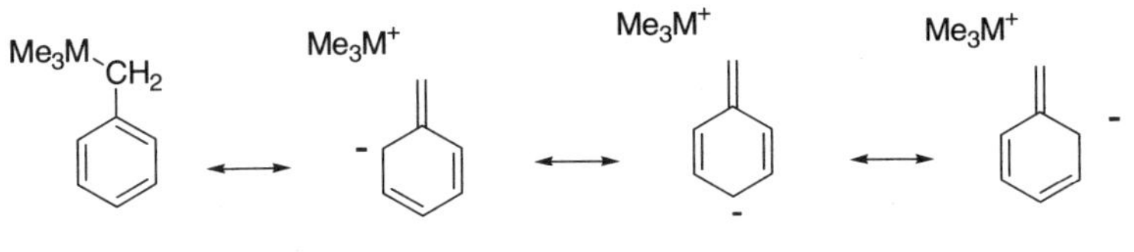

201a **201b** **201c** **201d**

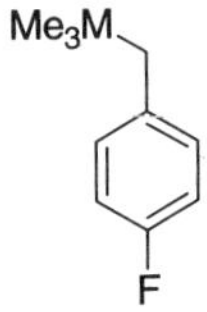

202

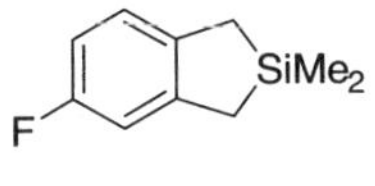

203

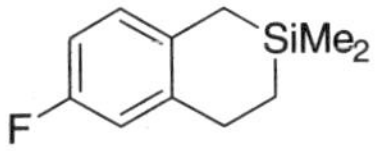

204

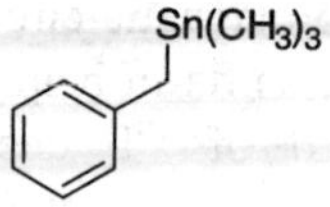

205

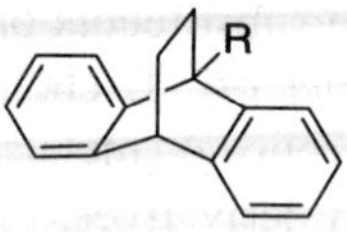

206: R = $SnMe_3$
207: R = H

six-membered rings constrain the C–Si bond to be close to the nodal plane of the π system, the effects are significantly diminished.

A similar dependence of σ–π conjugation on stereochemistry is evident from the charge-transfer spectroscopy of benzyltrimethylstannane **205** and the bicyclic stannane **206** (where the C–Sn bond is constrained to lie in the nodal plane of the π system)[175]; whereas the charge-transfer frequency of **206** is identical with the parent hydrocarbon **207**, in benzyltrimethylstannane **205**, in which σ–π conjugation is possible, the charge-transfer frequency is lowered by 6800 cm^{-1} compared with toluene.

The stereoelectronic requirements of the σ–π interaction are also reflected in the barriers to rotation about the $Ar–CH_2MR_3$ bond.[178, 179] These barriers have been determined for a number of benzyl group 4 metal derivatives from the long-range coupling constants between the methylene protons and the *para* ring protons, and by ab initio calculations. These rotational barriers are shown to be dominated by a twofold term that is consistent with the σ–π interaction; the magnitudes of the barriers are in the order M = Si < Ge < Sn.

The σ_{C-M}–π interaction raises the HOMO level of the aromatic π system, as evidenced by a lowering of the ionization potential of group 4 metal benzyls relative to the corresponding carbon analogs. Vertical ionization potentials for group 4 metal benzyl derivatives have been determined by a number of techniques including photoelectron spectroscopy,[180–185] mass spectroscopy,[186] and charge-transfer spectroscopy.[11, 22, 23, 25, 175, 176] Ionization potentials for compounds $PhCH_2MMe_3$ have been determined to be 8.40–8.42 eV for Si, 8.36–8.40 eV for Ge, 8.08–8.21 eV for Sn, and 7.87 eV for Pb, which may be compared with toluene (8.82 eV) and neopentyl benzene (8.77 eV).[183, 184] Schweig et al. showed that the destabilization of the π molecular orbitals in allyl and benzyl compounds of silicon, germanium, and tin can be quantitatively accounted for on the basis of the σ–π interaction.[154] The difference in the ionization potentials for silicon-substituted indane derivative **208** and the related silaindane **209**, which are 8.13 and 8.42 eV, respectively, reflect the differing degrees of overlap between the C–Si bond and the aromatic π system in **208** and **209**.[187] Multiple substitution of benzene with trimethylsilylmethyl substituents lowers the ionization potential further; for example, 1,4-bis(trimethylsilylmethyl)benzene (**210**) and 1,2,4,5-tetrakis-(trimethylsilylmethyl)benzene (**211**) have remarkably low ionization potentials of 7.45 and 7.10 eV, respectively.[180]

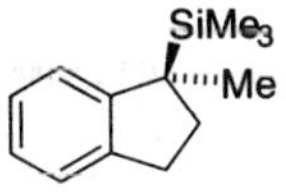

208

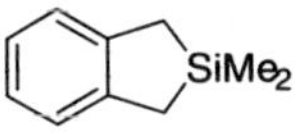

209

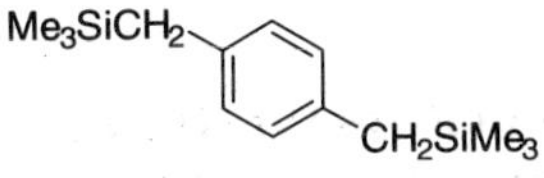

210

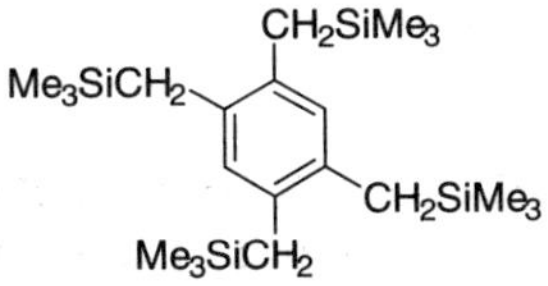

211

The observation of an intense absorption at 43,300 cm^{-1} (ε = 10,900) in the electronic spectra of phenylpentamethyldisilane **212** and a red shift of this ^1L_a band of 6000 cm^{-1} relative to benzene has been suggested to be due to σ–π interaction between the Si–Si bond and the aromatic ring.[188] Support for this interpretation was provided by the electronic spectra of the stereochemically constrained disilanes **213** and **214**. The disilane **214**, which can achieve a conformation allowing overlap between the Si–Si bond and the π system, displays an intense absorption at 42,900 cm^{-1} (ε = 11,000) similar to **212**. However, disilane **213**, in which the constraints of the five-membered ring place the Si–Si bond close to the nodal plane of the benzene ring, did not display this absorption.

Further spectroscopic evidence for neutral hyperconjugation between Me_3MCH_2 (where M = Si, Ge, Sn) and an adjacent aromatic ring has been claimed[189] from the one-bond ^{13}C–M (M = ^{29}Si, ^{119}Sn) and ^{13}C–^{13}C coupling constants in the series of *4*-substituted benzyl-silanes, -germanes, and -stannanes **215**, **216**, **217** (see also Ref. 190). When the substituent was held constant, there was an increase in the $^{13}CH_2$–$^{13}C_{ipso}$ coupling constant for the series M = Si < Ge < Sn consistent with increasing σ–π conjugation from Si to Sn. For the 4-substituted benzyltrimethylsilanes **215**, the ^{29}Si–$^{13}CH_2$ coupling constant decreased along the series X = OMe > Me > H > CN, indicating the increasing importance of the resonance form **215a** as the electronegativity of the substituent X increases.

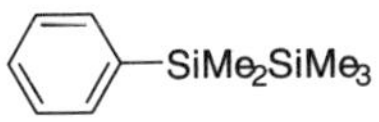

212

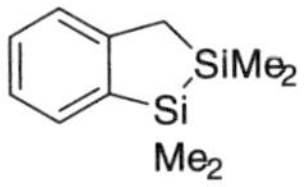

213

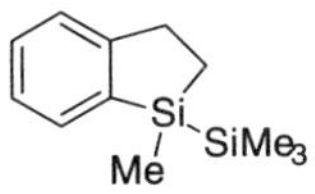

214

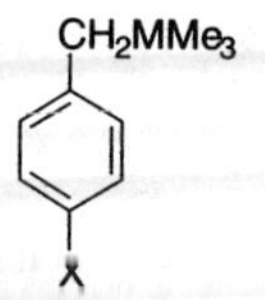

215: M = Si
216: M = Ge
217: M = Sn

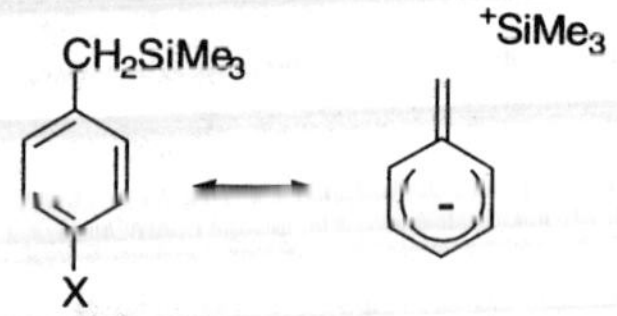

215 215a

VIII. INTERACTION OF C–MR3 BONDS WITH ELECTRON RICH ORBITALS AT THE β POSITION

A. Interaction of C–MR3 with Heteroatom NonBonding Orbitals at the β Position

The interaction of C–M bonding orbitals with heteroatom lone-pair orbitals is a two-center–four-electron interaction. This is destabilizing and raises the energy of the HOMO relative to the unperturbed orbitals, and hence lowers the ionization potential of the heteroatom lone-pair electrons. The interaction is qualitatively represented by the molecular orbital interaction diagram shown in Figure 11. The strength of the interactions depends on a number of factors but most importantly on the orientation of the C–M bond with respect to the lone-pair orbital and on the

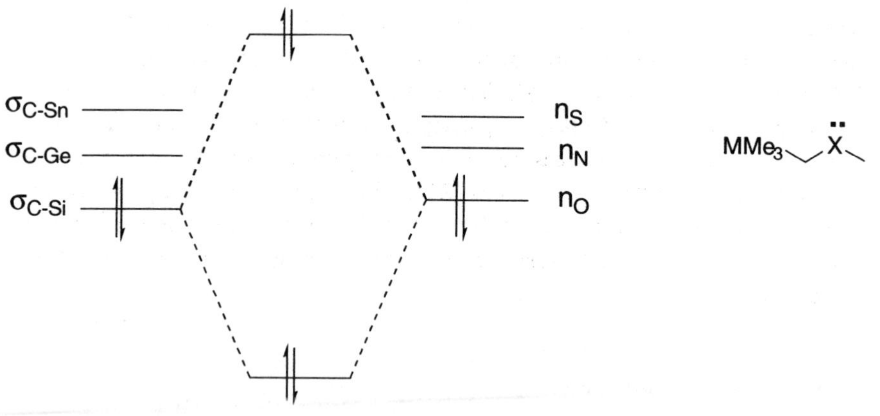

Figure 11. Energy-level diagram for the interaction of the C–M σ orbital and a heteroatom lone-pair orbital.

energy difference between the interacting orbitals. Orbital energies of C–M (M = Si, Ge, Sn)[16–18] bonds and of various heteroatom lone-pair orbitals[19–21] have been obtained from photoelectron spectroscopy on model compounds. These suggest that a C–Si bond should interact most strongly with an ether oxygen lone-pair orbital, for in this case there is a close energy match between the σ_{C-Si} and n_O orbitals. Interaction of the σ_{C-Si} orbital with nonbonded orbitals on heteroatoms such as nitrogen or sulfur is expected to be weaker, as these lie at higher energies. The lower group 4 metals, Ge and Sn, for which the σ_{C-M} orbitals lie at higher energies, interact more strongly with n_N and n_S lone-pair orbitals than silicon. The effects that the interaction between a C–M bond and a heteroatom lone-pair orbital have on the properties of O,[191–194] N,[191, 195] S,[196, 197] and P[155] containing compounds have been studied by a number of techniques, including photoelectron spectroscopy,[180, 193, 194, 196, 198] electrochemistry,[192, 199] electron spin resonance spectroscopy.[200]

1. *Organosilicon Compounds with Heteroatom Lone Pairs at the β Position*

Block et al.[194] examined the effects of trimethylsilyl substitution on the first vertical ionization potentials by photoelectron and Penning ionization electron spectroscopy studies of a range of cyclic and noncyclic sulfides and ethers. It was shown that substitution of oxirane **218** with a trimethylsilyl substituent as in **219** lowered the ionization potential by 0.90 eV (20.8 kcal/mol), while similar substitution of dimethyl ether **220** in **221** lowered the ionization potential by 0.64 eV (14.8 kcal/mol). By comparison, the effects of silyl substitution on sulfur lone-pair ionization potentials was found to be smaller; thus the ionization potential of dimethyl sulfide **222** is lowered by 0.37 eV upon trimethylsilyl substitution in **223**, and the trimethylsilyl-substituted thiirane **225** is lowered by 0.59 eV relative to thiirane **224**. The raising of the energy of the sulfur lone-pair electrons in the thiirane **225** is also apparent from its UV spectrum, where there is a bathochromic shift in the absorption maximum compared to the parent **224**.

Yoshida et al.[191] determined the electrochemical oxidation potentials of a range of silyl-substituted ethers and alcohols and found that silyl substitution lowered the oxidation potentials relative to the corresponding nonsilylated compounds. In the ethers **226–230** the effect of the silicon was shown to depend strongly on the Si–C–O–C dihedral angle. Maximum effects were found for **229**, where the C–Si bond was close to eclipsed with the oxygen *p* lone-pair orbital; much smaller effects were seen for **227** and **230**, where the C–Si bond overlaps less effectively with the oxygen *p* lone-pair orbital. Further evidence for the dependence of the oxidation potential on the conformation of the Si–C–O–C fragment was provided by the temperature dependence of the oxidation potential of the acyclic ether **231**. At low temperature the most stable conformation of **231** has the Si–C–O–C dihedral angle 180°; however, this conformation has the C–Si bond orthogonal to

218

IP 10.57

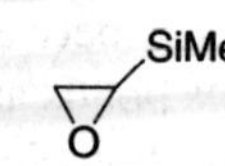

219

IP 9.67

220

IP 10.04

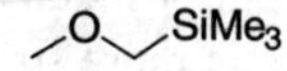

221

IP 9.40

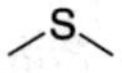

222

IP 8.72

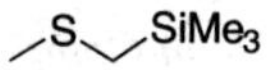

223

IP 8.35

224

IP 9.03

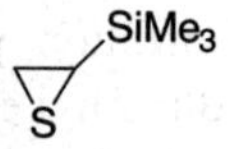

225

IP 8.44

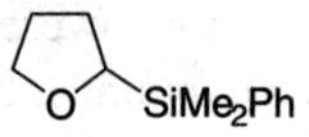

226

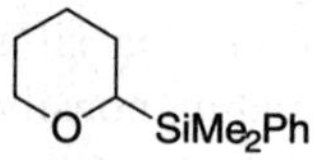

227

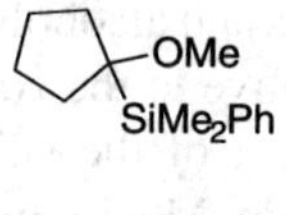

228

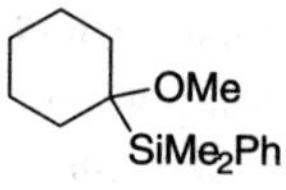

229

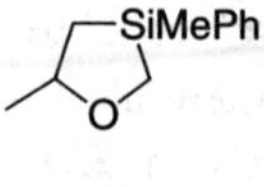

230

231

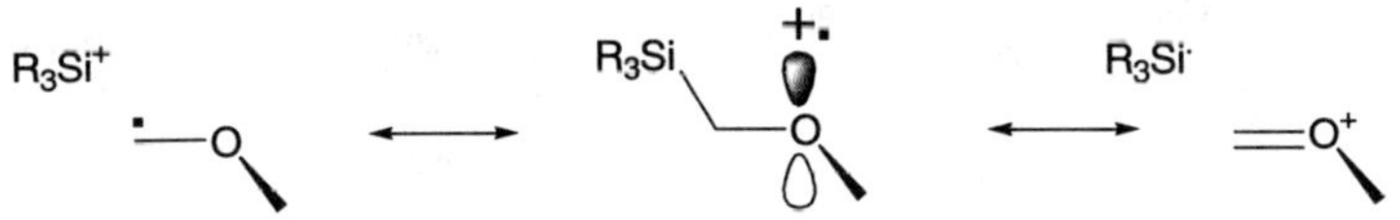

233

Figure 12. Stabilization of a heteroatom radical cation by hyperconjugation with the C–Si σ orbital.

the oxygen *p* lone-pair orbital and is unfavorable for electron transfer. As the temperature is raised, the oxidation potential of **231** decreased due to increasing population of a less favored conformation that is energetically more favorable for oxidation.

Theoretical calculations on the simple model compound **232** showed that the HOMO energy varied with the Si–C–O–H dihedral angle and was at a maximum when the Si–C–O–H angle was 90° (where the C–Si bond eclipses the oxygen *p* lone-pair orbital) and minimum at 180° (where the C–Si bond and oxygen *p* lone-pair orbitals are orthogonal). These calculations also suggested that the lowering of the oxidation potentials for **226–230** was due in most part to the raising of the energy of the oxygen lone-pair electrons by the destabilizing four-electron interaction with the C–Si bond (see Figure 11). Stabilization of the radical cation **233** by hyperconjugation (Figure 12), which would also contribute to a lowering of the oxidation potential, appeared to be of less significance. The C–Si bond was calculated to stabilize the β-radical cation by only 0.12 eV at the STO-3G* level of theory, which is similar to the stabilization of carbon radicals by β-silicon substituents[201]; in contrast, the HOMO is destabilized by 0.9 eV due to the two-center–four-electron interaction. Theoretical calculations on the analogous sulfur-containing model compound **234** also showed that the HOMO energy level varied with Si–C–S–C angle but that the effect was much less dramatic than was the case for **232**. This outcome is consistent with the much poorer energy match between a σ_{C-Si} orbital and a sulfur *p* lone-pair orbital. Sakurai and Kira[200] determined the ESR spectra of the radical cations **235** and **236** generated from their corresponding neutral precursors by ^{60}Co γ-irradiation in frozen $CFCl_3$ matrices. From analysis of the fine structure of the spectra of **235** and **236**, it was deduced that the preferred conformation of these radicals had the C–Si bond eclipsing the heteroatom singly occupied *p* orbital. For the sulfur radical cation **236** the SOMO

232 234 235 236

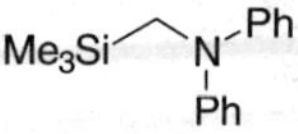

237

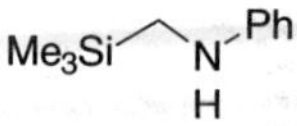

238

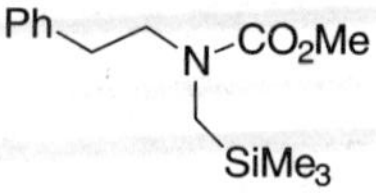

239

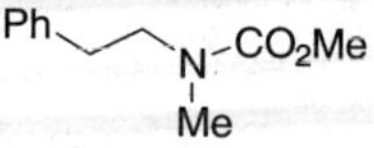

240

was described as an S-centered nonbonded orbital, perturbed moderately by σ_{C-Si}–n_S orbital mixing. However, for the O radical cation **235** the SOMO could not be regarded as an oxygen nonbonded orbital, but as one including large contributions from the σ_{C-Si} bonding orbital, this result is consistent with better mixing of the similar energy orbitals, σ_{C-Si} and n_O.

Interaction of the C–Si bond with the nitrogen lone-pair orbital in **237** and **238**[191, 195] causes only a moderate decrease of the oxidation potential (0.33 eV for **237**[195] and 0.14 eV for **238**[191]) relative to the parent amines. Interestingly, the effects of silyl substitution on the oxidation potentials of carbamate nitrogen lone-pair electrons is larger; for example, the oxidation potential of **239** is lowered by 0.5 eV relative to the parent **240**. This can be rationalized by the lowering of the energy of the nitrogen p orbital by conjugation with the carbonyl group, resulting in a better energy match with the σ_{C-Si} orbital.

The interaction of the σ_{C-M} bonding orbital with the carbonyl oxygen p lone-pair orbital is proposed largely to account for the bathochromic shift for the n–π^* transition in the UV–VIS spectra of acylsilanes, -germanes, and -stannanes compared with simple alkyl ketones.[198, 202–204] The σ_{C-M}–n_O interaction raises the energy of the lone-pair electrons on the carbonyl group and consequently decreases the n–π^* energy gap. Evidence for the raising of the energy of the oxygen nonbonded electrons in these compounds has been provided by photoelectron spectroscopy[198] and electrochemistry.[202] For example, Bock et al.[198] measured the vertical ionization potentials of the acylsilane **241** and the acylgermane **242** and found these to be 8.6 and 8.5 eV, respectively, which represents a decrease of 1.1 and 1.2 eV, respectively, relative to acetone (9.7 eV). The very broad photoelectron bands (0.5 eV) for **241** and **242** were seen to be consistent with significant mixing of the σ_{C-M} (M = Si, Ge) bonds with the oxygen p lone pair, and in fact the longest-wavelength transitions in acylmetalloids are better regarded as σ–π^* transitions rather than n_O–π^* transitions. It has been proposed that the

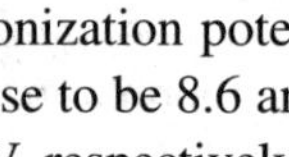

241: M = Si
242: M = Ge

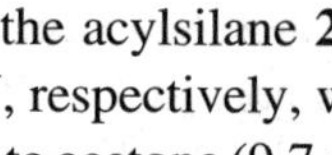

243

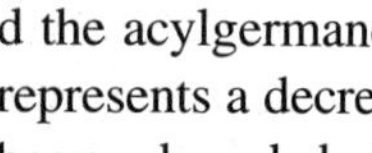

244

bathochromic shifts in the UV–VIS spectra of acylmetalloids is also due in part to the lowering of the energy of the LUMO of acyl metalloids from mixing of the carbonyl π^* orbital with the vacant d orbitals (or σ^*_{Si-C}) on the silicon.[204, 205] For example, Bock et al. have measured the half-wave reduction potentials of various acylsilanes and found stabilization of the LUMO of ca. 0.2–0.3 eV due to the presence of the silicon. The oxidation potentials of the acylsilanes **243** (1.23 V) and **244** (1.24 V) determined by rotating disk techniques[202] were dramatically lower than octanal and 2-octanone, both of which have oxidation potentials greater than 2.0 V.

2. *Organogermanium Compounds with Heteroatom Lone Pairs at the β Position*

Electrochemical studies of α-alkoxygermanes and acylgermanes[202, 203] show that these compounds are more easily oxidized than their parent germanium-free compounds and the corresponding silicon-containing compounds. For example, the oxidation potential of (menthoxylmethyl)trimethylgermane (**245**)[203] occurred at 1.67 eV, while that of the parent **247** occurred at 2.16 eV, and the corresponding silane **246** at 1.83 V. The oxidation potential of the acylgermane **248**[202] was determined to be 1.31 V compared 1.62 V for the corresponding acylsilane **249** and the parent, which is > 2.0 V. Mixing between the σ_{C-Ge} orbital and n_O orbitals, raising the energy of the HOMO, is proposed to account for these observations.

3. *Organostannane Compounds with Heteroatom Lone Pairs at the β Position*

Electrochemical studies of (menthyloxymethyl)trimethylstannane (**250**)[206] reveals an oxidation potential of 1.09 V, which is lower than the corresponding silane **246** and germane **245** and the parent **247**. The oxidation potential of tetrabutylstannane **251** was determined to be 1.52 V, which is higher than the alkoxystannane **250**; thus the decrease in the oxidation potential for **250** was attributed to the σ_{C-Sn}–n_O interaction. It can be argued that the σ_{C-Sn}–n_O interaction raises the energy of the

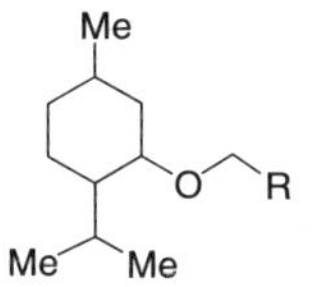

245: R = $GeMe_3$
246: R = $SiMe_3$
247: R = H
250: R = $SnMe_3$

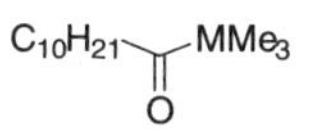

248: M = Ge
249: M = Si

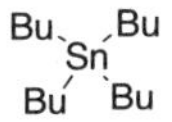

251

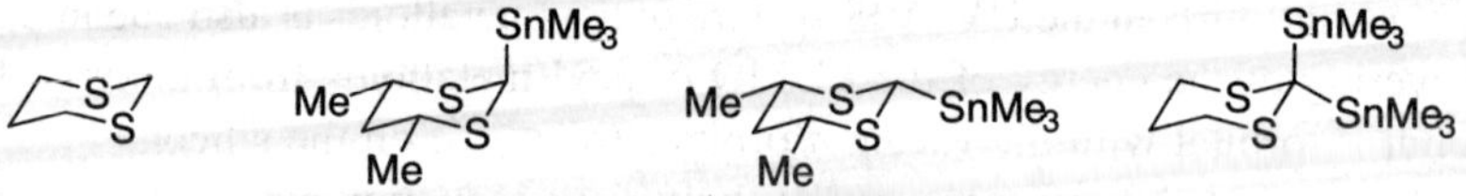

252 **253** **254** **255**

HOMO in **250** by ca. 0.4 V (the difference between the oxidation potentials of **250** and **251**); this is a much smaller effect than the corresponding σ_{C-Si}–n_O interaction, which lowers the oxidation potential of α-alkoxysilanes by as much as 0.9 V. This is consistent with the better energy match between the σ_{C-Si} and n_O orbitals.

The effects of stannyl substituents on the ionization potentials and oxidation potentials of the (higher-energy) sulfur lone pairs at the β position is much larger than the effects of silicon. For example, introduction of a trimethylstannyl substituent onto the 2-position of the 1,3-dithiane **252**[197] results in a dramatic decrease in the oxidation potential, and this effect was shown to be geometry dependent. The dithiane **253** in which the trimethystannyl substituent is constrained to be axial (with efficient overlap between the C–Sn bond and the sulfur *p* lone pairs) has an oxidation potential of 0.4 V, which represents a decrease in oxidation potential of 0.78 V compared with the parent dithiane **252** (1.18 V), whereas the isomer **254** in which the trimethylstannyl substituent is equatorial (and overlaps less effectively with the sulfur *p* lone-pair orbitals) displays an oxidation potential of 0.75 V, which represents a much smaller effect. The oxidation potential of the 2,2-bis(trimethylstannyl)dithiane **255** is lowered by a remarkable 1.0 V relative to the parent dithiane **252**. The vertical ionization energy of **255** was also investigated by measuring its photoelectron spectrum, which revealed a lowest ionization band for removal of a sulfur *p* lone-pair electron that was lowered by ca. 1 eV compared to dithiane **252**.

B. Interactions between Adjacent C–MR$_3$ Bonds

Interactions between vicinal C–Si and C–Sn bonds in the fragment R_3M–C–C–MR_3 raises the energy of the HOMO relative to reference compounds containing a single equivalent metal substituent. These interactions, and their dependence upon stereochemistry, have been revealed by spectroscopic studies on model compounds.[207] For example, in the photoelectron spectrum of 1,2-bis-(trimethylstannyl)ethane[207] (**256**) the first ionization energy [which corresponds to ionization of the antibonding combination of the vicinal σ_{C-Sn} orbitals (Figure 13)] occurs at 8.0 eV compared with ethyltrimethylstannane (**257**) of 9.1 eV; ionization of the bonding combination for **256** occurs at 10.6 eV. This splitting indicates an interaction between the two C–Sn σ orbitals of 2.6 eV. In

Figure 13. In-phase and out-of-phase combinations of vicinal C–Sn orbitals.

contrast, 1,3-bis-(trimethylstannyl)propane (**258**) shows a single ionization band corresponding to the $Me_3Sn–CH_2$ σ orbital at 9.4 eV. The photoelectron spectrum of trans-2,3-bis(trimethylstannyl)[2.2.1]bicycloheptane (**259**) reveals a first ionization band at 8.0 eV. The splitting between the in-phase and out-of-phase combinations of the σ_{C-Sn} orbitals was estimated to be 1.2 eV. The much smaller splitting observed for **259** compared with **256** is consistent with the reduced overlap between the σ_{C-Sn} orbitals in the former; Sn–C–C–Sn –120° for **258** and 180° in **256**. The σ–σ interactions in **256** and **259** were compared with the *n–n* interaction in the bicyclic tetraalkylhydrazines **260** and **261**, shown in Figure 14. The obvious similarity in the σ–σ interaction between vicinal C–Sn bonds and vicinal nitrogen lone-pair orbitals suggests that polarized C–metal bonds behave like nonbonded electrons in delocalization. Interactions between vicinal C–Si bonds are apparent in the photoelectron spectrum of 1,2-bis(trimethylsilyl)ethane (**262**), which has a first ionization band at 8.7 eV compared with that of 1,3-bis(trimethylsilyl)propane (**263**), which occurs at 9.4 eV.

Yoshida et al.[208] determined the oxidation potentials of a range of 1,2-bis (trialkylsilyl)-1,2-diphenylethane derivatives using rotating disk voltammetry. They found that the oxidation potentials of the two isomers **264** and **265**, which were 1.23 and 1.32 V relative to the reference Ag/AgCl electrode, were less positive than for benzyltrimethylsilane (**201**: M = Si) (σ–π interaction system) and 1,2-bis(trimethylsilyl)ethane (**262**) (a σ–σ interaction system). This result suggested some conjugation between the σ–π interaction system and the σ–σ interaction system in **264** and **265**. This was supported by molecular orbital calculations, which showed that the HOMO for **264** and **265** is delocalized through the two C–Si bonds and the two benzene rings.

260 **261**

ΔE 2.6 eV 1.8 eV

Figure 14. Lone-pair–lone-pair interactions in tetraalkyl hydrazines.

256

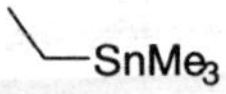

257

258

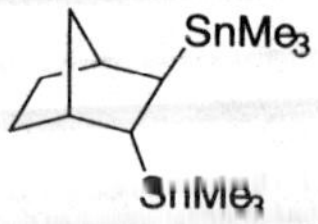

259

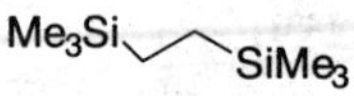

262

263

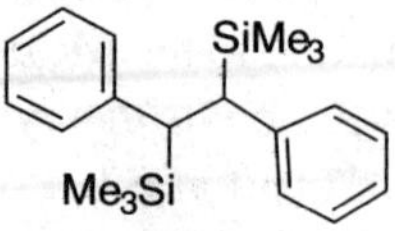

264 (*meso*)
265 (*dl*)

IX. COMMENT: INTERACTION OF GROUP 4 METAL SUBSTITUENTS WITH SINGLY OCCUPIED ORBITALS

Interaction of group 4 metal substituents with singly occupied orbitals will not be dealt with in any depth in this chapter. It is worth pointing out, however, that it has been well established that these substituents stabilize carbon-centered radicals at the α and β positions,[30, 38, 201, 209] by ca. 1–5 kcal/mol, the former by *p–d* conjugation, and the latter by σ–π conjugation and/or *p–d* homoconjugation.[210–215] Interaction with singly occupied orbitals on oxygen[191,200,216] and sulfur[200] have been studied by electron spin resonance and by theoretical calculations. The stabilizing effect of silicon on both α- and β-carbon-centered radicals has been used to advantage to direct the course of organic reactions involving free radical intermediates.[217, 218]

ACKNOWLEDGMENT

Our sincere thanks goes to John Lawlor of this department for his critical comments and proofreading in the preparation of this chapter.

REFERENCES

1. Bowden, K.; Gibbs, E. J. *Chem. Soc. Rev.* **1996**, 171.
2. Hofmann, R. *Acc. Chem. Res.* **1971**, *4*, 1.
3. Birnbaum, K. B. *Acta Cryst.* **1972**, *B28*, 2825.
4. Grob, C. A. *Angew. Chem. Int. Ed.* **1969**, *8*, 535.

5. Kirby, A. J. *The Anomeric Effect and Related Stereoelectronic Effects at Oxygen*; Springer-Verlag: New York, 1983.
6. *The Anomeric Effect and Associated Stereoelectronic Effects*; ACS Symposium Series 539, Thatcher, G. R. J., Ed.; ACS: Washington, 1993.
7. Juaristi, E.; Cuevas, G. *Tetrahedron* **1992**, *48*, 5019.
8. Wolfe, S. *Acc. Chem. Res.* **1972**, *5*, 102.
9. Juaristi, J. *J. Chem. Ed.* **1979**, *56*, 439.
10. Jarvie, A. W. P. *Organometallic Chemistry Reviews A* **1970**, *6*, 153.
11. Egorochkin, A. N. *Russ. Chem. Rev.* **1984**, *53*, 445.
12. Lambert, J. B. *Tetrahedron* **1990**, *46*, 2677.
13. Fleming, I. *Organic Silicon Chemistry*, In *Comprehensive Organic Chemistry*; Barten, D. H. R.; Ollis, D. W.; Eds.; Pergamon Press: Oxford, 1979; Vol. 3; p. 541.
14. Koopman, T. *Physica* **1933**, *1*, 104.
15. Bassindale, A. R.; Eaborn, C.; Walton, D. R. M.; Young, D. J. *J. Organomet. Chem.* **1969**, *20*, 49.
16. Bock, H.; Ensslin, W. *Angew. Chem. Int. Ed. Engl.* **1971**, *10*, 404.
17. Schweig, A.; Weidner, U.; Manuel, G. *J. Organomet. Chem.* **1973**, *54*, 145; Carlson, T. A.; McGuire, G. E.; Jonas, A. E.; Cheng, K. L.; Anderson, C. P.; Lu, C. C.; Pullen, B. P. *Electron Spectroscopy*; North-Holland: Amsterdam, 1972, p. 207.
18. Evans, S.; Green, J. C.; Joachim, P. J.; Orchard, A. F.; Turner, D. W.; Maier, J. P. *J. Chem. Soc. Faraday Trans. 2* **1972**, *68*, 905.
19. Bischof, P.; Hashmall, J. A.; Heilbronner, E.; Hornung, V. *Tetrahedron Lett.* **1969**, 4025.
20. Worley, S. D. *Chem. Rev.* **1971**, *71*, 295.
21. Bock, H.; Mollere, P.; Becker, G.; Fritz, G. *J. Organomet. Chem.* **1973**, *61*, 113.
22. Hanstein, W.; Berwin, H. J.; Traylor, T. G. *J. Am. Chem. Soc.* **1970**, *92*, 829.
23. Sennikov, P. G.; Skobeleva, S. E.; Kuznetsov, V. A.; Egorochkin, A. N. *J. Organomet. Chem.* **1980**, *201*, 213.
24. Davis, D. D. *J. Organomet. Chem.* **1981**, *206*, 21; Cook, M. A.; Eaborn, C.; Walton, D. R. M. *J. Organomet. Chem.* **1970**, *24*, 293.
25. Hanstein, W.; Berwin, H. J.; Traylor, T. G. *J. Am. Chem. Soc.* **1970**, *92*, 7476.
26. Exner, O. *Correlation Analysis in Chemistry*; Chapman, N. B.; Shorter, J., Eds., Plenum Press: New York, 1978.
27. Adcock, W.; Cox, D. P.; Kitching, W. *J. Organomet. Chem.* **1977**, *133*, 393.
28. Cuttress, N. C.; Katritzky, A. R.; Eaborn, C.; Walton, D. R. M. *J. Organomet. Chem.* **1972**, *43*, 131.
29. White, J. M. *Aust. J. Chem.* **1995**, *48*, 1227.
30. Colvin, E. *Silicon in Organic Synthesis*; Butterworths Monographs in Chemistry, Butterworth: Boston, 1981.
31. Ushakov, S. N.; Itenberg, A. M. *Zh. Obshch. Khim.* **1937**, *7*, 2495.

32. Sommer, L. H.; Braughman, G. A. *J. Am. Chem. Soc.* **1961**, *83*, 3346.

33. Jarvie, A. W. P.; Holt, A.; Thompson, J. *J. Chem. Soc. B* **1970**, 746.

34. Cook, M. A.; Eaborn, C.; Walton, D. R. M. *J. Organomet. Chem.* **1970**, *24*, 301

35. Hudrlik, P. F.; Peterson, D. *J. Am. Chem. Soc.* **1975**, *97*, 1464.

36. Connil, M-F.; Jousseaume, B.; Noiret, N.; Saux, A. *J. Org. Chem.* **1994**, *59*, 1925.

37. Wierschke, S. G.; Chandrasekhai, J.; Jorgensen, W. L. *J. Am. Chem. Soc.* **1985**, *107*, 1496.

38. Ibrahim, M. R.; Jorgensen, W. L. *J. Am. Chem. Soc.* **1989**, *111*, 819.

39. Wang, G.; Li, D.; Chelius, E. C.; Lambert, J. B. *J. Chem. Soc. Perkin Trans. 2* **1990**, 331.

40. Lambert, J. B.; Wang, G.; Finzel, R. B.; Teramura, D. H. *J. Am. Chem. Soc.* **1987**, *109*, 7838.

41. Lambert, J. B.; Wang, G.; Teramura, D. H. *J. Org. Chem.* **1988**, *53*, 5422.

42. Fujijama, R.; Munechika, T. *Tetrahedron Lett.* **1993**, *34*, 5907.

43. Davis, D. D.; Janocks, H. M. *J. Organomet. Chem.* **1981**, *206*, 33.

44. Traylor, T. G.; Koener, G. S. *J. Org. Chem.* **1981**, *46*, 3651.

45. Lambert, J. B.; Embledge, R. W.; Malany, S. *J. Am. Chem. Soc.* **1993**, *115*, 1317.

46. Shimizu, N.; Watanabe, S.; Tsuno, Y. *Bull. Chem. Soc. Jpn.* **1991**, *64*, 2249.

47. Fleming, I. *Chemtracts* **1993**, 113.

48. Olah, G. A.; Berrier, A. L.; Field, L. D.; Prakash, G. K. *J. Am. Chem. Soc.* **1982**, *104*, 1349.

49. Lew, C. S. Q.; McClelland, R. A. *J. Am. Chem. Soc.* **1993**, *115*, 11516.

50. Lambert, J. B.; Zhao, Y. *J. Am. Chem. Soc.* **1996**, *118*, 7867.

51. Steinberger, H.; Müller, T.; Auner, N.; Maerker, C.; von R. Schleyer, P. *Angew. Chem. Int. Ed. Engl.* **1997**, *36*, 626.

52. Lambert, J. B.; Chelius, E. C. *J. Am. Chem. Soc.* **1990**, *112*, 8120.

53. Lambert, J. B.; Zhao, X. *J. Organomet. Chem.* **1996**, *521*, 203.

54. Inagaki, S.; Iwase, K.; Mori, Y. *Chemistry Lett.* **1986**, 417.

55. Jousseaume, B.; Noiret, N.; Pereyre, M.; Francès, J.; Pétraud, M. *Organometallics* **1992**, *11*, 3910.

56. Shimizu, N.; Kinoshita, C.; Osajima, E.; Hayakawa, F.; Tsuno, Y. *Bull. Chem. Soc. Jpn.* **1991**, *64*, 3280.

57. Hill, E. A. *J. Org. Chem.* **1972**, *37*, 4008.

58. Jerkunica, J. M.; Traylor, T. G. *J. Am. Chem. Soc.* **1971**, 6278.

59. Traylor, T. G.; Berwin, H. J.; Jerkunica, J.; Hall, M. L. *Pure Appl. Chem.* **1972**, 599.

60. Hannon, S. J.; Traylor, T. G. *J. Org. Chem.* **1981**, *46*, 3645.

61. Seyferth, D.; Washburne, S. S.; Attridge, C. J.; Yamamoto, K. *J. Am. Chem. Soc.* **1970**, *92*, 4405.

62. Seyferth, D., Damrauer, R.; Washburne, S. S. *J. Am. Chem. Soc.* **1967**, *89*, 1538.

63. Seyferth, D.; Damrauer, R.; Andrews, S. B.; Washburne, S. S. *J. Am. Chem. Soc.* **1971**, *93*, 3709.

64. Frey, H. M.; Walsh, R.; Watts, I. M. *J. Chem. Soc. Chem. Comm.* **1989**, 284.

65. Weidenbruch, M.; Rankers, R. *J. Organomet. Chem.* **1980**, *198*, 29.

66. Seyferth, D.; Cheng, Y. M.; Traficante, D. D. *J. Organomet. Chem.* **1972**, *46*, 9.

67. Hosomi, A.; Mikami, M.; Sakurai, H. *Bull. Chem. Soc. Jpn.* **1983**, *56*, 2784.

68. Sarkar, T. K.; Ghorai, B. K. *J. Chem. Soc. Chem. Comm.* **1992**, 1184.

69. Furth, P. S.; Hwu, J. R. *J. Org. Chem.* **1989**, *54*, 3404.

70. Fleming, I.; Pearce, A. *J. Chem. Soc. Perkin Trans. 1* **1981**, 251.

71. Fleming, I.; Michael, J. P. *J. Chem. Soc. Perkin Trans. 1* **1981**, 1549.

72. Tanino, K.; Hatanaka, Y.; Kuwajima, I. *Chem. Lett.* **1987**, 385.

73. Taylor, R. T.; Paquette, L. A. *J. Org. Chem.* **1978**, *43*, 242.

74. Hwu, J. R.; Wetzel, J. M. *J. Org. Chem.* **1992**, *57*, 922.

75. Fleming, I.; Michael, J. P. *J. Chem. Soc. Chem. Comm.* **1978**, 245.

76. Fleming, I.; Patel, S. K. *Tet. Lett.* **1981**, *22*, 2321.

77. Green, A. J.; Kuan, Y. L.; White, J. M. *J. Chem. Soc. Chem. Comm.* **1994**, 2023.

78. Denmark, S. E.; Klix, R-C. *Tetrahedron* **1988**, *44*, 4043.

79. Denmark, S. E.; Jones, T. K. *J. Am. Chem. Soc.* **1982**, *104*, 2642.

80. Smith, D. A.; Ulmer, C. W. *J. Org. Chem.* **1993**, *58*, 4118.

81. Hosomi, A. *Acc. Chem. Res.* **1988**, *21*, 200.

82. Chan, T. H.; Fleming, I. *Synthesis* **1979**, 761.

83. Hayashi, T.; Konishi, M.; Ito, H.; Kumada, M. *J. Am. Chem. Soc.* **1982**, *104*, 4962.

84. Weidner, U.; Schweig, A. J. *J. Organomet. Chem.* **1972**, *39*, 261.

85. Young, D.; Kitching, W. *J. Org. Chem.* **1983**, *48*, 615.

86. Furth, P. S.; Hwu, J. R. *J. Org. Chem.* **1989**, *54*, 3404.

87. Hudrlik, P. F.; Hudrlik, A. M.; Nagendrappa, G.; Yimena, T.; Zellers, E. T.; Chin, E. *J. Am. Chem. Soc.* **1980**, *102*, 6894.

88. Aida, T.; Asaoka, M.; Sonoda, S.; Takei, H. *Heterocycles* **1993**, *36*, 427.

89. Hudrlik, P. F.; Waugh, M. A.; Hudrlik, A. M. *J. Organomet. Chem.* **1984**, *271*, 69.

90. Nishiyama, H.; Sakuta, K.; Osaka, N.; Itoh, K. *Tetrahedron Lett.* **1983**, 4021.

91. White, J. M.; Horvat, S. *unpublished material.*

92. Issa, W.; Green, A. J.; White, J. M. *Aust. J. Chem.* **1997**, *50*, 927.

93. Amos, R. D.; Handy, N. C.; Jones, P. G.; Kirby, A. J.; Parker, J. K.; Percy, J. M.; Su, M. D. *J. Chem. Soc. Perkin Trans. 2* **1992**, 549.

94. White, J. M.; Robertson, G. B. *J. Org. Chem.* **1992**, *57*, 4638.

95. Green, A. J.; Kuan, Y. L.; White, J. M. *J. Org. Chem.* **1995**, *60*, 2734.

96. Chan, V. Y.; Clark, C. I.; Giordano, J.; Green, A. J.; Karalis, A.; White, J. M. *J. Org. Chem.* **1996**, *61*, 5227.

97. Green, A. J.; Giordano, J.; White, J. M. *manuscript in preparation.*

98. White, J. M.; Robertson, G. B. *Acta Crystallogr. Sect. C* **1994**, *49*, 347.
99. White, J. M.; Green, A. J. *Acta Crystallogr. Sect. C* **1997**, *53*, 1084.
100. White, J. M.; Green, A. J. *Acta Crystallogr. Sect. C* **1995**, *51*, 1905.
101. Albright, T. A.; Burdett, J. K., Whangbo, W. *Orbital Interactions in Chemistry*; Wiley: New York, 1985.
102. Allen, F. H.; Kirby, A. J. *J. Am. Chem. Soc.* **1984**, *106*, 6197.
103. Briggs, A. J.; Glenn, R.; Jones, P. G.; Kirby, A. J.; Ramaswamy, P. *J. Am. Chem. Soc.* **1984**, *106*, 6200.
104. Kuan, Y. L.; White, J. M. *J. Chem. Soc. Chem. Comm.* **1994**, 1195.
105. Eliel, E. L. *Angew. Chem. Int. Ed. Engl.* **1965**, *4*, 761.
106. White, J. M.; Green, A. J. *unpublished data.*
107. White, J. M.; Giordano, J. *manuscript in preparation.*
108. Horner, J. H.; Newcomb, M. *Organometallics* **1991**, *10*, 1732; Kitching, W.; Doddrell, D.; Grutzner, J. B. *J. Organomet. Chem.* **1976**, *107*, C5; Moder, T. I.; Hsu, C. C. H.; Jensen, F. R. *J. Org. Chem.* **1980**, *45*, 1009.
109. Pilz, M.; Michel, H.; Berndt, A. *Angew. Chem. Int. Ed. Engl.* **1990**, *29*, 401.
110. Allwohn, J.; Hunold, R.; Pilz, M.; Müller, R-G.; Massa, W.; Berndt, A. *Z. Naturoforsch.* **1990**, *45b*, 290.
111. Sommer, L. H.; Dorfman, E.; Goldberg, G. M.; Whitmore, F. C. *J. Am. Chem. Soc.* **1946**, *68*, 488.
112. Shiner, Jr., V. J.; Ensinger, M. W.; Kris, G. S. *J. Am. Chem. Soc.* **1986**, *108*, 842.
113. Shiner, Jr., V. J.; Ensinger, M. W.; Kriz, G. S.; Halley, K. A. *J. Org. Chem.* **1990**, *55*, 653.
114. Shiner, Jr., V. J.; Ensinger, M. W.; Rutkowske, R. D. *J. Am. Chem. Soc.* **1987**, *109*, 804.
115. Ensinger, M. W.; Shiner, Jr., V. J. *Physical Organic Chemistry 1986, A Collection of the Invited Lectures Presented at the 8th IUPAC Conference on Physical Organic Chemistry, Tokyo, Japan, 24–29 August 1986,* Studies in Organic Chemistry, Volume 31, pp. 41–58, Elsevier Science Publishers B. V.: Amsterdam, 1987.
116. Davidson, E. R.; Shiner, V. J. *J. Am. Chem. Soc* **1986**, *108*, 3135.
117. Lambert, J. B.; Salvador, L. A.; So, J-H. *Organometallics* **1993**, *12*, 697.
118. Grob, C. A.; Gründel, M.; Sawlewicz, P. *Helv. Chim. Acta* **1988**, *71*, 1502.
119. Kirmse, W.; Söllenböhmer, F. *J. Am. Chem. Soc.* **111**, 4127.
120. Grob, C. A.; Sawlewicz, P. *Tetrahedron Lett.* **1987**, *28*, 951.
121. Davis, D. D.; Black, R. H. *J. Organomet. Chem.* **1974**, *82*, C30.
122. Green, A. J.; Pigdon, T.; White, J. M.; Yamen, J. *J. Org. Chem.* **1998**, *63*, 3943.
123. Nickson, A.; Werstiuk, N. H. *J. Am. Chem. Soc.* **1967**, *89*, 3914.
124. Adcock, W.; Clark, C. I.; Schiesser, C. H. *J. Am. Chem. Soc.* **1996**, *118*, 11541.
125. Fleming, I.; Urch, C. J. *J. Organomet. Chem.* **1985**, *285*, 173.
126. Fleming, I.; Urch, C. J. *Tetrahedron Lett.* **1983**, *24*, 4591.

127. Fleming, I.; Patel, S. K.; Urch, C. J. *J. Chem. Soc. Perkin Trans. 1* **1989**, 115.

128. Davis, D. D.; Chambers, R. L.; Johnson, H. T. *J. Organomet. Chem.* **1970**, *25*, C13.

129. Davis, D. D.; Johnson, H. T. *J. Am. Chem. Soc.* **1974**, *96*, 7576.

130. McWilliam, D. C.; Balasubramanian, T. R.; Kuivila, H. G. *J. Am. Chem. Soc.* **1978**, *100*, 6407.

131. Ueno, Y.; Ohta, M.; Okawara, M. *Tetrahedron Lett.* **1982**, *23*, 2577.

132. Peterson, D. J.; Robbins, M. D.; Hansen, J. R. *J. Organomet. Chem.* **1974**, *73*, 237.

133. Hwu, J. R.; Gilbert, B. A. *J. Am. Chem. Soc.* **1991**, *113*, 5917.

134. Adcock, W.; Kok, G. B. *J. Org. Chem.* **1987**, *52*, 356.

135. Adcock, W.; Trout, N. A. *unpublished data.*

136. White, J. M.; Giordano, J.; Green, A. J. *Acta Crystallogr. Sect. C* **1996**, *52*, 3204.

137. Adcock, W.; Krystic, A.; Duggan, P. J.; Shiner, V. J.; Coope, J.; Ensinger, M. W. *J. Am. Chem. Soc.* **1990**, *112*, 3140.

138. Hrovat, D.; Borden, W. T. *J. Org. Chem.* **1992**, *57*, 2519.

139. Adcock, W.; Coope, J.; Shiner, Jr., V. J.; Trout, N. A. *J. Org. Chem.* **1990**, *55*, 1411.

140. Hoffmann, R.; Imamura, R. A.; Hehre, W. J. *J. Am. Chem. Soc.* **1968**, *90*, 1499.

141. Adcock, W.; Trout, N. A. *J. Org. Chem.* **1991**, *56*, 3229.

142. Adcock, W. A.; Abeywickrema, A. N. *J. Org. Chem.* **1982**, *47*, 2957.

143. White, J. M.; Green, A. J.; Van, V. *Aust. J. Chem.* **1998**, *51,* 555.

144. Lambert, J. B.; Salvador, L. A.; Stern, C. L. *J. Org. Chem.* **1993**, *58*, 5428.

145. Riches, B. H.; Kitching, W. *J. Chem. Soc. Chem. Comm.* **1996**, 1907.

146. Imachi, M.; Nakagawa, J.; Hayashi, M. *J. Mol. Struct.* **1983**, *102*, 403.

147. Ohno, K.; Kaga, T.; Murata, H. *Bull. Chem. Soc. Jpn.* **1977**, *50*, 2870.

148. Kitching, W.; Penman, K. G.; Valle, G.; Tagliavini, G.; Ganis, P. *Organometallics* **1989**, *8*, 785.

149. Marton, D.; Tagliavini, G.; Valle, G.; Ganis, P. *J. Organomet. Chem.* **1989**, *362*, 281.

150. Ganis, P.; Valle, G.; Tagliavini, G.; Kitching, W.; Penman, K. G.; Jones, M. A. *J. Organomet. Chem.* **1988**, *349*, 57.

151. Ganis, P.; Furlani, D.; Marton, D.; Tagliavini, G.; Valle, G. *J. Organomet. Chem.* **1985**, *293*, 207.

152. Weidner, U.; Schweig, A. *Angew. Chem. Int. Ed. Engl.* **1972**, *11*, 146.

153. Ponec, R.; Chvalovsky, V.; *Collect. Czech. Chem. Comm.* **1973**, *38*, 3845.

154. Schweig, A.; Weidner, U.; Manuel, G. *J. Organomet. Chem.* **1974**, *67*, C4.

155. Cauletti, C.; Furlani, C.; Grandinetti, F. *J. Organomet. Chem.* **1986**, *315*, 287.

156. Weidner, U.; Schweig, A. *J. Organomet. Chem.* **1972**, *39*, 261.

157. Egorochkin, A. N.; Vyazankin, N. S.; Skobeleva, S. E.; Khorshev, S. Ya.; Mironov, V. F.; Gar, T. K. *Zh. Obsh. Khim.* **1972**, *42*, 643.

158. Hagen, G.; Mayr, H. *J. Am. Chem. Soc.* **1991**, *113*, 4954.

159. Kahn, S. D.; Pau, C. F.; Chamberlin, A. R.; Hehre, W. J. *J. Am. Chem. Soc.* **1987**, *109*, 650.

160. Fleming, I.; Sarkar, A. K.; Thomas, A. P. *J. Chem. Soc. Chem. Comm.* **1987**, 157 and references cited therein.

161. Pitt, C. G. *J. Organomet. Chem.* **1973**, *61*, 49.

162. Egorochkin, A. N.; Razuvaev, G. A.; Lopatin, M. A. *J. Organomet. Chem.* **1988**, *344*, 49.

163. Symons, M. C. R. *Tetrahedron* **1962**, *18*, 333.

164. Adcock, W.; Hegarty, B. F.; Kitching, W.; Smith, A. J. *J. Organomet. Chem.* **1968**, *12*, 21.

165. Smith, A. J.; Adcock, W.; Kitching, W. *J. Am. Chem. Soc.* **1970**, *92*, 6140.

166. Adcock, W.; Rizvi, S. Q. A.; Kitching, W. *J. Am. Chem. Soc.* **1972**, *94*, 3657.

167. Kitching, W.; Drew, G.; Adcock, W.; Abeywickrema, A. N. *J. Org. Chem.* **1981**, *46*, 2252.

168. Rizvi, S. Q. A.; Gupta, B. D.; Adcock, W.; Doddrell, D. D.; Kitching, W. *J. Organomet. Chem.* **1973**, *63*, 67.

169. Bullpitt, M.; Kitching, W.; Adcock, W.; Doddrell, D. *J. Organomet. Chem.* **1976**, *116*, 187.

170. Cutress, N. C.; Grindley, T. B.; Katritzky, A. R.; Eaborn, C.; Walton, D. R. M.; Topsom, R. D. *J. Organomet. Chem.* **1974**, *65*, 17.

171. Egorov, Y. P.; Leites, L. A.; Chernyshev, E. A. *Izv. Akad. Nauk. SSSR, Otd. Khim., Nauk* **1961**, 45.

172. Egorov, Y. P.; Leites, L. A.; Mironov, V. F. *Zhur. Strukt. Khim.* **1961**, *2*, 562.

173. Egorov, Y. P.; Kirei, G. G.; Leites, L. A.; Mironov, V. F.; Petrov, A. D. *Izv. Akad. Nauk. SSSR, Otd. Khim., Nauk* **1962**, 1880.

174. Egorov, Y. P.; Kirei, G. G. *Zhur. Obshch. Khim.* **1964**, *34*, 3615.

175. Traylor, T. G.; Berwin, H. J.; Jerkunica, J.; Hall, M. L. *Pure Appl. Chem.* **1972**, *30*, 599.

176. Jerkunica, J. M.; Traylor, T. G. *J. Am. Chem. Soc.* **1971**, *93*, 6278.

177. Weeks, G. H.; Adcock, W.; Klingensmith, K. A.; Waluk, J. W.; West, R.; Vasay, M.; Downing, J.; Michl, J. *Pure Appl. Chem.* **1986**, *58*, 39.

178. Schaefer, T.; Sebastian, R.; Penner, G. H. *Can. J. Chem.* **1991**, *69*, 496.

179. Schaefer, T.; Penner, G. H.; Takeuchi, C. S.; Beaulieu, C. *Can. J. Chem.* **1989**, 1283.

180. Bock, H.; Solouki, B. *The Chemistry of Organic Silicon Compounds*; Patai, S.; Rappoport, Z., Eds., Ch. 9, John Wiley: Chichester, 1989.

181. Cauletti, C.; Stranges, S. *The Chemistry of Organic Germanium, Tin and Lead Compounds*; Patai, S., Ed., Ch. 6, John Wiley: Chichester, 1995.

182. Distefano, G.; Modelli, A.; Guerra, M.; Jones, D.; Rossini, S. *J. Mol. Struct.* **1988**, 177.

183. Distefano, G.; Pignataro, S.; Ricci, A.; Colonna, F. P.; Pietropaolo, D. *Annalimdi Chimica* **1974**, *64*, 153.

184. Bischof, P. K.; Dewar, M. J. S.; Goodman, D. W.; Jones, T. B. *J. Organomet. Chem.* **1974**, *82*, 89.

185. Sakurai, H.; Kira, M.; Ochiai, M. *Chem. Lett.* **1972**, 87.

186. Rakita, P. E.; Hoffman, M. K.; Andrews, M. N.; Bursey, M. M. *J. Organomet. Chem.* **1973**, *49*, 213.

187. West, R.; Barton, T. J. *J. Chem. Ed.* **1980**, *57*, 165.

188. Sakurai, H.; Tasaka, S.; Kira, M. *J. Am. Chem. Soc.* **1972**, *94*, 9285.

189. Lambert, J. B.; Singer, R. A. *J. Am. Chem. Soc.* **1992**, *114*, 10246.

190. Kamien'ska-Trela, K. *Mag. Res. in Chem.* **1995**, *33*, 406.

191. Yoshida, J.; Maekawa, T.; Murata, T.; Matsunaga, S.; Isoe, S. *J. Am. Chem. Soc.* **1990**, *112*, 1962.

192. Yoshida, J. *Top. Curr. Chem.* **1994**, *170*, 40.

193. Bock, H.; Meuret, J. *J. Organomet. Chem.* **1993**, *459*, 43.

194. Block, E.; Yencha, A. J.; Aslam, M.; Eswarakrishnan, V.; Luo, J.; Sano, A. *J. Am. Chem. Soc.* **1988**, *110*, 4748.

195. Cooper, B. E.; Owen, W. J. *J. Organomet. Chem.* **1971**, *29*, 33.

196. Nyulászi, L.; Veszprémi, T.; Réffy, J. *J. Organomet. Chem.* **1993**, *445*, 29.

197. Glass, R. S.; Radspinner, A. M.; Singh, W. P. *J. Am. Chem. Soc.* **1992**, *114*, 4921.

198. Ramsey, B. G.; Brook, A.; Bassindale, A. R.; Bock, H. *J. Organomet. Chem.* **1974**, *74*, C41.

199. Michman, M.; in *The Chemistry of Organic Germanium, Tin and Lead Compounds*; Patai, S., Ed., Ch. 13, John Wiley: Chichester, 1995.

200. Kira, M.; Nakazawa, H.; Sakurai, H. *Chem. Lett.* **1986**, 497.

201. Auner, N.; Walsh, R.; Westrup, J. *J. Chem. Soc. Chem. Comm.* **1986**, 207.

202. Yoshida, J.; Itoh, M.; Matsunaga, S.; Isoe, S. *J. Org. Chem.* **1992**, *57*, 4877.

203. Yoshida, Y.; Morita, Y.; Itoh, M.; Ishichi, Y.; Isoe, S. *Synlett* **1992**, 843.

204. Bock, H.; Alt, H.; Seidl, H. *J. Am. Chem. Soc.* **1969**, *91*, 355.

205. Grev, R. S.; Schaefer, H. F. *J. Am. Chem. Soc.* **1989**, *111*, 6137.

206. Yoshida, J.; Ishichi, Y.; Nishiwaki, K.; Shiozawa, S.; Isoe, S. *Tetrahedron Lett.* **1992**, *33*, 2599.

207. Hosomi, A.; Traylor, T. G. *J. Am. Chem. Soc.* **1975**, *97*, 3682.

208. Nishiwaki, K.; Yoshida, J. *Chem. Lett.* **1996**, 787.

209. Jackson, R. A.; Ingold, K. U.; Griller, D.; Nazran, A. S. *J. Am. Chem. Soc.* **1985**, *107*, 208.

210. Krusic, P. K.; Kochi, J. K. *J. Am. Chem. Soc.* **1969**, *91*, 6161.

211. Gruller, D.; Ingold, K. U. *J. Am. Chem. Soc.* **1974**, *96*, 6715.

212. Kawamura, T.; Kochi, J. K. *J. Organomet. Chem.* **1971**, *30*, C8.

213. Kawamura, T.; Kochi, J. K. *J. Am. Chem. Soc.* **1971**, *93*, 846.

214. Kawamura, T.; Kochi, J. K. *J. Am. Chem. Soc.* **1972**, *94*, 648.

215. Kira, M.; Sugiyama, H.; Sukurai, H. *J. Am. Chem. Soc.* **1983**, *105*, 6436.

216. Sakurai, H.; Uchida, T.; Kira, M. *J. Organomet. Chem.* **1976**, *107*, 15.
217. Hwu, J. R.; Gilbert, B. A.; Lin, L. C.; Liaw, B. R. *J. Chem. Soc. Chem. Comm.* **1990**, 161.
218. Hwu, J. R.; Chen, B-L.; Shiao, S-S. *J. Org. Chem.* **1995**, *60*, 2448.

Asymmetric Catalysis with Chiral Lanthanoid Complexes

MASAKATSU SHIBASAKI† AND HIROAKI SASAI‡

†*Graduate School of Pharmaceutical Sciences, The University of Tokyo, Hongo, Bunkyo-ku, Tokyo 113-0033, Japan*
‡*The Institute of Scientific and Industrial Research, Osaka University, Mihogaoka, Ibaraki-shi, Osaka 567-0047, Japan*

Topics in Stereochemistry, Volume 22, Edited by Scott E. Denmark.
ISBN 0-471-25316-2

I. INTRODUCTION

The lanthanoids are composed of a series of 15 analogous elements from La (atomic number 57) to Lu (atomic number 71). These are amphoteric elements, which are stable in the +3 oxidation state, are widely recognized as strong Lewis acids on the basis of their high oxophilicity. The ionic radius of these octa-coordinated trivalent lanthanoids decreases from 1.18 Å in La to 0.97 Å in Lu as the atomic number increases, demonstrating the so-called lanthanoid contraction. Meanwhile, the ionic radius increases as the coordination number increases. In recent years, catalytic asymmetric synthesis using chiral lanthanoid complexes has received much attention. It is envisioned that by varying the lanthanoid element in a chiral catalyst, it would be possible to modify slightly the metal–oxygen bond characteristics in an appropriate asymmetric environment without changing the fundamental structure of the catalyst. Therefore, utilization of chiral lanthanoid complexes may provide a unique opportunity for creating tailor-made asymmetric catalysts. Moreover, some of the lanthanoid elements are abundant and nontoxic, which also make them desirable as the key component of an asymmetric catalyst.

The first part of this chapter describes recent advances in the use of novel, chiral, alkali metal free-lanthanoid–BINOL derivative complexes for a variety of efficient, catalytic, asymmetric reactions. For example, using a catalytic amount of chiral Ln–BINOL derivative complexes, asymmetric Michael reactions and asymmetric epoxidations of enones proceed in a highly enantioselective manner.

The second part of the chapter deals with several kinds of asymmetric reactions catalyzed by unique heterobimetallic complexes. These reagents are lanthanoid–alkali metal hybrids which form BINOL derivative complexes (LnMB, where Ln = lanthanoid, M = alkali metal, and B = BINOL derivative). These complexes efficiently promote asymmetric aldol-type reactions as well as asymmetric hydrophosphonylations of aldehydes (catalyzed by LnLB, where L = lithium), asymmetric Michael reactions (catalyzed by LnSB, where S = sodium), and asymmetric hydrophosphonylations of imines (catalyzed by LnPB, where P = potassium) to give the corresponding desired products in up to 98% ee. Spectroscopic analysis and computer simulations of these asymmetric reactions have revealed the synergistic cooperation of the two different metals in the complexes. These complexes are believed to function as both Brønsted bases and as Lewis acids may prove to be applicable to a variety of new asymmetric catalytic reactions.[1, 2]

The third part of this chapter reviews previously described catalytic asymmetric reactions that can be promoted by chiral lanthanoid complexes. Transformations such as Diels–Alder reactions, Mukaiyama aldol reactions, several types of reductions, Michael addition reactions, hydrosilylations, and hydroaminations proceed under asymmetric catalysis in the presence of chiral lanthanoid complexes.

II. CHIRAL, ALKALI METAL FREE-LANTHANOID–BINOL DERIVATIVE COMPLEXES

A. Basicity of Lanthanoid Alkoxides

The evolution of lanthanide–BINOL complexes emerged from the desire to develop a new class of basic asymmetric catalysts from early transition metal alkoxides. Early investigations focused on the use of the amphoteric Zr(O-*t*-Bu)$_4$, a group 4 metal alkoxide for the direct preparation of zirconium enolates from ketones. With Zr(O-*t*-Bu)$_4$, various reactions of ketones and aldehydes proceeded smoothly to give aldol products, indicating that the direct formation of zirconium enolates was occurring (Figure 1).[3] These reactions, however, proceeded efficiently only when stoichiometric quantities of Zr(O-*t*-Bu)$_4$ were used. Attempts to carry out aldol reactions using catalytic amounts of Zr(O-*t*-Bu)$_4$ were unsuccessful. Clearly, metal alkoxides more basic than group 4 metal alkoxides were needed. Lanthanoid alkoxides were expected to exhibit enhanced basicity due to their lower ionization potentials (ca. 5.4–5.6 eV) and lower electronegativities (1.1–1.2) compared to group 4 metals. Moreover, this increased basicity was expected to increase the utility of lanthanoid alkoxides for catalysis of organic reactions. This expectation was borne out in fact. As shown in Figure 2, lanthanoid and group 3 metal alkoxides were suitable for use as a base in aldol-type reactions,

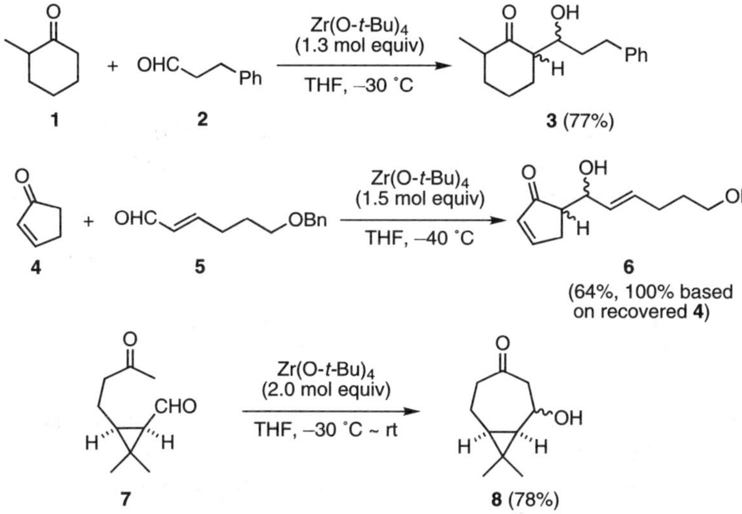

Figure 1. Aldol reactions with Zr(O-*t*-Bu)$_4$.

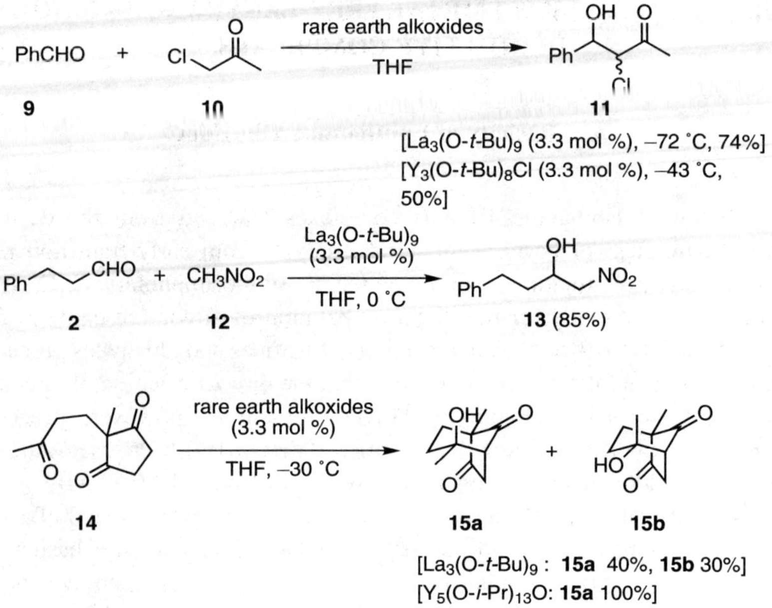

Figure 2. Typical aldol-type reactions catalyzed by lanthanoid and/or yttrium alkoxides.

and moreover that the *catalytic* use of these alkoxides was feasible, suggesting the possibility for developing basic asymmetric catalysts from lanthanoid alkoxides.

B. Preparation of Ln–BINOL Complexes

The first basic asymmetric catalysts of type A were prepared from $LnCl_3$ (Ln = Y or La) and enantiopure bidentate diol derivatives such as **16**[4] under the conditions outlined in Figure 3. Although the structures of the type A catalytic species were still open to conjecture, they nonetheless proved effective for catalytic asymmetric reactions. Thus treatment of **14** with the yttrium catalyst of type A in THF at –52 °C for 6 days gave a 60% yield of **15a** with 52% ee.[5] However, it was shown that the enantiomeric purity of isolated **15a** *decreased* when further exposed to the type A catalyst at –30 °C, suggesting that the product had a tendency to undergo a retro aldol reaction.

The nitroaldol (Henry) reaction[6] was next examined, and it too proceeded efficiently in the presence of a catalytic amount of lanthanoid alkoxides. Specifically, by using type A catalysts the reaction of **2** with nitromethane (**12**) gave the desired nitroaldol product, however, in racemic form. The stereorandom course of this reaction could be understood by NMR spectroscopic analysis of the

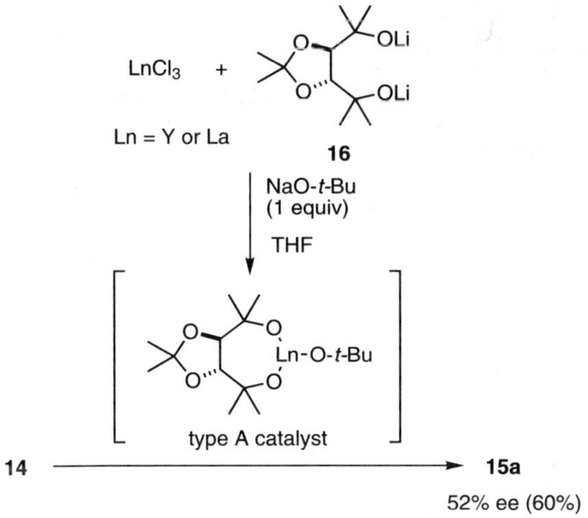

Figure 3. Preparation of type **A** asymmetric catalysts and a catalytic asymmetric intramolecular aldol reaction.

reaction mixture. A ^{13}C NMR spectrum of a mixture of the asymmetric catalyst and nitromethane showed clearly that an undesired exchange of the asymmetric ligand for (acidic) nitromethane was occurring, producing an achiral nitronate. This result suggested the preparation of a chiral lanthanoid complex from commercially available Ln(O-*i*-Pr)$_3$ by exchanging an isopropoxide moiety with an acidic chiral BINOL (**17**). The inexpensive La metal was chosen as the lanthanoid. Thus to a stirred solution of La(O-*i*-Pr)$_3$ (1.0 mmol) in anhydrous THF was added 1 mol equiv of (*S*)-**17** at 0 °C, which resulted in the formation of the La–**17** complex. The preparation of the La–**17** complex is found to be highly dependent on the lot of La(O-*i*-Pr)$_3$ employed. Occasionally a suspension is obtained that is also an efficient catalyst for the asymmetric Michael reaction described in the following section. The complex La–**17** (albeit of unknown structure) displays excellent catalytic activity in a number of asymmetric reactions.

C. Catalytic, Asymmetric Michael Reaction Promoted by the La–BINOL Complex

Michael reactions have been found to proceed in a highly enantioselective manner with a catalytic amount of the La–**17** complex.[7] For example, dibenzyl methylmalonate (**18**) reacted with cyclopentenone (**4**) in the presence of 10 mol % of the La–**17** complex (–20 °C for 48 h) to give the Michael adduct **19** in 70% ee in 86% yield. Tetrahydrofuran (THF) was found to be the solvent of choice. A

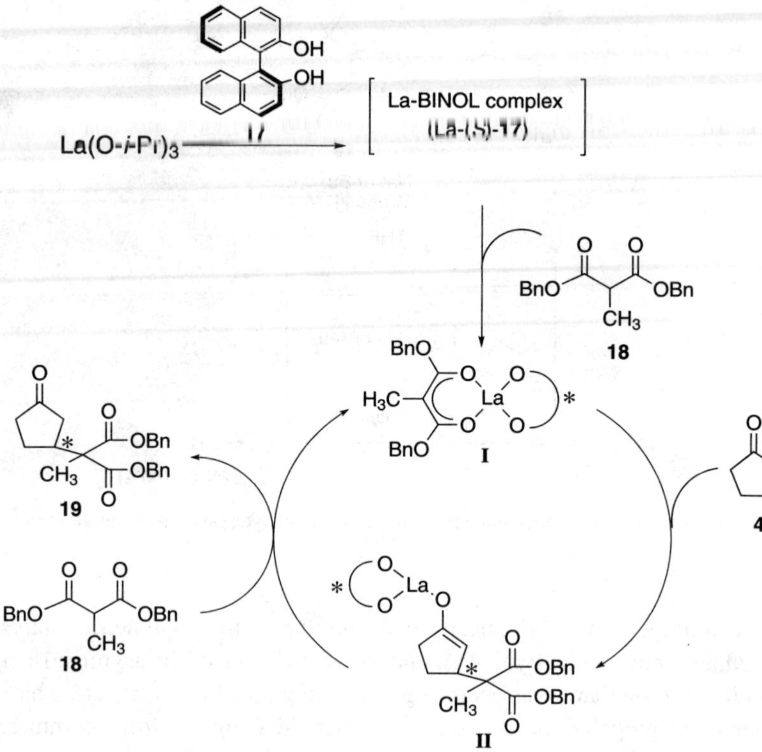

Figure 4. Proposed mechanism for the La–BINOL (La–**17**)-catalyzed asymmetric Michael reaction.

plausible mechanism for this reaction is shown in Figure 4. In the first step, dibenzyl methylmalonate (**18**) is deprotonated with La–**17** to give the BINOL–lanthanum ester enolate **I**. This enolate would then react with cyclopentenone (**4**) to give the lanthanum enolate **II** in the stereochemistry-determining step. A critical turnover event is the proton transfer to the enolate **II** from **18**, which should be favorable due to the difference in their pK_a values. This key step forms the Michael adduct **19** and simultaneously regenerates the lanthanum ester enolate **I**, thus making the catalytic cycle possible. These mechanistic speculations suggested that the intermediate lanthanum ester enolate **I** is the real asymmetric catalyst and could be more directly prepared starting with La(O-*i*-Pr)$_3$, dibenzyl methylmalonate (**18**), and BINOL (**17**). This indeed proved to be the case.[7]

The independent preparation of the catalytic complex involved combination of equimolar amounts of **18** and La(O-*i*-Pr)$_3$ in THF at 0 °C, followed by addition of (*S*)-**17** (1 mol equiv). The residue obtained after removal of the solvent was redissolved in THF and used as a catalyst. For example, this complex (10 mol %) was used to catalyze the reaction between dibenzyl methylmalonate (**18**) and cyclopentenone (**4**) (–10 °C, 84 h) to afford the Michael adduct **19** in 91% ee in 91% yield. Representative results of Michael reactions with this catalyst complex and various substrates are shown in Table 1. Although the reactions are slow (60–84 h), the yields and enantioselectivities are extremely high for both cyclopentenone (**4**) and cyclohexenone (**22**).

Previous studies on asymmetric Michael reactions with a chiral lanthanoid catalyst are on record. Scettri et al. have reported the use of (+)-Eu(tfc)$_3$[tris(3-(trifluoromethylhydroxymethylene)-(+)-camphorato)europium(III)] as a catalyst with modest enantioselectivities.[8]

Table 1
Catalytic, Asymmetric Michael Reactions Promoted by La–(*S*)-**17** Enolate **I** (10 mol %)

Entry	Enone	Michael donor	Product	Temp. (°C)	Time (h)	Yield (%)	ee (%)
1	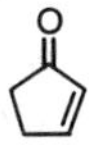 **4**	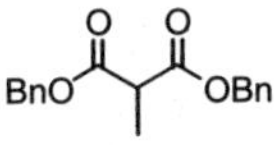 **18**	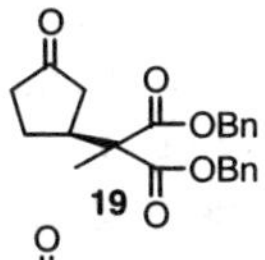 **19**	–10 –20	84 60	91 97	91 95
2	**4**	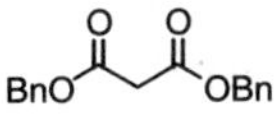 **20**	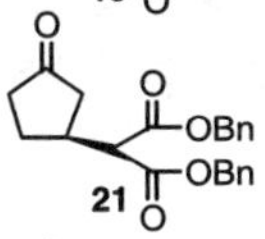 **21**	–20	72	96	92
3	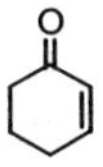 **22**	**18**	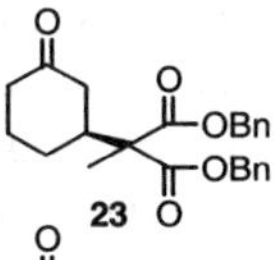 **23**	0	84	83	87
4	**22**	**20**	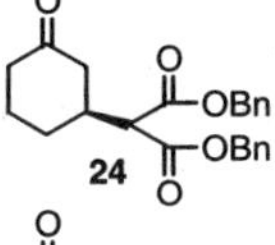 **24**	–10	84	94	92
5	**22**	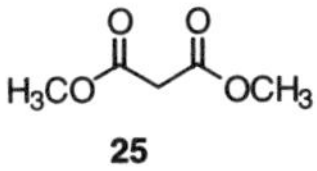 **25**	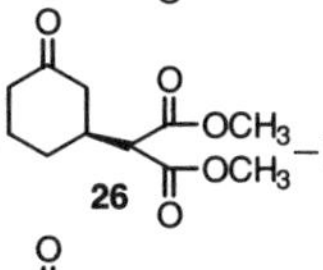 **26**	–10	84	100	75
6	**22**	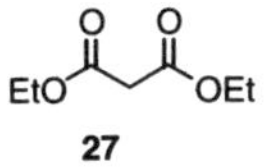 **27**	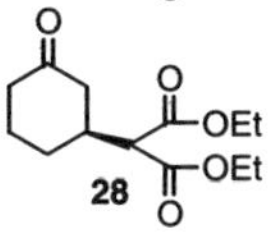 **28**	–10	84	97	78
7	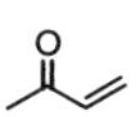 **29**	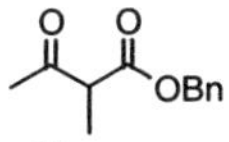 **30**	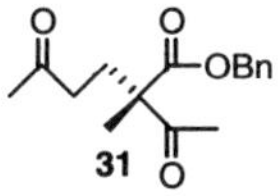 **31**	–50	48	86	62

D. Catalytic, Asymmetric Epoxidation of Enones Promoted by Ln–BINOL Derivative Complexes

Catalytic, asymmetric epoxidations are one of the most important asymmetric processes. In 1980 Katsuki and Sharpless reported a stoichiometric asymmetric epoxidation of allylic alcohols, a method that was later improved to become a catalytic process.[9] Moreover, catalytic asymmetric epoxidations of unfunctionalized olefins using salen–manganese complexes have been reported independently by several groups.[10–12] In striking contrast to these successful achievements, an efficient catalytic asymmetric epoxidation of enones *with broad generality* has not been developed.[13–22]

In view of the successful demonstration of La–**17** as an efficient asymmetric catalyst for Michael reactions, it is reasonable that an Ln–**17** complex might be useful for the asymmetric epoxidation of enones using hydroperoxides such as *tert*-butyl hydroperoxide (TBHP). Since this type of epoxidation proceeds through a Michael-type addition of hydroperoxides to enones, the mechanistic analogy is clear. The catalyst for this purpose [La–(*R*)-**17**] was prepared by combining La(O-*i*-Pr)$_3$ with (*R*)-**17** in the presence of powdered, activated 4A molecular sieves (MS) in THF (Figure 5). The reaction of chalcone (**33**) with TBHP (1.5 equiv) at rt for 0.5 h catalyzed by La–(*R*)-**17** (10 mol %) afforded **34** with 62% ee in 90% yield. Moreover, the use of cumene hydroperoxide (CMHP) instead of TBHP improved the asymmetric epoxidation, giving **34** with 83% ee in 93% yield (5 mol % of the La-**17** complex).[23] These reactions are clearly accelerated by the addition of MS 4A, though the origin of the effect is not clear at present.

This type of chiral lanthanum catalyst was found to be applicable for epoxidation of a range of enone substrates. Thus **35** was converted to **36** with 86% ee and in 93% yield, and **37** was transformed to **38** with 85% ee in 85% yield (Table 2, entries 1, 4, and 6). The enantioselectivity of the asymmetric epoxidations could be substantially improved by the use of (*R*)-3-hydroxymethyl-BINOL (**32**) instead of **17** (Table 2, entries 2, 3, 5, and 7). Namely, **34**, **36**, and **38** were obtained in excellent yields with 91, 94, and 83% ee, respectively.

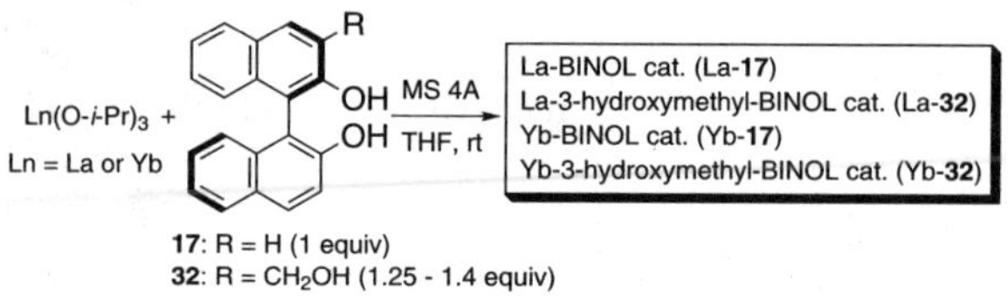

Figure 5. Preparation of chiral lanthanoid–BINOL derivative catalysts.

Table 2
Catalytic, Asymmetric Epoxidations Using Alkali Metal–Free Lanthanoid Complexes

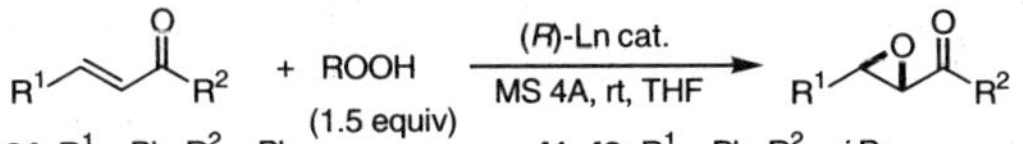

33, 34: R^1 = Ph, R^2 = Ph
35, 36: R^1 = *i*-Pr, R^2 = Ph
37, 38: R^1 = Ph, R^2 = *o*-MOMOC$_6$H$_4$
39, 40: R^1 = Ph, R^2 = CH$_3$
41, 42: R^1 = Ph, R^2 = *i*-Pr
43, 44: R^1 = Ph(CH$_2$)$_2$, R^2 = CH$_3$
45, 46: R^1 = CH$_3$(CH$_2$)$_4$, R^2 = CH$_3$

Entry	Enone	Epoxide	Ln cat. (mol %)	ROOH	Time (h)	Yield (%)	ee (%)[a]
1	**33**	**34**	La–**17** (5)	CMHP	6	93	83
2	**33**	**34**	La–**32** (5)	CMHP	7	93	91
3	**33**	**34**	La–**32** (1)	CMHP	44	95	89
4	**35**	**36**	La–**17** (5)	CMHP	12	93	86
5	**35**	**36**	La–**32** (5)	CMHP	7	95	94
6	**37**	**38**	La–**17** (5)	CMHP	20	85	85
7	**37**	**38**	La–**32** (5)	CMHP	96	78	83
8	**39**	**40**	Yb–**32** (5)	TBHP	96	83	94
9	**41**	**42**	Yb–**32** (8)	TBHP	159	55	88
10	**43**	**44**	Yb–**32** (8)	TBHP	118	91	88
11	**45**	**46**	Yb–**32** (8)	TBHP	67	71	91

[a] Absolute configurations were determined to be ($\alpha S,\beta R$).

In contrast to the results presented above, the enones shown in Table 2 bearing aliphatic groups at R^2 (entries 8, 9, 10, and 11) were best converted to the corresponding epoxides by using the ytterbium complex generated from Yb(O-*i*-Pr)$_3$, (*R*)-**32**, and MS 4A in THF at 40 °C for 1 h (Figure 5, Yb–**32**). Treatment of **39** with TBHP (1.5 equiv) in the presence of 5 mol % of Yb–**32** in THF at rt for 96 h was found to give **40** with 94% ee in 83% yield. Gratifyingly, **42**, **44**, and **46** were obtained in excellent yields with 88, 88, and 91% ee, respectively (Table 2). In contrast, the use of the La–CMHP system afforded **40** with less satisfactory results. These variations may be due to the difference in ionic radii between lanthanum and ytterbium, as well as the difference in their Lewis acidities.

Empirical experimentation has revealed that the catalyst formed from a 1:1–1:1.4 ratio of Ln(O-*i*-Pr)$_3$ (Ln = La or Yb) and **17** provides the maximum enantiomeric excesses for epoxidation (Figure 6). The ^{13}C NMR spectrum of La–**17** was not interpretable, suggesting that both the chiral Yb–**17** catalyst and the La–**17** catalyst exist as oligomers. Moreover, the catalytic epoxidation of **39** with Yb–**17** displays a pronounced nonlinear effect (Figure 7). (For a treatment of nonlinear effects, see Chapter 5 in this volume.) Thus the oligomeric structure of these lanthanoid–BINOL catalysts may play a key role in the catalytic asymmetric

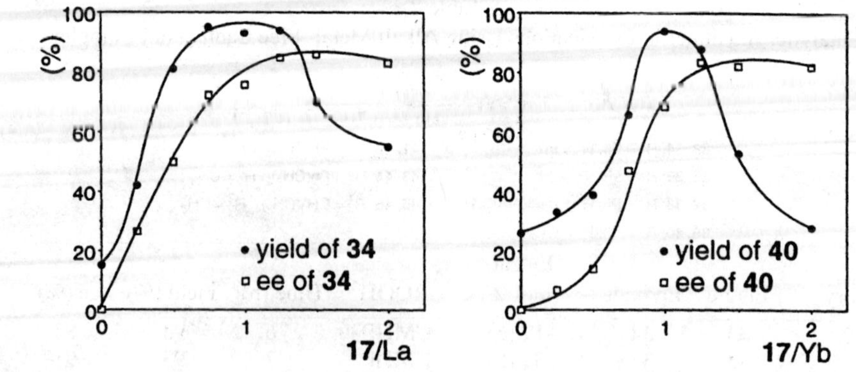

Figure 6. Influence of the ratio of Ln(O-*i*-Pr)$_3$ and **17** on yield and ee. Left: epoxidation of **33** catalyzed by the La–**17** catalyst. Right: epoxidation of **39** catalyzed by the Yb–**17** catalyst.

epoxidations of these enones. In other words, the Ln–alkoxide moiety in the catalysts appears to act as a Brønsted base, to activate the hydroperoxide moiety for the Michael reaction. Simultaneously, another Ln metal ion may act as a Lewis acidic site, both activating and controlling the orientation of the enone. However, other interpretations, such as a "reservoir-effect" have not been ruled out.

The Ln–BINOL derivative complexes are efficient asymmetric catalysts for Michael reactions and the epoxidations of enones. However, as was mentioned above, almost racemic products are obtained in the case of the asymmetric nitroaldol reaction of **2** with **12**. For this transformation, a new class of catalysts, heterobimetallic species, have been developed.

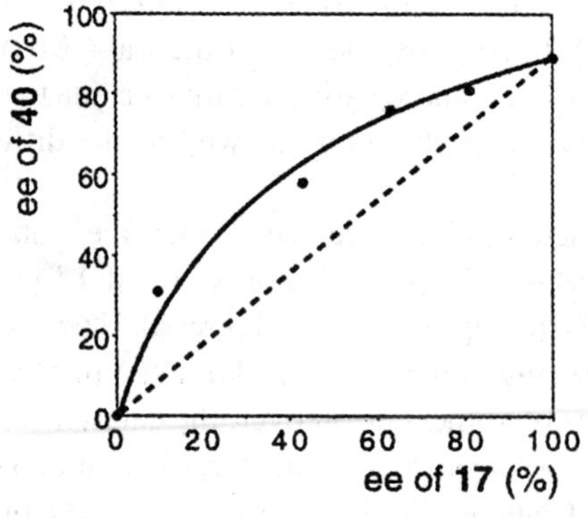

Figure 7. Asymmetric amplification in the epoxidation of **39** catalyzed by the Yb–**17** catalyst.

III. CHIRAL LANTHANOID–ALKALI METAL–BINOL COMPLEXES (LnMB)

A. Catalytic, Asymmetric Nitroaldol Reaction Promoted by the Lanthanoid–Lithium–BINOL Complex (LnLB)

This class of hybrid lanthanoid complexes is prepared from inexpensive lanthanoid trichlorides. Remarkably, the reaction of *anhydrous* $LaCl_3$ in THF with the dilithium salt of BINOL **47** or the disodium salt of BINOL **48** did not proceed even at reflux. However, the reaction of anhydrous $LaCl_3$ with the dialkali metal salts of BINOL proceeded smoothly in the presence of a small amount of water! Accordingly, an excellent catalyst **49** for nitroaldol reactions was prepared from **47** and $LaCl_3$, as shown in Figure 8.[5] This catalyst enabled the first examples of a catalytic asymmetric nitroaldol reaction to be realized. The results summarized in Figure 9 do not specify the loading of the catalyst **49**, since its structure is not known. The ee's are high (73–90%) for aliphatic and branched aldehydes.

B. Structural Determination of the Heterobimetallic Lanthanoid Complexes

LDI-TOF mass spectral analysis of **49** revealed that the structure was a heterobimetallic complex consisting of one lanthanum, three lithiums, and three BINOL moieties.[24] The relevant spectra are shown in Figure 10. LDI-TOF mass

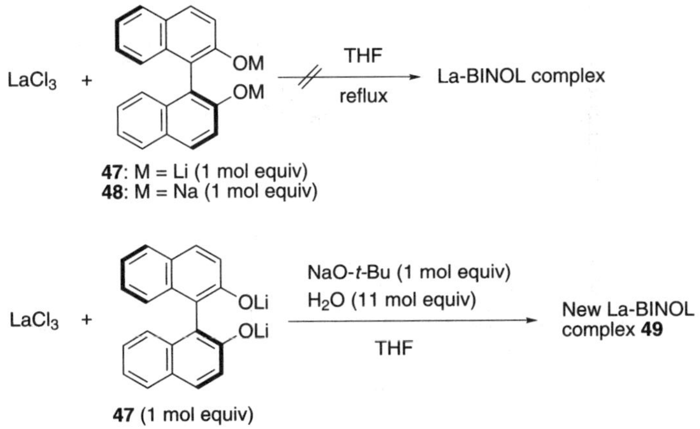

Figure 8. Preparation of a new chiral La–BINOL catalyst **49**.

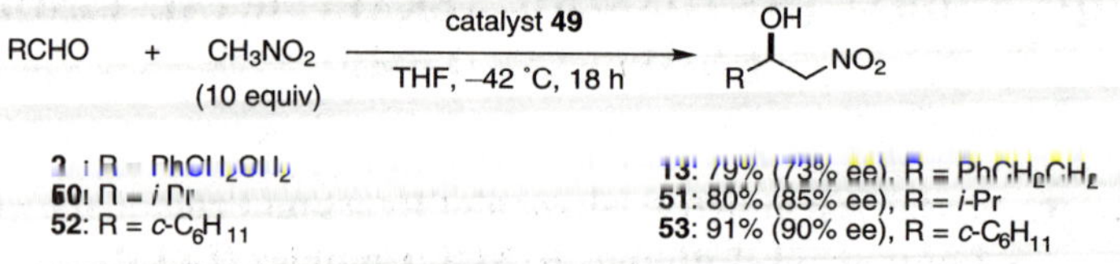

Figure 9. Catalytic asymmetric nitroaldol reactions promoted by catalyst **49**.

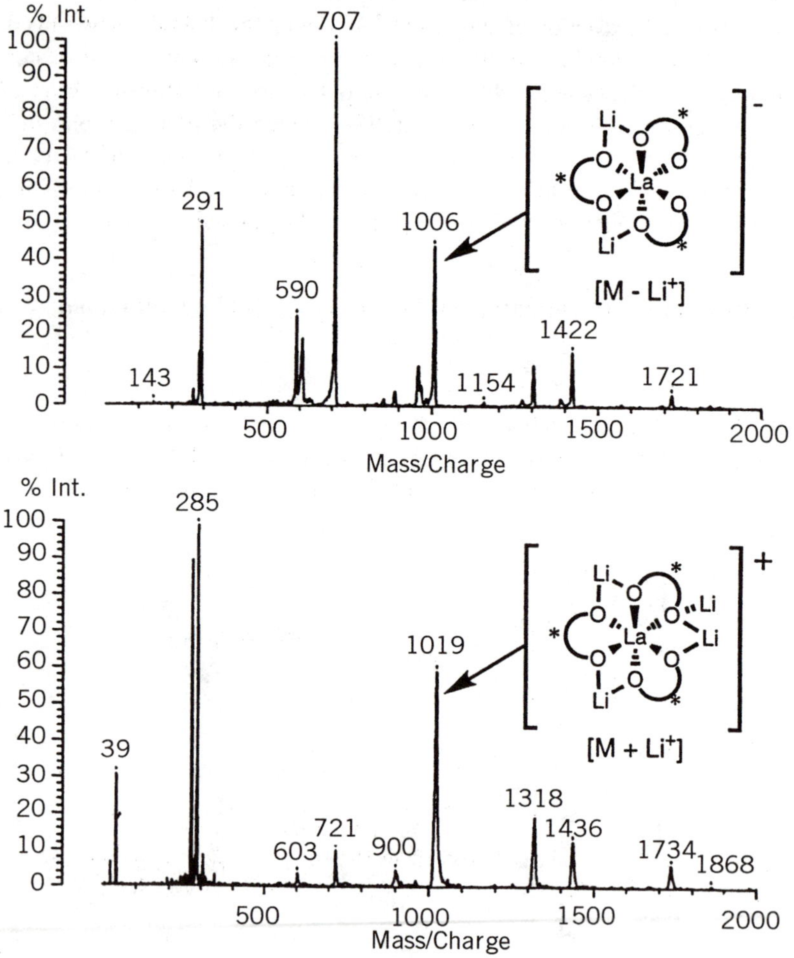

Figure 10. Laser desorption/ionization time-of-flight mass spectra of La–Li–BINOL complex in anionic (top) and cationic (bottom) modes.

spectrometry has a mass accuracy of approximately ±0.1%. However, the proposed structural framework was strongly supported by the similarity of the mass spectra of various other lanthanoid complexes, since lanthanoid elements have their own atomic weight and isotope abundance distribution.

Early attempts at obtaining an X-ray-quality crystal of a lanthanoid complex were largely unsuccessful. One reason for the low crystalline character of these lanthanoid complexes might be contamination due to the presence of LiCl in the catalyst solution. Accordingly, a more successful effort was the preparation of crystals under Li-free conditions. From the combination of lanthanoid trichlorides, disodium (*R*)-binaphthoxide (**48**), NaO-*t*-Bu, and H_2O, several lanthanoid complexes could be crystallized from THF. Elemental analyses suggested a stoichiometry of $LnNa_3$tris(binaphthoxide)–6THF–$2H_2O$, which was confirmed by X-ray crystallographic analyses of La, Pr, Nd, and Eu complexes.[24, 25] Figure 11 contains the X-ray structures of four lanthanoid–sodium–BINOL complexes (LnSB, S = sodium). The diffraction analyses showed that these substances are structurally homologous, differing only in the distance of the atoms around the central lanthanoid metal. In each lanthanoid–(*S*)-BINOL complex the lanthanoid

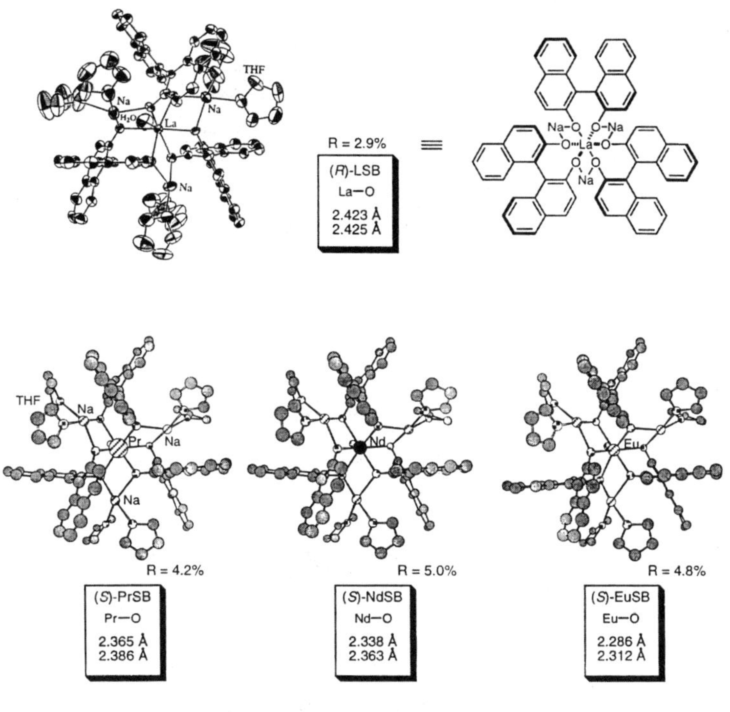

Figure 11. Structure of lanthanoid–sodium–BINOL complexes (LnSB).

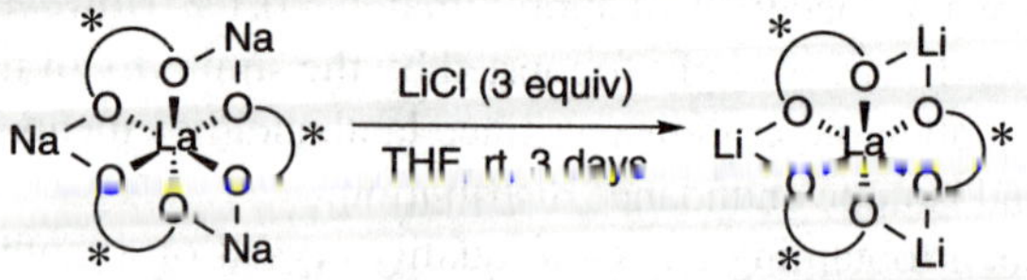

Figure 12. Preparation of LLB from LSB.

metal is a stereogenic center, so that the complexes can exist as metal-centered diastereomers. Nevertheless, it is interesting to note that La, Pr, Nd, and Eu complexes starting from (*S*)-BINOL exist only in the Λ configuration, most likely due to the greater thermodynamic stability of this configuration compared to the Δ form.[24]

All the above-mentioned LnSB crystals showed similar mass patterns to those of the Ln–Li–BINOL complexes, although they contained sodium instead of lithium. LnSB complexes were basic and acted as catalysts for the nitroaldol reaction. However, the nitroaldols thus obtained were mostly racemic mixtures. Interestingly, it is possible to generate the lithium-containing heterobimetallic complexes from the LnSB complexes by treatment with 3 equiv. of LiCl in THF at room temperature for 3 days (Figure 12). Under these conditions, the complete exchange of sodium for lithium was achieved, as was detected by LDI-TOF MS. These lithium-containing heterobimetallic catalysts gave almost identical results in catalytic asymmetric nitroaldol reactions to those obtained from the same catalyst prepared directly from $LnCl_3$. The role of NaOH in the catalyst preparation appears to be simply that of a buffer, neutralizing the HCl generated by the reaction of $LaCl_3$ with H_2O. Furthermore, it seems likely that the presence of H_2O accelerates the catalyst formation (Figure 13).[26]

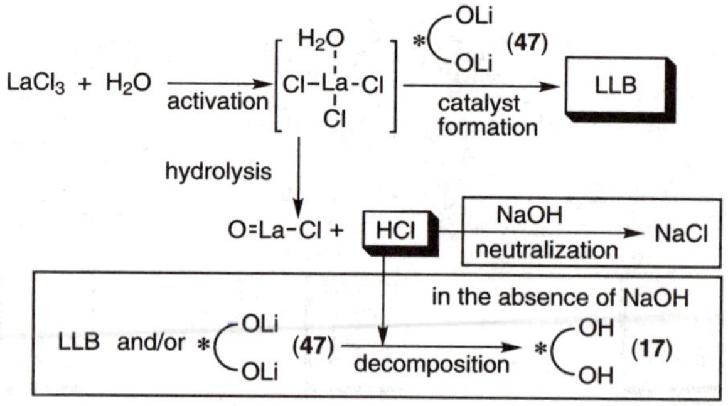

Figure 13. Schematic view of the role of H_2O and NaOH in the formation of the optically active La complex.

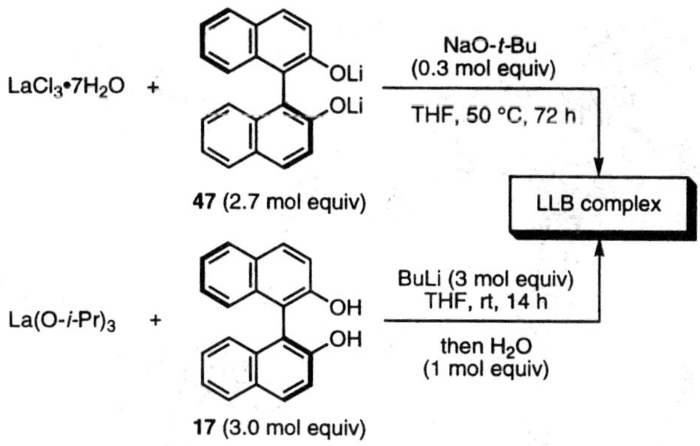

Figure 14. Best preparative procedures for LLB.

With indirect structural information on the LnLB complex and an understanding of the role of H_2O in catalyst formation, it was possible to formulate the conditions for the preparation of the lanthanum–lithium–BINOL complex (LLB) (Figure 14). Treatment of $LaCl_3 \cdot 7H_2O$ with 2.7 mol equiv. of dilithium salt of BINOL (**47**), and NaO-*t*-Bu (0.3 mol equiv.) in THF at 50 °C for 50 h gave the LLB complex most efficiently.[27] The second procedure shown in Figure 14 combines La(O-*i*-Pr)$_3$, 3 mol equiv. of BINOL, and 3 mol equiv. of butyllithium. Interestingly, the addition of 1 equiv. of water to LLB was found to improve the activity of the catalyst. By these two procedures a variety of heterobimetallic asymmetric complexes such as PrLB, NdLB, SmLB, EuLB, GdLB, and YbLB were also efficiently prepared. A crystal of the SmLB complex was recently obtained, and the preliminary X-ray crystallographic analysis of the SmLB complex is shown in Figure 15. This structure also suggests that the effective asymmetric catalyst for nitroaldol reactions is the LnLi$_3$tris(binaphthoxide) complex (LnLB).[28] It is noteworthy that these heterobimetallic asymmetric complexes, including LLB, are stable in organic solvents such as THF, CH_2Cl_2, and toluene that contain small amounts of water, and that these complexes are also insensitive to oxygen.

The lanthanoid and group 3 metals, the so-called rare earth elements, are generally regarded as a group of 17 elements with similar properties, especially with respect to their chemical reactivity. However, in the above-mentioned catalytic asymmetric nitroaldol reactions, pronounced differences were observed both in the reactivity and in the enantioselectivity of the various rare earth metals used.[29] For example, when benzaldehyde (**54**) and nitromethane (**12**) were used as starting materials, the EuLB complex gave **55** in 72% ee (91% yield) compared to 37% ee (81% yield) in the case of LLB (–40 °C, 40 h). The unique relationship

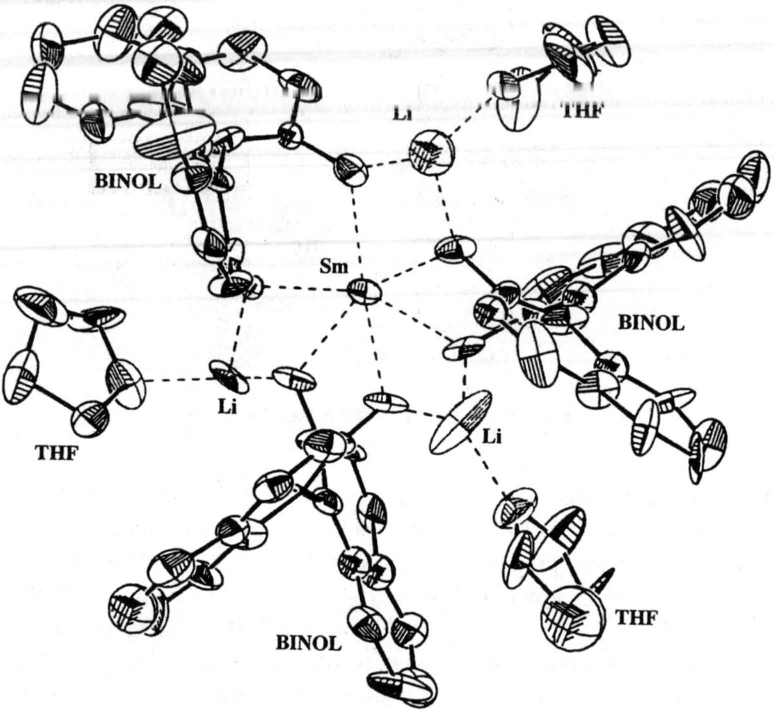

Figure 15. X-ray structure of SmLB.

between the ionic radii of rare earth metals and the enantiomeric purities of the nitroaldols is depicted in Figure 16. These results show that small changes in the structure of the catalyst (ca. 0.1 Å in ionic radius of the rare earth cation) cause drastic changes in the enantiomeric purity of the nitroaldols produced. Although, in general, nitroaldol reactions are regarded as equilibrium processes, in the LLB-catalyzed, asymmetric nitroaldol reactions no detectable retro-nitroaldol reaction was observed.

C. Application of Catalytic, Asymmetric Nitroaldol Reactions Using Nitromethane

The nitroaldol (Henry) reaction[6] is a powerful synthetic transformation and has been utilized in the construction of numerous natural products and other useful compounds.[30, 31] As shown in Figure 9, as little as 3.3 mol % of the LLB complex is a general and effective catalyst for the asymmetric nitroaldol reaction. The

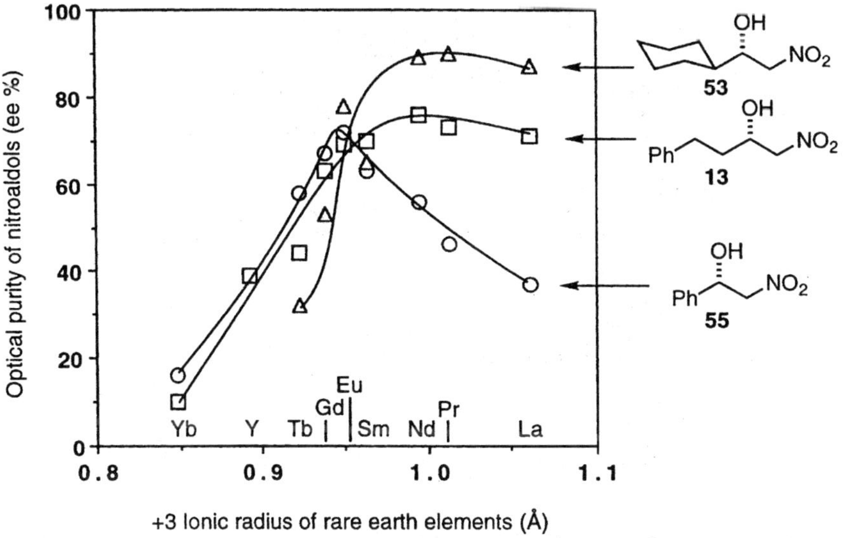

Figure 16. Effects of the ionic radii of rare earth elements on the optical purities of nitroaldol derivatives.

product nitroaldols are readily converted into β-amino alcohols and/or α-hydroxy carbonyl compounds. The utility of the LLB-catalyzed nitroaldol reaction is illustrated by the convenient syntheses of three kinds of enantiomerically enriched β-blockers, as presented in Figure 17.[29, 32–34] For example, by combining 10 mol equiv. of nitromethane (**12**) and **62** at –50 °C in the presence of 3.3 mol % of the (*R*)-LLB catalyst, a 76% yield of nitroaldol **63** in 92% ee was obtained. Reductive alkylation of the nitroaldol **63** to **64** was accomplished in 88% yield by a PtO_2-catalyzed hydrogenation in the presence of 5 mol equiv. of acetone in methanol. (*S*)-(–)-Pindolol **64** was thus synthesized in only four steps from 4-hydroxyindole.[33] Interestingly, the nitroaldols **57**, **60**, and **63** were found to have the *S*-absolute configuration when (*R*)-LLB was used. The nitronates thus appear to react preferentially with the *Si* face of the aldehydes, in contrast to the enantiofacial selectivity that might have been expected on the basis of previous results (cf. Figure 9). These results suggested that the presence of an oxygen atom at the β position had a great influence on the enantiofacial selectivity.

Nitroaldol reactions of enantiomerically pure α-amino aldehydes with nitromethane using a catalytic amount of LLB were found to proceed in a highly diastereoselective manner.[35] The adducts (3-amino-2-hydroxy-1-nitro derivatives) are attractive intermediates for the synthesis of unnatural *erythro*-amino-2-hydroxy acids, which are important components of several biologically active compounds. As an example, the promising HIV-protease inhibitor KNI-272 (**65**)[36, 37]

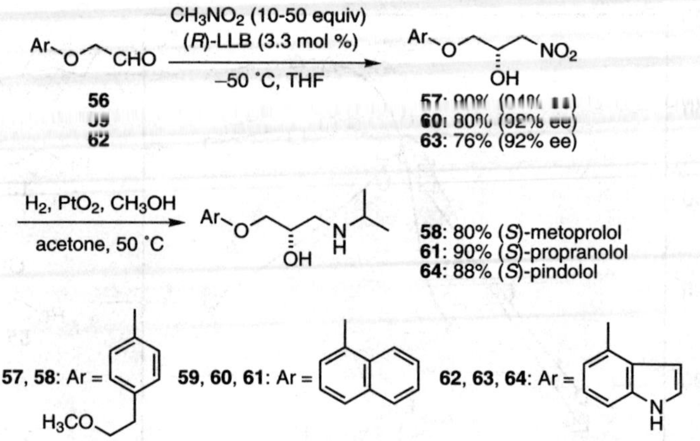

Figure 17. Catalytic, asymmetric syntheses of β-blockers using (*R*)-LLB as a catalyst.

contains (2*S*,3*S*)-3-amino-2-hydroxy-4-phenyl-butanoic acid (*erythro*-AHPA, **66**) as a subunit (Figure 18). As shown in Table 3, treatment of *N*-phthaloyl-L-phenylalanal (**67**) with nitromethane at –40 °C in the presence of (*R*)-LLB (3.3 mol %) gave almost exclusively a single stereoisomer of (2*R*,3*S*)-2-hydroxy-4-phenyl-3-phthaloylamino-1-nitrobutane (**68**) in 92% yield (>99:1 *erythro*-selectivity). The enantiofacial selectivity for the C(2) hydroxyl group of **68** agreed with results previously observed in enantioselective nitroaldol reactions for non-β-oxy aldehydes using LLB. Interestingly, the reaction of the aldehyde (*S*)-**67** with nitromethane, using the (*S*)-LLB complex as a catalyst, led to a reduced diastereoselectivity. Conversion of the nitroaldol adduct **68** into carboxylic acid **66** was achieved in 80% yield (Figure 19).

LLB-type catalysts were also able to promote diastereoselective and enantioselective nitroaldol reactions starting from substituted nitroalkanes. In preliminary work, however, LLB itself gave unsatisfactory results in terms of both diastereoselectivity (*syn*/*anti* ratio 63:37 to 77:23) and enantioselectivity (<78% ee).[32] To address the problem of modest enantio- and diastereoselectivities with

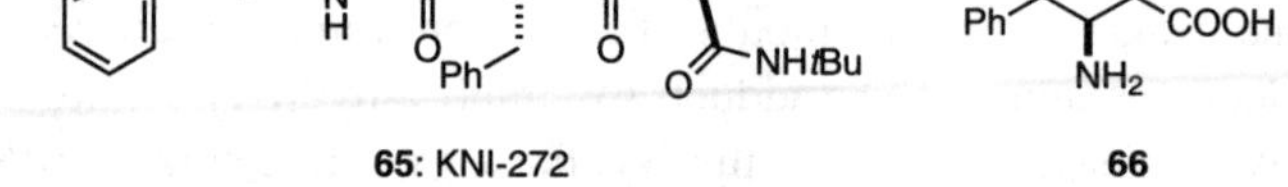

Figure 18. Structures of HIV protease inhibitor KNI-272 and its component (2*S*,3*S*)-3-amino-2-hydroxy-4-phenylbutanoic acid (*erythro*-AHPA, **66**).

Table 3
Diastereoselective Nitroaldol Reaction Catalyzed by LLB

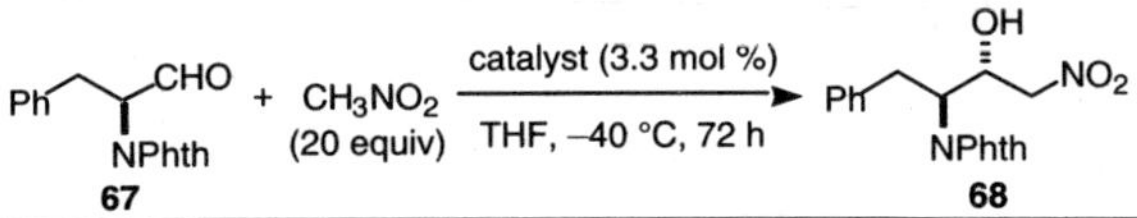

Entry	Catalyst	Yield (%)	*erythro* (% ee):*threo*
1	(*R*)-LLB	92	99 (96):1
2	(*S*)-LLB	96	74 (90):26

substituted nitroalkanes, new catalysts that bore substituents on the BINOL skeleton were investigated. The preparation of these complexes followed by analogy with LLB from La(O-*i*-Pr)$_3$; 3 mol equiv. of butyllithium and 3 mol equiv. of various (*R*)-BINOL derivatives in which certain positions were substituted by alkyl, alkenyl, alkynyl, aryl, and/or cyano groups. The utility of these new catalysts was first assessed in a nitroaldol reaction with nitromethane (**12**) and hydrocinnamaldehyde (**2**). The 3,3′-dimethyl-BINOL[38]- and 3,3′-bis(trimethylsilyl)BINOL[39]-derived complexes gave racemic **13**, and the BIPOL[40]-derived catalyst gave **13** in only 39% ee. Surprisingly, however, substitution at the 6,6′ position of BINOL proved effective in obtaining a superior asymmetric catalysts.[41] The structures of these lanthanoid–lithium–6,6′-disubstituted-BINOL complexes (as elucidated by ^{1}H and ^{13}C NMR) are similar to those of the (*R*)-LLB complex depicted in Figure 20. Results obtained for the nitroaldol reaction (–40 °C for 91 h) in THF using 3.3 mol % of the various catalysts and 10 equiv. of nitromethane are shown in Table 4. Using catalysts **69e–69i**, which bear silylalkynyl groups, nitroaldols were routinely obtained with enantiomeric purity *superior* to that of adducts obtained using LLB.

Another advantage was conferred by introducing 6,6′-substituents to BINOL. In general, catalytic asymmetric syntheses of fluorine-containing compounds are rather difficult.[42] However, an effective asymmetric nitroaldol reaction of the rather unreactive α,α-difluoro aldehydes proceeded satisfactorily when using the heterobimetallic asymmetric catalysts generated from 6,6′-bis[(triethylsilyl)-ethynyl]BINOL, as shown in Table 5.[43] The *S*-configuration of the nitroaldol adduct **71** showed that the nitronate reacted preferentially on the *Si* face of the

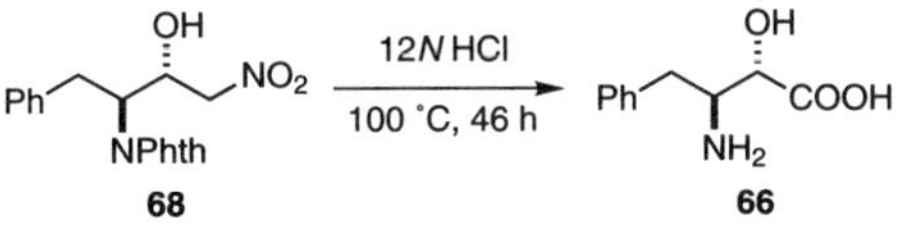

Figure 19. Synthesis of *erythro*-AHPA (**66**).

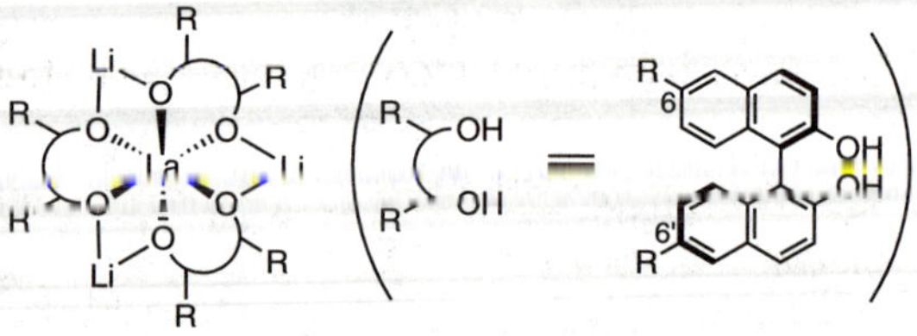

LLB: R = H
69a: R = Br
69b: R = CH_3
69c: R = C≡N
69d: R = C≡CH

69e: R = C≡CPh
69f: R = C≡CSi$(CH_3)_3$
69g: R = C≡CSiEt$_3$
69h: R = C≡CTBS
69i: R = C≡CSi$(CH_3)_2$Ph

Figure 20. Structural modification of LLB.

aldehyde in the presence of (*R*)-LLB. In contrast, (*R*)-LLB generally causes the nitronate to attack the aldehydes preferentially on the *Re* face; so it is noteworthy that the enantiotopic face selection for α,α-difluoro aldehydes is reversed from that of the nonfluorinated aldehydes. The stereoselectivity for α,α-difluoro aldehydes is identical to that of the β-oxy aldehydes, suggesting that the fluorine atoms at the α position have a great influence on the enantioface selection.

D. Diastereoselective and Enantioselective Nitroaldol Reactions

In all cases described up to this point, nitromethane has been employed as the nucleophile. Consequently the nitroaldol adducts bear a single stereogenic center.

Table 4
Catalytic, Asymmetric Nitroaldol Reactions Promoted by Various LLB-type Complexes

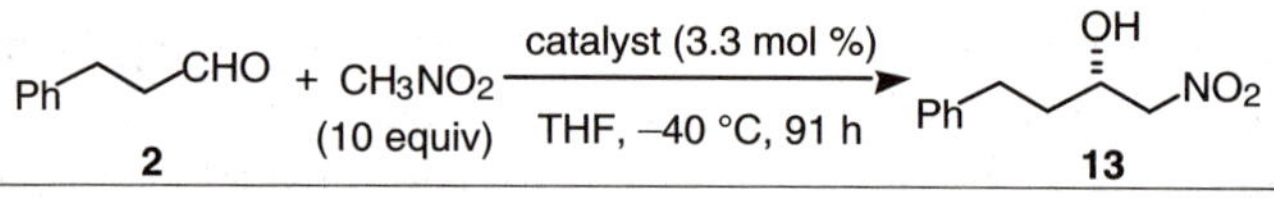

Entry	Catalyst	Yield (%)	ee (%)
1	LLB	79	73
2	**69a**	80	67
3	**69b**	84	63
4	**69c**	67	55[a]
5	**69d**	69	71
6	**69e**	74	79
7	**69f**	85	88
8	**69g**	84	85
9	**69h**	59	85
10	**69i**	54	86

[a]6,6′-Dicyano-BINOL of 93% ee was used.

Table 5
Catalytic, Asymmetric Nitroaldol Reactions of α,α-difluoro Aldehydes

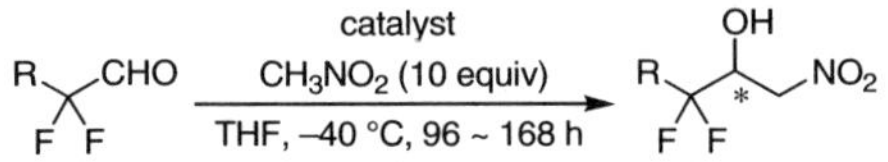

70: R = $PhCH_2CH_2CH_2$ — **71**: R = $PhCH_2CH_2CH_2$
72: R = $CH_3(CH_2)_6$ — **73**: R = $CH_3(CH_2)_6$
74: R = $PhCH_2O(CH_2)_2$ — **75**: R = $PhCH_2O(CH_2)_2$
76: R = *i*-$PrSCH_2$ — **77**: R = *i*-$PrSCH_2$
78: R = *c*-C_6H_{11} — **79**: R = *c*-C_6H_{11}

Entry	Aldehyde	Catalyst (mol %)	Product	Yield (%)	ee (%)[a]
1	**70**	**A** (20)	**71**	74	55
2	**70**	**B** (20)	**71**	77	87
3	**70**	**B** (5)	**71**	67	74
4	**70**	**C** (20)	**71**	58	94
5	**70**	**C** (5)	**71**	55	92
6	**72**	**C** (5)	**73**	73	70
7	**74**	**C** (5)	**75**	52	80
8	**76**	**C** (5)	**77**	55	85
9	**78**	**C** (5)	**79**	58	95

Catalyst: **A** = LLB; **B** = SmLB; **C** = $SmLi_3$tris((*R*)-6,6′-bis((triethylsilyl)ethynyl)binaphthoxide).
[a]Absolute configuration of **71** was determined to be (*S*) by X-ray crystallography of a derivative.

With substituted nitroalkanes, the formation of diastereomers becomes a possibility.

The new complexes **69f** and **69g** were found to be effective catalysts for these diastereoselective nitroaldol reactions; in all cases, high *syn*-selectivity and enantioselectivity were obtained using 3.3 mol % of the catalysts **69**.[41] Representative results are shown in Table 6. As expected, much higher *syn*-selectivity and enantioselectivity were obtained using the new catalysts than with LLB (entry 1). It is noteworthy that structurally simple aldehydes such as hexanal (which has no neighboring group to assist asymmetric induction) produced nitroaldols in high enantiomeric purity. The enantiomeric purities of the minor *anti*-adducts were lower than those of the *syn*-adducts, indicating that the former were *not* generated by an epimerization of the nitro group. In fact, treatment of the *syn*-adducts with catalysts such as LLB and its derivatives resulted in the near-quantitative recovery of the unchanged nitroaldols without detectable epimerization. It appears that the *syn*-selectivity in the nitroaldol reaction can best be explained as arising from steric hindrance in the bicyclic transition state—as can readily be seen in the Newman projection (Figure 21). In the favored transition state (leading to the *syn*-adduct), the catalyst is believed to act simultaneously at

Table 6

Diastereoselective and Enantioselective Nitroaldol Reactions

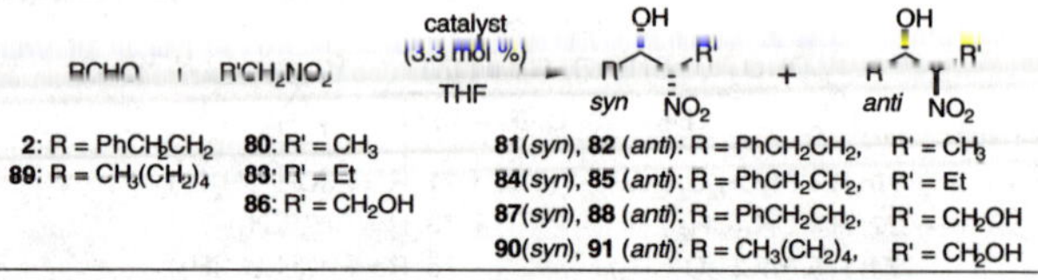

Entry	Alde-hyde	Nitro-alkane	Catalyst	Time (h)	Temp. (°C)	Nitroaldols	Yield (%)	*syn*:*anti*	ee of *syn* (%)
1	**2**	**80**	**LLB**	75	−20	**81 + 82**	79	74:26	66
2	**2**	**80**	**69b**	75	−20	**81 + 82**	80	74:26	65
3	**2**	**80**	**69d**	75	−20	**81 + 82**	77	84:16	90
4	**2**	**80**	**69f**	75	−20	**81 + 82**	72	85:15	92
5	**2**	**80**	**69g**	75	−20	**81 + 82**	70	89:11	93
6	**2**	**80**	**69g**	115	−40	**81 + 82**	21	94:6	97
7	**2**	**83**	**LLB**	138	−40	**84 + 85**	89	85:15	87
8	**2**	**83**	**69g**	138	−40	**84 + 85**	85	93:7	95
9	**2**	**86**	**LLB**	111	−40	**87 + 88**	62	84:16	66
10	**2**	**86**	**69g**	111	−40	**87 + 88**	97	92:8	97
11	**89**	**86**	**LLB**	93	−40	**90 + 91**	79	87:13	78
12	**89**	**86**	**69g**	93	−40	**90 + 91**	96	92:8	95

different sites as both a Lewis acid and Brønsted base. In contrast, the non-chelation-controlled transition state appears to afford *anti*-adducts with lower enantiomeric purities.

The *syn*-selective asymmetric nitroaldol reaction was successfully applied to the catalytic, asymmetric synthesis of *threo*-dihydrosphingosine (**94**), which elicits a variety of cellular responses by inhibiting protein kinase C (Figure 22).[44] The nitroaldol reaction of hexadecanal **92** with 3 equiv. of nitroethanol (**86**) in the

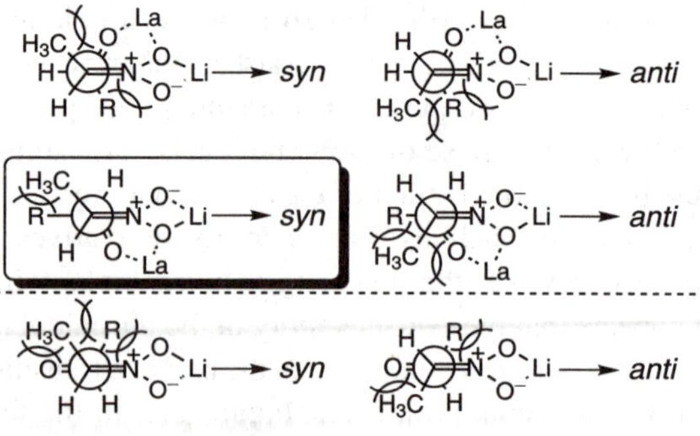

Figure 21. Proposed transition states of diastereoselective and enantioselective nitroaldol reactions.

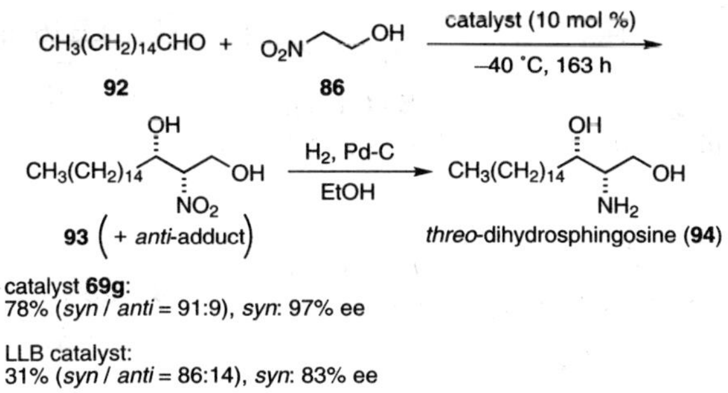

Figure 22. Catalytic, asymmetric synthesis of *threo*-dihydrosphingosine.

presence of 10 mol % of **69g** gave the corresponding nitroaldol adducts in high *syn/anti*-selectivity (91:9) and in 78% yield, with the *syn*-adduct **93** being obtained with up to 97% ee. The LLB-catalyzed reaction under similar conditions proceeded much more slowly to give an 86:14 ratio of the *syn* and *anti* adducts in 31% yield, and with lower enantiomeric purity (83% ee). Hydrogenation of **93** in the presence of 10% Pd on charcoal afforded *threo*-dihydrosphingosine (**94**) in 71% yield.[41]

The reasons for the salutary effects of the 6,6′-substituents on BINOL remain obscure. Nevertheless, it may be speculated that the introduction of 6,6′-bis-(trialkylsilyl)ethynyl substituents completely suppresses undesired ligand exchange between nitroalkane and BINOL, whereas this exchange does appear to occur in the case of LLB (albeit in only tiny amounts). The support for this explanation was provided by the following observation: In general an excess of nitroalkanes is required for efficient nitroaldol reactions. However, in the case of nitroethanol (**86**), only ca. 1 equiv. of the nitroalkane is sufficient, with the undesired ligand exchange clearly being noncompetitive and thus avoided. In this case, even using unmodified LLB gave high diastereoselectivity and enantiomeric excess, in a quite practical synthetic procedure (Figure 23).

E. Second-Generation LLB Catalyst (LLB-II)

The LLB catalysts described above served an important role in demonstrating the proof of principle for catalysis with lanthanide–BINOL complexes. In addition, they were the first catalysts for the enantioselective nitroaldol reaction and gave respectable selectivities in synthetically useful yields. However, the reactions required at least 3.3 mol % of the catalysts for efficient conversion, and at that loading the reactions are rather slow. Clearly, the need for more effective catalysts is indicated. Consideration of the mechanism for the catalytic asymmetric

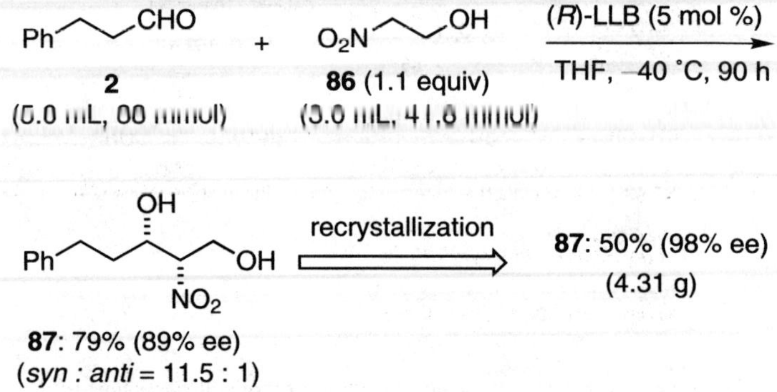

Figure 23. Practical procedure for the synthesis of **87**.

nitroaldol reaction suggested a possible strategy, Figure 24. In this proposal, a lithium nitronate is generated by deprotonation (either before or after aldehyde coordination). All attempts to detect the postulated intermediate **I** using various methods proved to be unsuccessful, probably owing to the low concentrations of the intermediate. The low concentration of **I** might be ascribable to the presence of an acidic OH group in close proximity to the nitronate. To remove a proton from **I**, almost 1 equiv. of base was added to the LLB catalyst. Optimally, 1 mol equiv. of H_2O and 0.9 mol equiv. of butyllithium were used to generate LiOH in situ

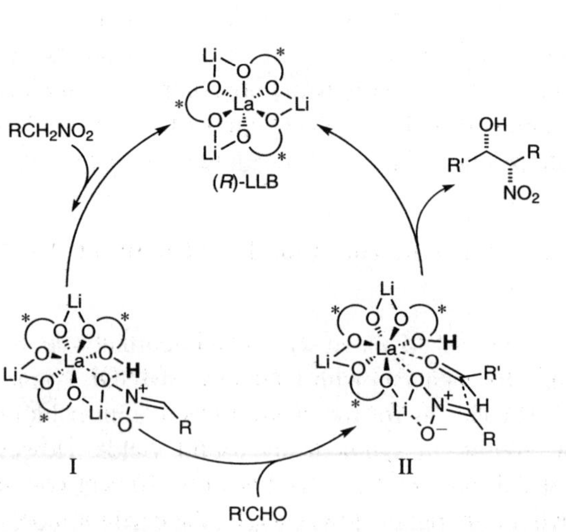

Figure 24. A possible mechanism for catalytic asymmetric nitroaldol reactions.

together with LLB. Indeed, the new "second-generation LLB" (LLB-II) efficiently promoted catalytic asymmetric nitroaldol reactions at loadings as low as 1 mol %, albeit still rather slowly (24 h at –50 °C; Table 7).[45] However, we also found that the use of 3.3 mol % of LLB-II accelerated catalytic asymmetric nitroaldol reactions within a shorter time span (4 h at –50 °C). However, to achieve optimal selectivities with substituted nitroalkenes low temperatures and consequently long reaction times were necessary (entries 6, 8 and 10). The use of other bases such as NaO–*t*-Bu, KO–*t*-Bu and $Ca(O\text{-}i\text{-}Pr)_2$ gave less satisfactory results.

The structure of LLB-II is still not secure. However, a reasonable proposal is that it is composed of a complex of LLB and LiOH. A proposed reaction course for an improved catalytic, asymmetric nitroaldol reaction is shown in Figure 25. The key feature of this proposal is the formulation of a discrete complex (**II**) between LLB and the lithio nitronate. This species is proposed to be a faster-acting reagent than the protio-form. In support of this hypothesis is the observation that treatment of the lithium nitronate (0.9 mol %) generated from nitropropane and butyllithium, with **69g** (1 mol %), **2**, and nitropropane (**83**) under similar conditions as those described above, gave comparable results (59% yield, *syn/anti*, 94/6, 94%

Table 7
Comparisons of Catalyst Activity Between Either LLB and Second-generation LLB (LLB-II) or **69g** and **69g-II**

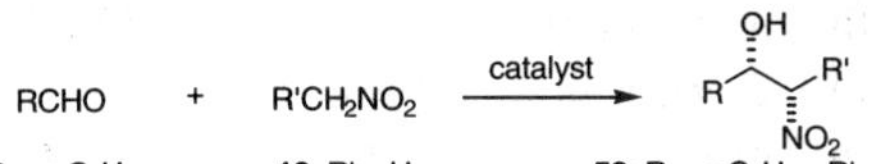

52: R = *c*-C_6H_{11} **12**: R' = H **53**: R = *c*-C_6H_{11}, R' = H
2: R = $PhCH_2CH_2$ **80**: R' = CH_3 **81**: R = $PhCH_2CH_2$, R' = CH_3
83: R' = Et **84**: R = $PhCH_2CH_2$, R' = Et
86: R' = CH_2OH **87**: R = $PhCH_2CH_2$, R' = CH_2OH

Entry	Substrate	Catalyst (mol %)	Time (h)	Temp. (°C)	Product	Yield (%) (*syn*/*anti*)	ee (%) of *syn*
1	**52 + 12**	LLB (1)	24	–50	**53**	5.6	88
2	**52 + 12**	LLB-II (1)	24	–50	**53**	73	89
3	**52 + 12**	LLB-II (3.3)	4	–50	**53**	70	90
4	**52 + 12**	**A** (1)	42	–50	**53**	86	51
5	**2 + 80**	**69g** (1)	113	–30	**81**	25 (70/30)	62
6	**2 + 80**	**69g-II** (1)	113	–30	**81**	83 (89/11)	94
7	**2 + 83**	**69g** (1)	166	–40	**84**	trace	—
8	**2 + 83**	**69g-II** (1)	166	–40	**84**	84 (95/5)	95
9	**2 + 86**	**69g** (1)	154	–50	**87**	trace	—
10	**2 + 86**	**69g-II** (1)	154	–50	**87**	76 (94/6)	96

LLB-II: LLB + H_2O (1 mol equiv.) + BuLi (0.9 mol equiv.).

Catalyst **A**: LLB + H_2O (1 mol equiv.) + BuLi (2 mol equiv.).

69g-II: **69g** + H_2O (1 mol equiv.) + BuLi (0.9 mol equiv.).

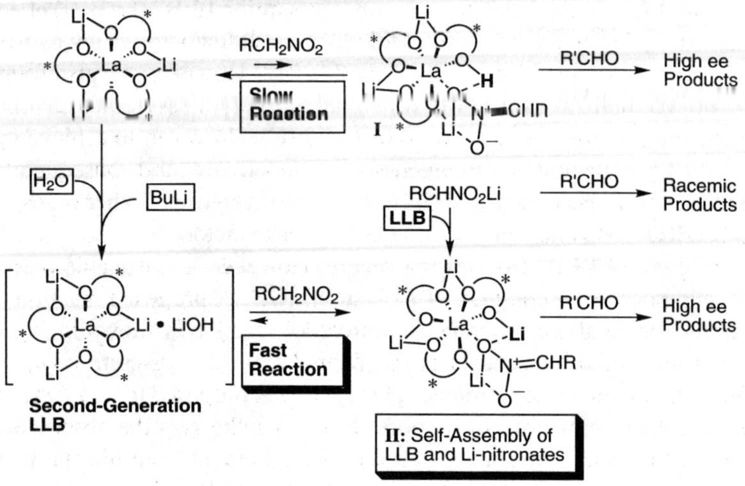

Figure 25. Proposed mechanism for the catalytic asymmetric nitroaldol reaction promoted by LLB, LLB-II, or LLB–Li–nitronate.

ee), suggesting the presence of a heteropolymetallic intermediate **II**, as shown in Figure 25. Although a molecular ion peak was not observed, a fragment ion peak for the complex of LLB and lithium nitronate was observed at 816 by FAB mass spectral analysis. The above-mentioned results suggest a quite interesting chemical phenomenon, namely, the tight complexation of LLB and LiOH and the high rate of aggregation between LLB and lithium nitronates.

The second-generation LnLB catalyst, which consists of 6,6′-bis[(trimethylsilyl)-ethynyl]- BINOL and Sm, as shown in Figure 26, successfully promoted an efficient asymmetric synthesis of arbutamine (**97**),[28] a useful β-agonist.[46, 47]

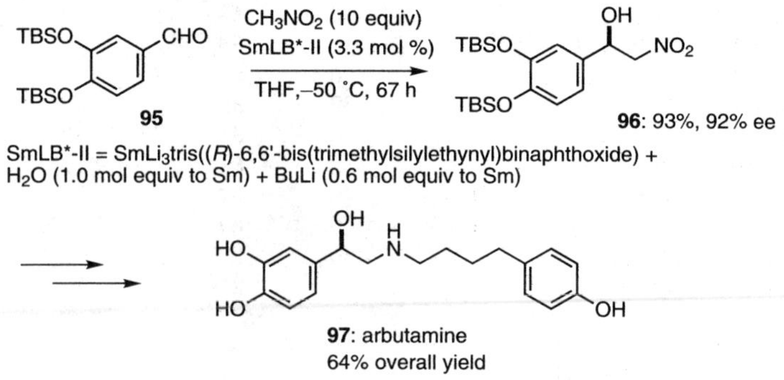

Figure 26. A catalytic, asymmetric synthesis of arbutamine.

F. Tandem Inter-Intramolecular Catalytic, Asymmetric Nitroaldol Reaction

Tandem, catalytic, asymmetric reactions are especially useful because enantiomerically enriched compounds containing multiple stereogenic centers can be constructed from simple achiral compounds using a chiral catalyst. The development of such methods has been recognized as one of the most challenging themes in organic synthesis.[48–50] One illustration of this process is the tandem, inter-intramolecular, catalytic asymmetric double nitroaldol reaction using **98** and nitromethane (**12**) (Figure 27).[51] In the presence of LLB, the reaction of **98** with **12** at –20 °C gave intermediates **99a**, **99b**, and one diastereoisomer of **100**. Complete consumption of **98** at room temperature led to the complete conversion of **100a** into another diastereoisomer, **100b**. The structure of the new compound **100b** was determined by X-ray structural analysis, as shown in Table 8. The enantiomeric purity of **100b** was found to be 39% ee. Crystallization of the reaction mixture from CH_2Cl_2 produced **100b** with 41% ee in 59% yield.

The use of many other heterobimetallic asymmetric catalysts such as LnLB, LnSB, and LnPB was examined under a variety of reaction conditions. Of these, LnLB was shown to be the most useful asymmetric catalyst for the present purpose. Entries 1 and 2 in Table 8 reveal that only modest results were obtained using LLB in THF. An increase in reaction temperature to 50 °C caused an expected decline in the ee of **100b** (entry 2). A survey of various lanthanoid metals led to an improvement in the ee of **100b** (entries 3–7). When PrLB was used as the catalyst, the enantiomeric purity of **100b** was improved to 48% ee. The initial temperature also appeared to be important. That is, when the initial temperature was –40 °C, the enantiomeric purity of **100b** increased to 58% ee (entry 8). Moreover, it was possible to reduce the amount of (*R*)-PrLB; when 5 mol % of PrLB was used, **100b** with 53% ee was formed (entry 9). An increase in the amount of nitromethane was surprisingly found to improve the ee of **100b**. That is, **100b** with 65% ee was produced by the reaction of **98** and 30 mol equiv. of nitromethane catalyzed by 5 mol % of (*R*)-PrLB (–40 °C for 112 h and then at room temperature for 24 h) and, after crystallization, **100b** with 79% ee was isolated in 41% yield (entry 10). The enantiomeric purity of **100b** (69% ee) increased to 96% ee (57% recovery) after recrystallization.[51]

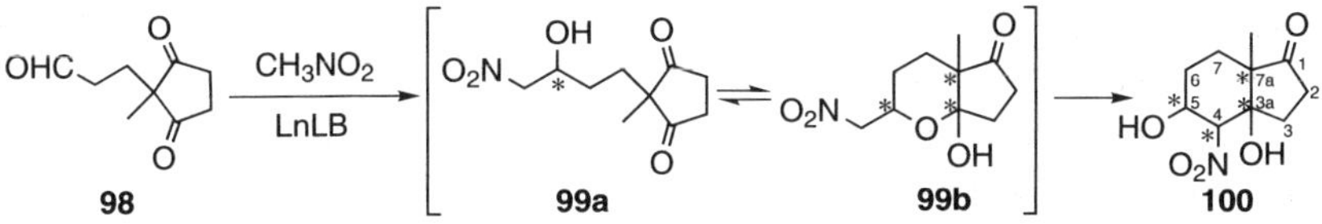

Figure 27. Tandem inter-intramolecular catalytic asymmetric nitroaldol reaction.

Table 8
Catalytic, Asymmetric Tandem Inter-Intramolecular Nitroaldol Reaction of **98** with Nitromethane (**12**)

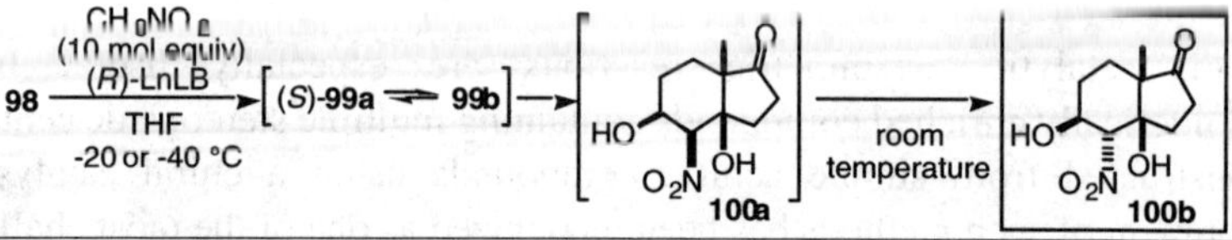

Entry	Catalyst (mol %)	Temp. (°C) [Time (h)]	ee of crude **100b**	Yield of **100b** (%)	ee of **100b** (%)
1	LLB (10)	−20 (115) to rt (24)	39	59	41
2	LLB (10)	−20 (115) to 50 (24)	~5	56	7
3	PrLB (10)	−20 (119) to rt (24)	48	47	60
4	SmLB (10)	−20 (119) to rt (24)	28	43	51
5	GdLB (10)	−20 (119) to rt (24)	36	47	41
6	DyLB (10)	−20 (119) to rt (24)	34	34	43
7	YbLB (10)	−20 (119) to rt (24)	7	37	12
8	PrLB (10)	−40 (115) to rt (24)	58	42	71
9	PrLB (5)	−40 (145) to rt (71)	53	45	66
10[a]	PrLB (5)	−40 (112) to rt (24)	65	41	79

[a]30 mol equiv. of nitromethane were used.

A mechanistic proposal for the production of **100b** is shown in Figure 28. Compound PrLB is a multifunctional heterobimetallic asymmetric catalyst; the Pr atom works as a Lewis acid, and the Li–naphthoxide portion functions as a Brønsted base. Thus **98** is activated by the Pr atom, and nitromethane is deprotonated by the Li–naphthoxide portion, resulting in the formation of **I**. Both **98** and **12** then react intermolecularly to give **II**. In addition, a lithium nitronate, again formed from **II**, reacts rapidly with one of the carbonyl groups (−40 or −20 °C) to give **III**, followed by the generation of **100a** and a regeneration of PrLB. At room temperature, however, an equilibrium appears to exist among **III**, **II**, and **IV**, and the reaction proceeds toward the generation of the thermodynamically more stable isomer **100b**.

In fact, the reaction of **100b** (42% ee) and 30 mol equiv. of nitromethane in the presence of 10 mol % of PrLB at room temperature for 7 days generated a trace amount of **99a** and **99b**, allowing for the recovery of **100b** with 42% ee, supporting the hypothesis that **100a** is the kinetic product and that an equilibrium exists among **III**–**II**–**IV** at room temperature. The subsequent reaction of a mixture of **99a** and **99b**, isolated after the reaction of **98** and nitromethane in the presence of PrLB at −40 °C, with nitromethane in the presence of PrLB at −40 °C for 59 h and at room temperature for 28 h, gave **100b** with 65% ee, whereas the reaction of a mixture of **99a** and **99b** obtained under similar conditions, in the presence of NaO–*t*-Bu

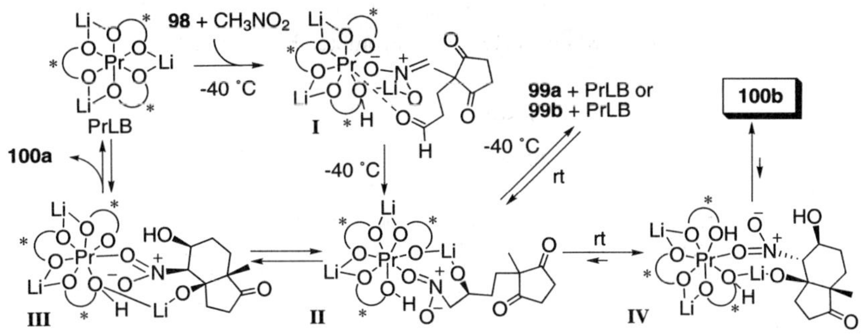

Figure 28. Proposed mechanism for the tandem inter- and intramolecular catalytic asymmetric nitroaldol reaction.

instead of PrLB, generated **100b** with 66% ee. These results suggest that the enantiomeric purity of **100b** depends only on the enantioselectivity of the intermolecular nitroaldol reaction and that, in an intramolecular reaction, there is no kinetic resolution of **99**. The cyclization of **99** to **100b** is controlled by the configuration of the **99a** hydroxy group.

Because **98** was easily prepared from 2-methyl-1,3-cyclopentanedione and acrolein in one step,[52] and because after recrystallization the adduct **100b** can be obtained in a highly enantiomerically pure form, this tandem inter-intramolecular catalytic asymmetric reaction could be useful for the preparation of chiral intermediates in the synthesis of natural products and/or chiral ligands.

G. Catalytic, Asymmetric Michael Reactions Promoted by LSB

Catalytic, asymmetric Michael reactions are among the most important carbon–carbon bond-forming reactions associated with the creation of new stereogenic centers.[53, 54] 1,3-Dicarbonyl compounds in particular are highly promising Michael donors for the enantioselective construction of carbon frameworks. Although LSB (L = lanthanum, S = sodium) was ineffective as an asymmetric catalyst for nitroaldol reactions,[24] it proved to be very useful for the catalytic asymmetric Michael reaction of various enones with malonates. The corresponding Michael adducts were produced in up to 92% ee and almost quantitative yield.[25] Representative examples are summarized in Table 9. In general, THF was found to be the best solvent. However, in the case of the LSB-catalyzed reaction of trans-chalcone (**101**) with **25**, the use of toluene was essential to afford **102** in good enantiomeric excess. Metal-center effects were also investigated for the Michael reaction of **101** with **25**. Lanthanum was shown to give superior results in LnSB-catalyzed, asymmetric Michael reactions.

Table 9
Catalytic, Asymmetric Michael Reactions Promoted by (*R*)-LnMB (10 mol %)

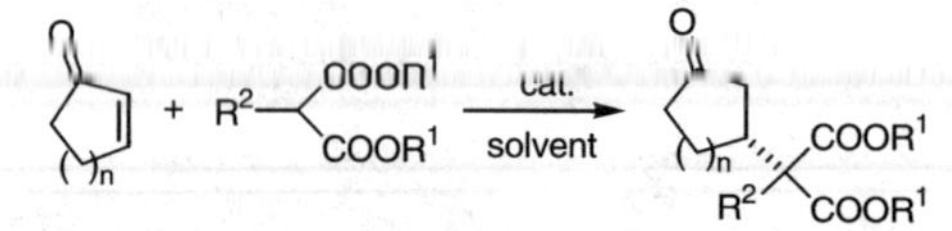

22: n = 2 **20**: R^1 = Bn, R^2 = H **24**: n = 2, R^1 = Bn, R^2 = H
4: n = 1 **18**: R^1 = Bn, R^2 = CH_3 **23**: n = 2, R^1 = Bn, R^2 = CH_3
25: R^1 = CH_3, R^2 = H **26**: n = 2, R^1 = CH_3, R^2 = H
27: R^1 = Et, R^2 = H **28**: n = 2, R^1 = Et, R^2 = H
19: n = 1, R^1 = Bn, R^2 = CH_3

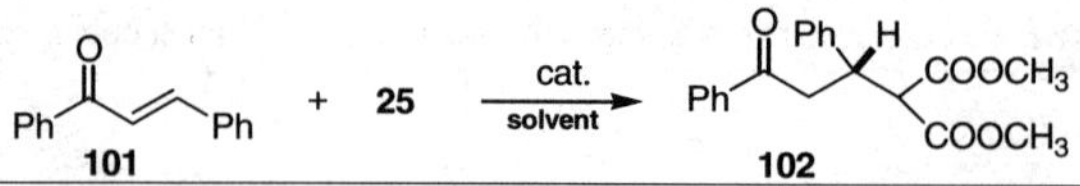

101 **102**

Entry	Enone	Michael donor	Product	Catalyst	Solvent	Temp. (°C)	Time (h)	Yield (%)	ee (%)
1	**22**	**20**	**24**	LSB	THF	0	24	97	88
2	**22**	**20**	**24**	LSB	THF	rt	12	98	85
3	**22**	**20**	**24**	LSB	toluene	rt	12	96	82
4	**22**	**20**	**24**	LLB	THF	rt	12	78	2
5	**22**	**20**	**24**	LPB	THF	rt	12	99	48
6	**22**	**18**	**23**	LSB	THF	0	24	91	92
7	**22**	**18**	**23**	LSB	THF	rt	12	96	90
8	**22**	**25**	**26**	LSB	THF	rt	12	98	83
9	**22**	**27**	**28**	LSB	THF	rt	12	97	81
10	**4**	**18**	**19**	LSB	THF	−40	36	89	72
11	**101**	**25**	**102**	LSB	THF	−50	36	62	0
12	**101**	**25**	**102**	LSB	toluene	−50	24	93	77
13	**101**	**25**	**102**	PrSB	toluene	−50	24	96	56
14	**101**	**25**	**102**	GdSB	toluene	−50	24	54	6

What is the origin of the catalytic activity and mode of enantioselectivity in the LSB-catalyzed Michael reactions? To clarify the nature of the interaction between the enone and the asymmetric catalyst, the complexation was studied by ^{1}H NMR spectroscopy. A mixture of cyclohexenone (**22**) and the asymmetric bimetallic complexes was examined, noting the chemical shift of the α-proton of the enone **22** (Figure 29). It is well known that, in general, praseodymium complexes induce an upfield shift, while europium complexes induce a downfield shift. In addition, ordinary Lewis acids such as $La(OTf)_3$ and Et_2AlCl also induce a downfield shift. Complexation with LSB induced a small downfield shift on the α-proton of **22**, and PrSB, a moderately effective asymmetric catalyst for Michael reactions, induced a large upfield shift. Interestingly, in the case of either EuSB or LLB, which gave

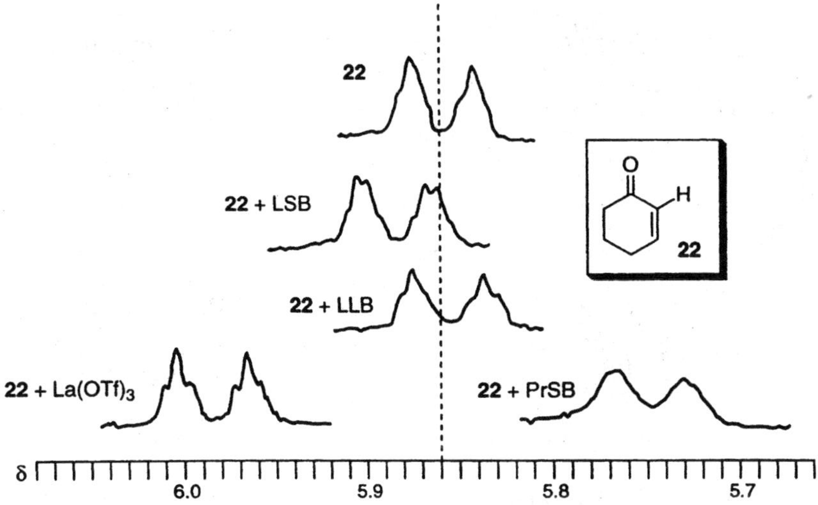

Figure 29. Chemical shift of α-proton on cyclohexenone (**22**).

only nearly-racemic Michael adducts, the ^{1}H NMR spectra showed no changes in the chemical shift of the α-proton of **22**.[25] These NMR studies indicated that the carbonyl group of the enone is coordinated to the lanthanum and/or praseodymium metal in the LnSB molecule, while the enone did not coordinate to either LLB or EuSB. These changes in chemical shift were observed even in the presence of **25**. The chemical phenomenon described above might be understood by considering the differing dihedral angles with which the BINOL moiety binds to the central metal in each case.

Taking the X-ray structure of LnSB as a reference point, computational simulations of the enantioselection process were carried out using Rappé's universal force field (UFF).[25, 55–58] As shown in Figure 30, with enone **22** coordinated to the lanthanum metal cation, the plane of the cyclohexenone ring should be almost parallel to the nearest of the naphthyl ring systems, facilitating an attack by the coordinated sodium enolate of **25** to give the Michael adduct **26**. UFF calculations and conformational searching for models of the pro-(*R*)- and pro-(*S*)-adduct systems clearly indicated that (*R*)-LSB complexes better as a pro-(*R*) adduct than as a pro-(*S*) adduct ($\Delta E = 4.9$ kcal/mol). LSB can thus generate an intermediate such as that shown at the top of Figure 30, giving high ee. The resultant sodium enolates of the enantiomerically enriched Michael adducts appear to abstract a proton from an acidic OH so as to regenerate the LSB catalyst. Thus the basic LSB complex also acts as a Lewis acid, controlling the orientation of the carbonyl function and so activating the enone for attack. It appears that the

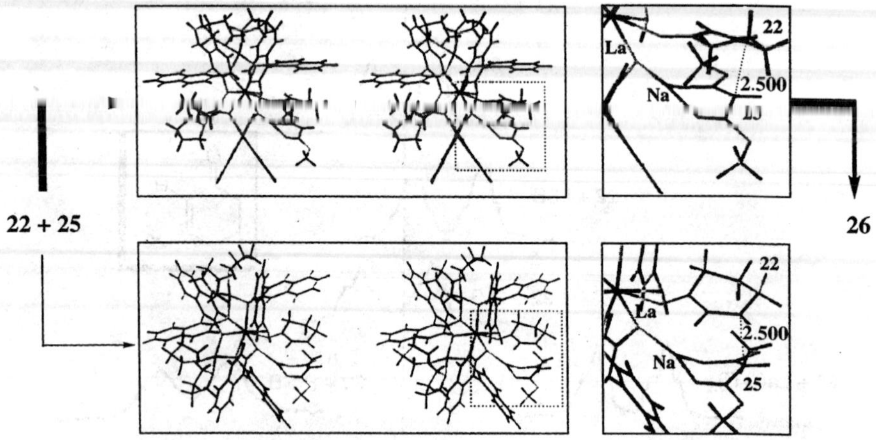

Figure 30. UFF computational simulations of the Michael reaction of **22** with **25** catalyzed by (*R*)-LSB. Top: Pro-(*R*) model (favorable model). Bottom: Pro-(*S*) model (unfavorable model). Enlarged pictures of the reaction sites are drawn on the each right side of the stereoviews.

multifunctional nature of the LSB catalyst makes possible the formation of Michael adducts with high enantiomeric excesses even at room temperature.

In both catalytic, asymmetric Michael reactions and nitroaldol reactions, enones and/or aldehydes appear to coordinate to the lanthanoid metal. Why, then, is LSB more effective for catalytic, asymmetric Michael reactions, whereas LLB is more effective for catalytic, asymmetric nitroaldol reactions? This disparity might arise from slight differences in bond lengths in the chelated intermediate, as well as slight differences in "bite" angle for the BINOL moiety caused by varying the alkali metal.

LSB was next applied to another type of catalytic, asymmetric Michael reaction, in which the new stereogenic center is induced on the side of the adduct originating from the Michael donor. In a preliminary study, it was found that the reaction of **103** with **29** in THF using 10 mol % of LSB gave **104** with 23% ee, while carrying out the reaction in toluene afforded **104** with 75% ee (Table 10). However, when the amount of LSB was reduced to 5 mol %, the enantiomeric excess of **104** declined to a more modest 25% ee. To offset this decline while still using a smaller amount of catalyst, the effect of the slow addition of **103** was examined. This proved effective to give **104** with high enantiomeric excess. In addition, the asymmetric Michael reaction catalyzed by 5 mol % of LSB in CH_2Cl_2 gave **104** in 89% yield and with 91% ee, without the need for the slow addition. Moreover, in this solvent it was found that the catalytic asymmetric Michael reaction for **104** was not so affected by the choice of lanthanoid metal.[39]

Table 10

LSB-Catalyzed Michael Reactions of **103** with **29** under Various Conditions

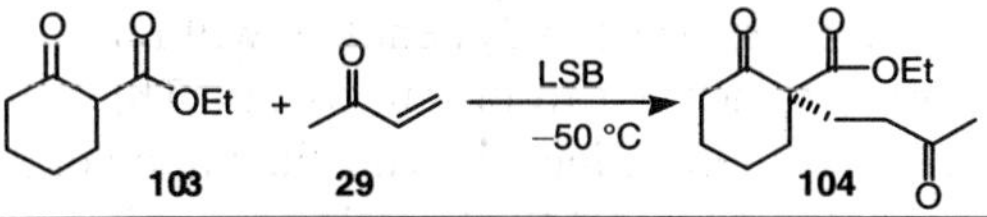

Entry	Solvent	Catalyst amount (mol %)	Time (h)	Yield (%)	ee (%)
1	THF	10	21	81	23
2	toluene	10	14	97	75
3	toluene	5	14	83	25
4[a]	toluene	5	22	76	89
5	CH_2Cl_2	10	19	85	93
6	CH_2Cl_2	5	19	89	91
7	Et_2O	10	14	85	71

[a]Compound **103** was added slowly using syringe pump methods over 8 h.

As shown in Table 11, the slow addition of a β-keto ester and the use of CH_2Cl_2 as the solvent are generally quite effective methods for suppressing the reduction in enantiomeric excess for the various Michael adducts. On the other hand, malonates give adducts with high enantiomeric excesses regardless of the solvent used.[25] These results can be explained by comparing the pK_a of a β-keto ester with

Table 11

Catalytic, Asymmetric Michael Reactions Promoted by LSB in CH_2Cl_2

Michael donor	Michael acceptor	Product	Catalyst amount (mol %)	Temp. (°C)	Time (h)	Yield (%)	ee (%)
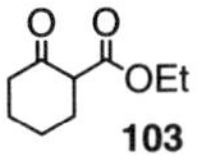 **103**	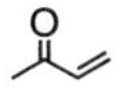 **29**	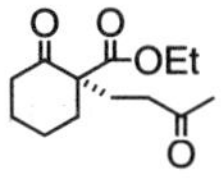 **104**	5	–50	19	89	91
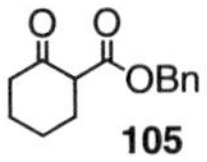 **105**	**29**	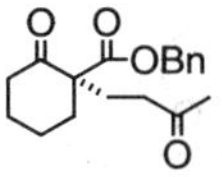 **106**	10	–50	12	73	91
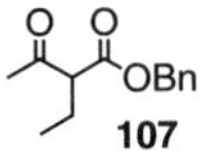 **107**	**29**	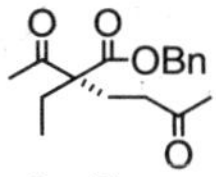 **108**	5	–50	16	98	89
103	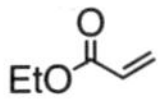 **109**	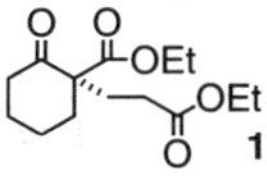 **110**	10	0 → rt	17	60	76

that of a malonate; the former is more acidic than the latter. Therefore, the concentration of the resulting Na-enolate can be expected to be greater in the case of the β-keto ester, and, moreover this Na-enolate will react with an enone more slowly than the Na-enolate derived from a malonate. This combination of more rapid formation and longer lifetime increases the likelihood of a dissociation of the Na-enolate from the chelated ensemble, thus giving a product with lower ee. On the other hand, in the less polar CH_2Cl_2, the Na-enolate would, even in this case, remain as part of the ensemble, thereby affording the product with high ee (Figure 31). Furthermore, the slow addition of the β-keto ester also acts to limit undesired ligand exchange between BINOL moieties and the Michael donor.

In both types of catalytic asymmetric Michael reactions, the use of 6,6′-substituted BINOL-derived LSB-type catalysts did not result in significantly improved results.

The heterobimetallic chiral catalysts such as LLB and LSB promote various asymmetric reactions efficiently by means of a synergistic cooperation between two different metals and a chiral template. In addition to the above-mentioned chiral lanthanoid complexes, another type of heterobimetallic complexes containing group 13 elements as the center metal have been developed. The aluminum–lithium–BINOL complex (ALB) and the gallium–sodium–BINOL complex (GaSB) are representative of this class and have been prepared as shown in Figure 32.[45, 60] Structures of these complexes were determined by either X-ray crystallography or NMR spectroscopy, which indicated that these complexes consist of one group 13 metal and one alkali metal, and two molecules of BINOL. These are efficient catalysts that afford Michael adducts in up to 99% ee (Figure 33). Moreover, tandem Michael–aldol reactions of cyclopentenone (**4**), diethyl methylmalonate (**111**), and hydrocinnamaldehyde (**2**) have been realized. In these cases, neither LLB, LSB, nor La–**17** gave satisfactory results.

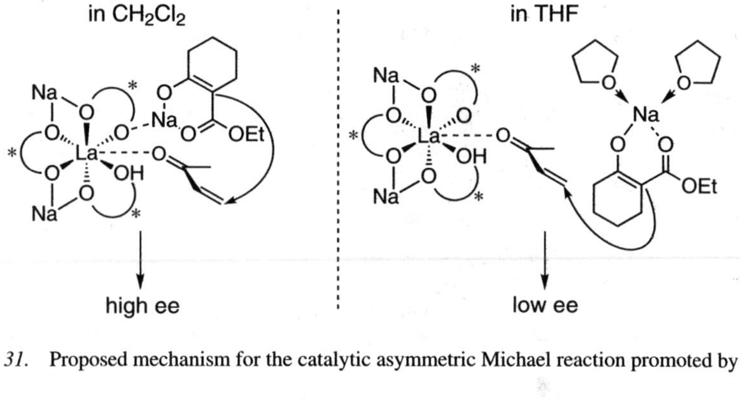

Figure 31. Proposed mechanism for the catalytic asymmetric Michael reaction promoted by LSB.

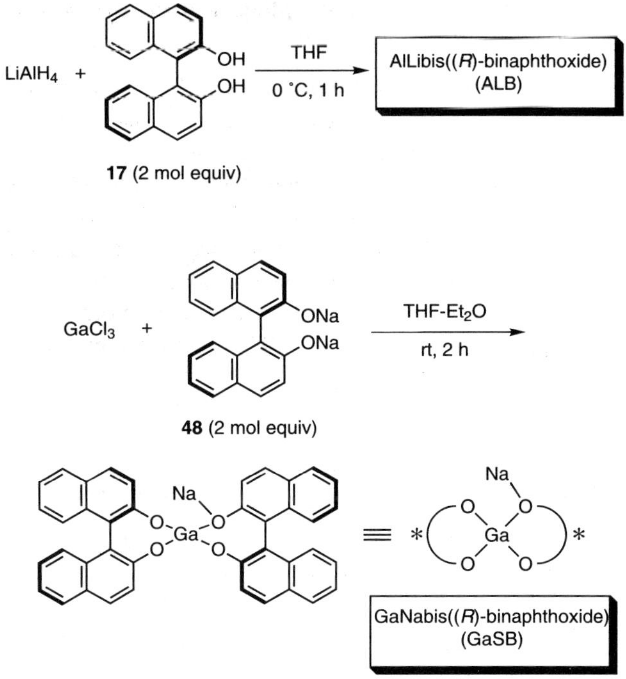

Figure 32. Preparation of AlLibis[(*R*)-binaphthoxide] (ALB) and GaNabis[(*R*)-binaphthoxide] (GaSB).

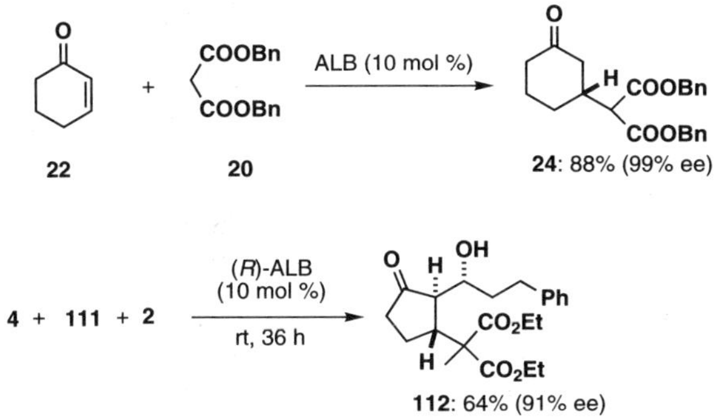

Figure 33. Three-component coupling reaction using hydrocinnamaldehyde (**2**) as an electrophile.

H. Catalytic, Asymmetric Hydrophosphonylations of Aldehydes

In recent years, chiral α-hydroxy phosphonates have attracted much attention due to their wide ranging biological activity and their usefulness as synthetic intermediates for other biologically important α-substituted phosphoryl compounds. Two groups have independently reported enantioselective hydrophosphonylations of aldehydes using LLB.[61–63] Because the purity of the LLB catalyst in these reports appears to be rather low, the reactions were reinvestigated. It was found that in the presence of 10 mol % of LLB, hydrophosphonylation of benzaldehyde (**54**) and cinnamaldehyde (**124**) with 1.3 equiv. of dimethyl phosphite in THF at –40 °C gave the corresponding α-hydroxy

Table 12
Catalytic, Asymmetric Hydrophosphonylation of Aldehydes

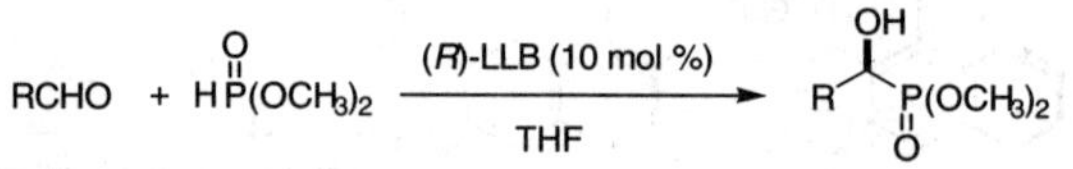

54: R = Ph
114: R = *p*-NO_2-Ph
116: R = *p*-Cl-Ph
118: R = *p*-Me-Ph
120: R = *p*-MeO-Ph
122: R = *p*-Me_2N-Ph
124: R = (*E*)-PhCH=CH
126: R = (*E*)-PhCH=C(CH_3)
128: R = (*E*)-$CH_3(CH_2)_2$CH=CH
130: R = $CH_3(CH_2)_4CH_2$

113: R = Ph
115: R = *p*-NO_2-Ph
117: R = *p*-Cl-Ph
119: R = *p*-Me-Ph
121: R = *p*-MeO-Ph
123: R = *p*-Me_2N-Ph
125: R = (*E*)-PhCH=CH
127: R = (*E*)-PhCH=C(CH_3)
129: R = (*E*)-$CH_3(CH_2)_2$CH=CH
131: R = $CH_3(CH_2)_4CH_2$

			LLB (–78 °C)		ALB (–40 °C)[a]	
Entry	Aldehyde	Product	Time (h)	Yield (%) [ee (%)]	Time (h)	Yield (%) [ee (%)]
1	**54**	**113**	8	88 (79)	51	95 (90)
2	**114**	**115**	12	85 (36)	40	85 (71)
3	**116**	**117**	8	80 (63)	38	80 (83)
4	**118**	**119**	7	93 (78)	92	82 (86)
5	**120**	**121**	9	83 (88)[a]	115	88 (78)
6	**120**	**121**	8	87 (93)		
7	**122**	**123**	12	88 (95)[a]	no reaction	
8	**122**	**123**	12	80 (95)		
9	**124**	**125**	8	90 (84)	83	85 (82)
10	**126**	**127**	8	94 (92)	61	47 (56)
11	**128**	**129**	8	63 (75)	39	53 (55)
12	**130**	**131**	8	88 (61)	41	95 (16)

[a]Aldehyde was added at one portion.

phosphonates **113** and **125** in 76% ee (79% yield) and 72% ee (78% yield), respectively (Table 12).[64] Shibuya[62] and Spilling[63] have reported that hydrophosphonylation of **54** using the LLB prepared in their groups gave the corresponding α-hydroxy phosphonates in less than 30% ee under similar conditions. It is noteworthy that with slow addition of the aldehydes, the enantiomeric excesses of both **113** and **125** at –40 °C were increased to 83% ee (73% yield) and 79% ee (88% yield), respectively. Furthermore, it was found that ALB is also effective at generating α-hydroxyphosphonates from aldehydes, especially ones with an electron-withdrawing substituent.[65] ALB and LLB can thus be used in a complementary manner for asymmetric catalysis of the hydrophosphonylation of a wide variety of aldehydes. Representative results obtained at –78 °C for LLB and –40 °C for ALB are summarized in Table 12. It is clear from the comparisons that while ALB is as selective as LLB in most cases, the reaction rates are significantly lower. The use of neither 6,6′-bis[(triethylsilyl)-ethynyl]BINOL nor the second-generation LLB catalyst produced a significant improvement.

The effects of the slow addition of the aldehydes on enantioselection can be best explained as follows. Heterobimetallic catalysts such as LLB are believed to activate both nucleophiles and electrophiles. For the hydrophosphonylation of comparatively unreactive aldehydes, the activated phosphite can react only with aldehydes that are precoordinated to lanthanum. However, in the case of reactive aldehydes such as **54** and **124**, the Li-activated phosphite may be able to undergo a competing reaction with the unactivated aldehyde. If such aldehydes are added in one portion, the ee of the product will thus be reduced. Slow addition of the

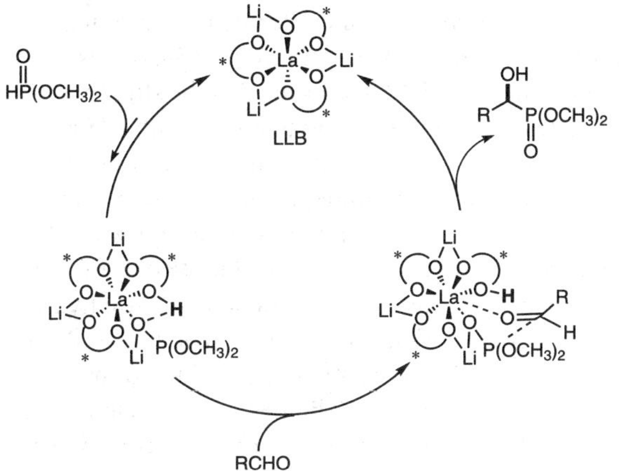

Figure 34. A proposed catalytic cycle for asymmetric hydrophosphonylation.

aldehyde, in contrast, has the effect of maximizing the ratio of activated to unactivated aldehyde present in the solution, by allowing time for the catalytic cycle to complete and regenerate the catalyst, thereby facilitating aldehyde activation. Reactive aldehydes should, therefore, be added slowly in order to avoid the side reaction that proceeds without activation of the aldehyde by LLB (Figure 34).

I. Catalytic, Asymmetric Hydrophosphonylation of Imines Promoted by the Lanthanoid–Potassium–BINOL Catalyst (LnPB)

α-Aminophosphonic acids are interesting compounds for use in the design of enzyme inhibitors. The concept of mimicking the tetrahedral transition structures of enzyme-mediated peptide bond hydrolyses has led to the successful design and synthesis of phosphonamide-containing peptides as a promising new class of proteinase inhibitors.[66, 67] Heterobimetallic complexes have also been successfully employed in catalytic, asymmetric hydrophosphonylation of imines using to produce enantiomerically enriched α-aminophosphonates in modest to high enantiomeric excesses.[68]

The catalytic, asymmetric hydrophosphonylation of imines was found to be highly dependent on the imine nitrogen substituent. For example, reaction of *N*-trityl imine **132** with dimethyl phosphite was found to proceeded at 60 °C only in the presence of a stoichiometric amount of the heterobimetallic complexes; LaK_3tris[(*R*)-binaphthoxide] (LPB) and LSB (Table 13). However, reaction of the modified benzhydryl imine **134** proceeded at room temperature in the presence of even a catalytic amount of the heterobimetallic complexes, in particular LPB. Thus treatment of imine **134** with 1.5 equiv. of dimethyl phosphite and 10 mol % of LPB in THF–toluene (1:7) at room temperature for 96 h gave α-aminophosphonate **135** with 96% ee in 70% yield. Moreover, with imine **136**, the use of 5 mol % of LPB gave **137** with 92% ee in 82% yield. It appears that the slightly weaker coordination of amine **137** to lanthanum with respect to that seen with **135** improves the catalyst turnover. As shown in Table 13, several imines were converted effectively to the corresponding optically active α-aminophosphonates. For the cinnamaldehyde-derived imine **144**, the GdK_3tris(binaphthoxide) complex (GdPB) gave better results. Finally, for the 4-methoxyaniline imine **146**, the use of PrPB gave the best results.

The proposed mechanism for this catalytic asymmetric hydrophosphonylation is shown in Figure 35. The first step of this reaction is the deprotonation of dimethyl phosphite by LPB to generate potassium dimethyl phosphite. This potassium phosphite immediately coordinates to a lanthanoid to give **I** due to the strong oxophilicity of lanthanoid metals. The complex **I** then reacts (in the stereochemistry-determining step) with an imine to give the potassium salt of the α-aminophosphonate. A proton-exchange reaction affords the product

Table 13
Catalytic, Asymmetric Hydrophosphonylation of Acyclic Imines

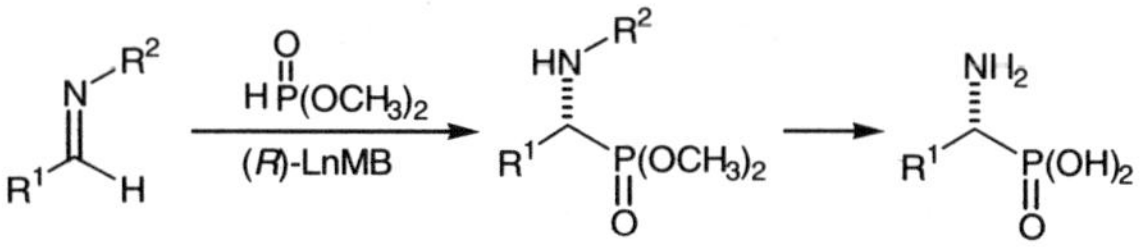

132: R^1 = Et, R^2 = Tr	**133**: R^1 = Et, R^2 = Tr
134: R^1 = *i*-Pr, R^2 = $CH(p\text{-}CH_3OC_6H_4)_2$	**135**: R^1 = *i*-Pr, R^2 = $CH(p\text{-}CH_3OC_6H_4)_2$
136: R^1 = *i*-Pr, R^2 = $CHPh_2$	**137**: R^1 = *i*-Pr, R^2 = $CHPh_2$
138: R^1 = CH_3, R^2 = $CHPh_2$	**139**: R^1 = CH_3, R^2 = $CHPh_2$
140: R^1 = Et, R^2 = $CHPh_2$	**141**: R^1 = Et, R^2 = $CHPh_2$
142: R^1 = C_5H_{11}, R^2 = $CHPh_2$	**143**: R^1 = C_5H_{11}, R^2 = $CHPh_2$
144: R^1 = (*E*)-PhCHCH, R^2 = $CHPh_2$	**145**: R^1 = (*E*)-PhCHCH, R^2 = $CHPh_2$
146: R^1 = *c*-C_6H_{11}, R^2 = $p\text{-}CH_3OC_6H_4$	**147**: R^1 = *c*-C_6H_{11}, R^2 = $p\text{-}CH_3OC_6H_4$
148: R^1 = $C_{12}H_{25}$, R^2 = $CHPh_2$	**149**: R^1 = $C_{12}H_{25}$, R^2 = $CHPh_2$

Entry	Imine	Cat. (mol %)	Conditions	Time (h)	Yield (%)	ee (%)
1	**132**	LSB (100)	A	18	**133**: 47	69
2	**134**	LSB (20)	B	18	**135**: 25	55
3	**134**	LPB (20)	B	18	**135**: 27	71
4	**134**	LPB (20)	C	21	**135**: 62	91
5	**134**	LSB (20)	C	21	**135**: 38	49
6	**134**	LLB (20)	C	21	**135**: 46	38
7	**134**	LPB (10)	C	96	**135**: 70	96
8	**136**	LPB (20)	C	63	**137**: 97	97
9	**136**	LPB (5)	C	143	**137**: 82	92
10	**138**	LPB (20)	C	70	**139**: 73	75
11	**140**	LPB (20)	C	63	**141**: 88	94
12	**142**	LPB (20)	C	63	**143**: 57	92
13	**144**	GdPB (20)	D	40	**145**: 86	66
14	**146**	PrPB (20)	C	68	**147**: 75	66
15	**146**	LPB (20)	C	89	**147**: 71	49
16	**148**	LPB (20)	C	84	**149**: 50	89

All reactions were performed in the presence of 5 equiv. of dimethyl phosphite except entries 7 and 8 (1.5 equiv.). Condition A: 60 °C in THF. Condition B: Room temperature in THF. Condition C: Room temperature in THF–toluene (1:7). Condition D: 50 °C in THF–toluene (1:7).

α-aminophosphonate and LPB, thereby completing the catalytic cycle. Although the mechanism is reasonable, it does not offer insight into the origin of stereoselection nor reveal why LPB is the best catalyst for asymmetric hydrophosphonylations of imines. The use of neither a second-generation LPB catalyst nor an LPB derivative derived from modified BINOLs gave better results.

The hydrophosphonylation of cyclic imines (thiazolines) has also been reported.[69] Representative results are shown in Table 14. For example, reaction of **150** with dimethyl phosphite in the presence of 20 mol % of LPB gave **151** in only modest yield (53%) and enantioselectivity (Table 14, entry 1). The rate of the

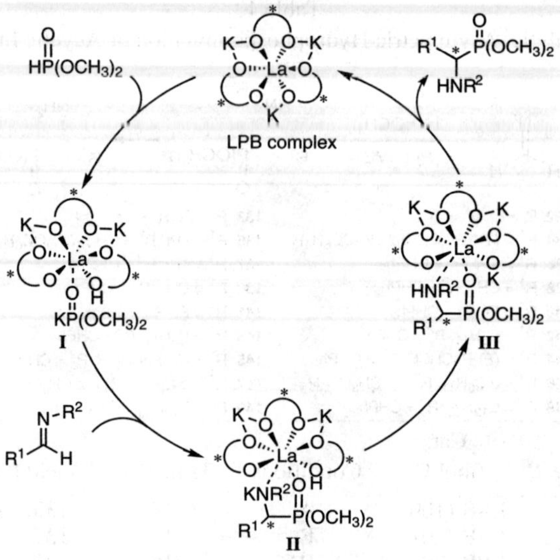

Figure 35. Proposed catalytic cycle for hydrophosphonylation of imines.

Table 14
LnPB-Catalyzed Asymmetric Hydrophosphonylation of **150**

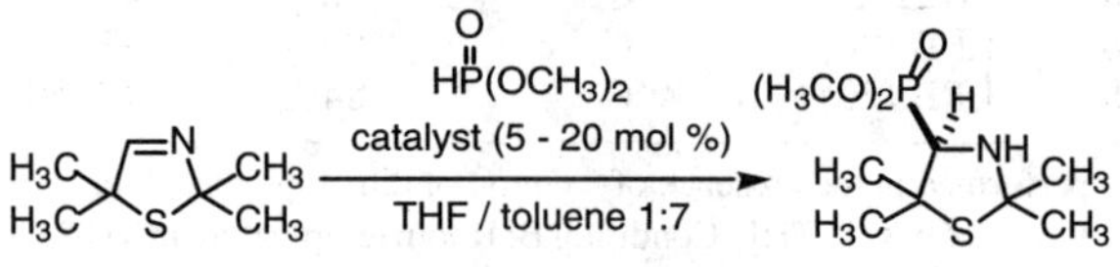

Entry	Catalyst (mol %)	Temp.	Time (h)	Yield (%)	ee (%)
1	LPB (20)	rt	144	53	61
2	LPB (20)	50 °C	50	55	64
3	PrPB (20)	50 °C	50	51	84
4	SmPB (20)	50 °C	40	97	93
5	GdPB (20)	50 °C	50	77	95
6	DyPB (20)	50 °C	50	76	97
7	YbPB (20)	50 °C	50	90	96
8	YbPB (20)	rt	50	86	98
9	YbPB (5)	50 °C	40	63	95

reaction could be improved by increasing the reaction temperature to 50 °C without adversely affecting the yield or selectivity (entry 2). Further optimization of the reaction using Sm, Gd, or Dy catalysts at 50 °C leads to a significant increase in enantioselectivity with ee values of up to 97% ee and good chemical yields being obtained (entries 4–6). In addition, the phosphite adduct (*R*)-**151** could be obtained with both excellent enantioselectivity (96% ee) and high chemical yield by using the (*R*)-YbPB complex as the heterobimetallic lanthanoid catalyst (entry 7). This catalyst could also effect the phosphite addition at room temperature; the product **151** was obtained with similarly high 86% chemical yield and in an excellent 98% ee (entry 8). These are among the highest enantioselectivities reported to date for catalytic, asymmetric hydrophosphonylation.

J. Direct Catalytic, Asymmetric Aldol Reaction

The aldol reaction is generally regarded as being among the most powerful carbon–carbon bond-forming reactions, and the development of a range of catalytic, asymmetric aldol-type reactions has thus proven to be a valuable contribution to asymmetric synthesis.[70–74] In all these asymmetric aldol-type reactions, however, preconversion of the ketone moiety to a more reactive species such as an enol silyl ether, enol methyl ether, or ketene silyl acetal is an unavoidable necessity.[75–86] Development of a direct catalytic, asymmetric aldol reaction, starting from aldehydes and *unmodified* ketones, is thus a worthwhile endeavor.[73, 87] Such reactions are well-documented enzymatic transformations,[88] with the fructose–1,6-bisphosphate and/or DHAP aldolases being characteristic examples. The mechanism of these reactions is thought to involve the cocatalysis by a Zn^{2+} cation and a basic functional group in the enzyme active site, with the latter abstracting a proton from a carbonyl compound and the former functioning as a Lewis acid so as to activate the other carbonyl component. These aldolases can thus be thought of as multifunctional catalysts displaying both Lewis acidity and Brønsted basicity, thus making possible efficient catalytic asymmetric aldol reactions under typically mild in vivo conditions. An analogous cooperative mode of action can be seen in reactions mediated by any of several heterobimetallic asymmetric catalysts having both Lewis acidity and Brønsted basicity, which have been discussed in this chapter.

The design for a direct catalytic asymmetric aldol reaction of aldehydes and unmodified ketones with bifunctional catalysts is shown in Figure 36. A Brønsted basic functionality (OM) in the heterobimetallic asymmetric catalyst (**I**) could deprotonate the α-proton of a ketone to generate the metal enolate (**II**), while at the same time a Lewis acidic functionality (LA) could activate an aldehyde to give (**III**), which would then react with the metal enolate (in a chelation-controlled fashion) in an asymmetric environment to afford a β-keto metal alkoxide (**IV**).

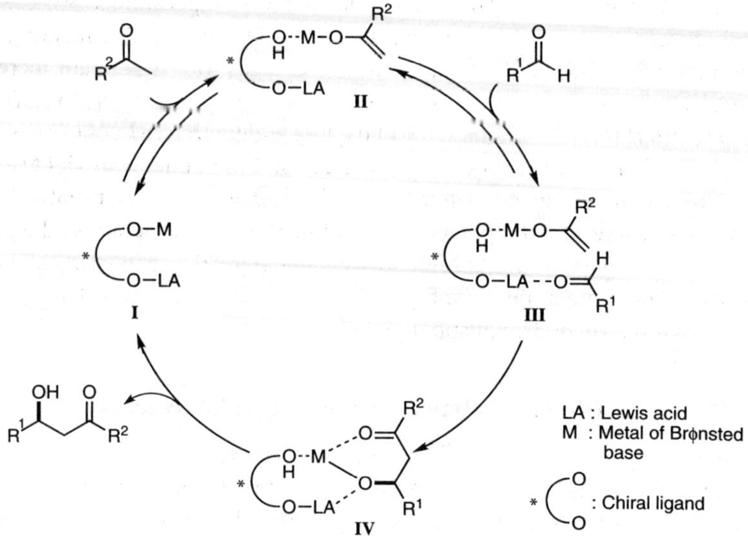

Figure 36. Catalytic cycle of direct catalytic asymmetric aldol reactions.

Proton exchange between the metal alkoxide moiety and an aromatic hydroxy proton or the α-proton of a ketone could then lead to the generation of an enantiomerically enriched aldol adduct and regeneration of the catalyst (**I**).

Several types of direct catalytic, asymmetric aldol reaction proceeded smoothly in the presence of 20 mol % (*R*)-LLB to afford aldol adducts with good to high enantiomeric purities.[89] Typical examples are shown in Table 15. Catalytic asymmetric aldol reactions of aldehydes with a single α-hydrogen also proceed without significant self-aldolization. Thus reaction of **52** with 8.0 equiv. of **153** in the presence of 20 mol % of LLB was found to give **159** with 44% ee in 72% yield, with no self-condensation products of **52** being detected (entry 7). For this example, other heterobimetallic catalysts were quite unsatisfactory (LSB: 16%, 13% ee; LPB: 49%, 0% ee; ALB: trace; LnLB: Ln = Pr, Sm, Gd, Dy, or Yb, low enantiomeric excesses). The reaction of isobutyraldehyde (**50**) with **153** also proceeded smoothly, to give **160** with 54% ee and in 59% yield (entry 8). The reaction of aldehydes that possess two α-hydrogens, for example, hydrocinnamaldehyde (**2**), proved more difficult; **161** was obtained with 52% ee; however, the yield was low (28%) due to the formation of self-condensation by-products (entry 9).

Aldol reactions that utilize acetone (**162**) as a starting material are generally difficult to control. However, in this case the reaction of aldehyde **157** and 10 equiv. of **162** with LLB was found to give **163** with 74% ee in 82% yield (entry 10). The reaction of **152** with 10 equiv. of **162** gives **164** with 73% ee in 53% yield

Table 15
Direct Catalytic, Asymmetric Aldol Reactions Promoted by (*R*)-LLB (20 mol %)

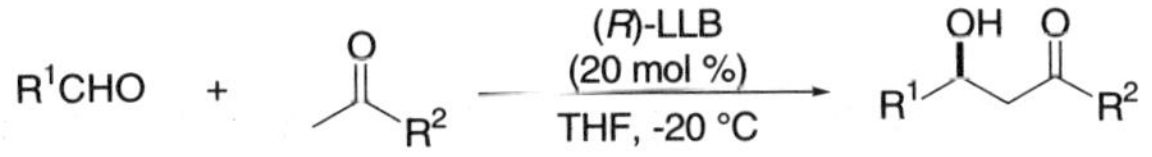

152: R^1 = *t*-Bu
157: R^1 = $PhCH_2C(CH_3)_2$
52: R^1 = *c*-C_6H_{11}
50: R^1 = *i*-Pr
2: R^1 = $Ph(CH_2)_2$

153: R^2 = Ph
155: R^2 = 1-naphthyl
162: R^2 = CH_3
165: R^2 = Et

154: R^1 = *t*-Bu, R^2 = Ph
156: R^1 = *t*-Bu , R^2 = 1-naphthyl
158: R^1 = $PhCH_2C(CH_3)_2$, R^2 = Ph
159: R^1 = *c*-C_6H_{11}, R^2 = Ph
160: R^1 = *i*-Pr, R^2 = Ph
161: R^1 = $Ph(CH_2)_2$, R^2 = Ph
163: R^1 = $PhCH_2C(CH_3)_2$, R^2 = CH_3
164: R^1 = *t*-Bu, R^2 = CH_3
166: R^1 = $PhCH_2C(CH_3)_2$, R^2 = Et

Entry	Aldehyde	Ketone	Product	Time (h)	Yield (%)	ee (%)
1[a]	**152**	**153** (5 equiv.)	**154**	88	43	89
2	**152**	**153** (5 equiv.)	**154**	88	76	88
3	**152**	**153** (1.5 equiv.)	**154**	135	43	87
4	**152**	**153** (10 equiv.)	**154**	91	81	91
5	**152**	**155** (8 equiv.)	**156**	253	55	76
6	**157**	**153** (7.4 equiv.)	**158**	87	90	69
7	**52**	**153** (8 equiv.)	**159**	169	72	44
8[b]	**50**	**153** (8 equiv.)	**160**	277	59	54
9	**2**	**153** (8 equiv.)	**161**	72	28	52
10	**157**	**162** (10 equiv.)	**163**	185	82	74
11	**152**	**162** (10 equiv.)	**164**	100	53	73
12	**157**	**165** (50 equiv.)	**166**	185	71	94

[a]Hydrated (*R*)-LLB by an addition of 1 mol equiv. of H_2O.
[b]The reaction was carried out at −30 °C.

(entry 11), and the reaction of **157** and 50 equiv. of 2-butanone (**165**) affords the adduct **166** with excellent ee (94%) in 71% yield (entry 12). Acetone (**162**) and 2-butanone (**165**) are widely used as solvents, and are much cheaper than the corresponding enol silyl ethers and/or methyl enol ethers, which are used as substrates in the catalytic asymmetric Mukaiyama–aldol reaction. The use of large excesses of the ketone can thus be justified in this particular case.

According to the hypothetical catalytic cycle (Figure 36), the lanthanum atom is believed to function as a Lewis acid and a lithium binaphthoxide moiety as a Brønsted base. The nature of the coordination of the aldehyde appears to be of first importance. This coordination provides activation of the aldehyde for reaction with the hypothetical LLB–enolate (**II**) (which on the basis of pK_a values can be present at most in low concentration), and also controls of the orientation of the aldehyde for enantioselective reaction. A ^{1}H NMR study also supports the existence of the coordination between aldehydes and the lanthanum cation.[89]

IV. PREVIOUS EXAMPLES OF ASYMMETRIC CATALYSIS WITH CHIRAL LANTHANOID COMPLEXES

A. Catalytic Asymmetric Diels–Alder-Type Reactions

The first report of the use of enantiomerically pure lanthanoid complexes as chiral Lewis acids for Diels–Alder-type reactions appeared in 1983.[90] As shown in Figure 37, in the presence of (+)-Eu(hfc)$_3$(tris(3-(heptafluoropropylhydroxy-methylene)-(+)-camphorato)europium(III)), usually utilized as a chiral shift reagent for NMR study, the hetero-Diels–Alder reaction of benzaldehyde (**54**) with compound **167** has been found to give **168** in 50% ee.

Since that early report, the enantioselectivity of catalytic asymmetric Diels–Alder reactions has been greatly improved. Kobayashi et al. have developed a quite efficient chiral lanthanoid catalyst for Diels–Alder reactions.[81, 91, 92] The catalyst is prepared from Yb(OTf)$_3$, (*R*)-BINOL, and a tertiary amine (Figure 38). Using the chiral Yb-catalyst, Diels–Alder adducts from **169** and cyclopentadiene (**170**) are obtained in high yields with high diastereo- and enantioselectivities. Representative examples are shown in Table 16. Interestingly, achiral additives greatly affect the enantiofacial selectivity. Therefore, both enantiomers are accessible from a single chiral source and different achiral additives.[91, 93] The proposed structure of the catalyst depicted in Figure 39 is based on spectroscopic analysis.[92]

Heterobimetallic catalysts have also been examined as Lewis acids in the asymmetric Diels–Alder reaction. Thus LaLi$_3$tris[(*R*)-6,6′-dibromobinaphthoxide] catalyzed the reaction of **172** with **170** to give **173** in high *endo* selectivity and in high enantiomeric excess (Figure 40). The same reaction promoted by LLB afforded **173** in 82% yield (*endo*:*exo* = 15:1) with 63% ee.[94]

As shown in Figure 41, Kobayashi's catalysis was also effective for asymmetric aza Diels–Alder reactions, and in this case 2.4 equiv. of DBU was used instead of 1,3,5-trimethylpiperidine. 2,6-Di-*tert*-butyl-4-methylpyridine appeared to be most effective additive.[95] Markó et al. have reported that these catalysts are also effective for asymmetric inverse electron-demand Diels–Alder reactions of 2-pyrone

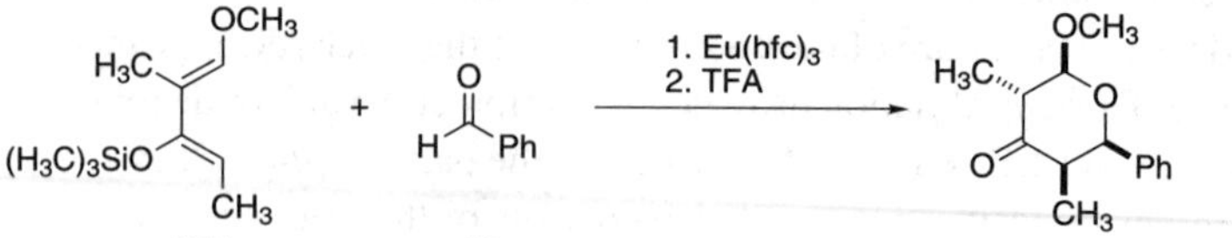

167 **54** **168**: 50% ee

Figure 37. Catalytic, asymmetric hetero-Diels–Alder reaction.

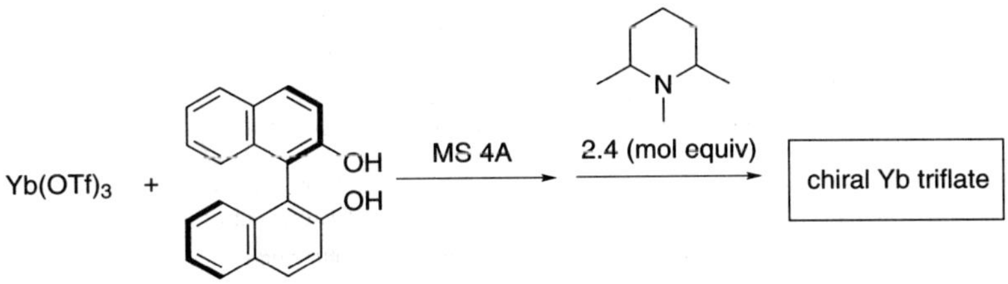

17 (1.2 mol equiv)

Figure 38. Preparation of a chiral Yb-catalyst for the Diels–Alder reaction.

Table 16
Catalytic, Asymmetric Diels–Alder Reaction

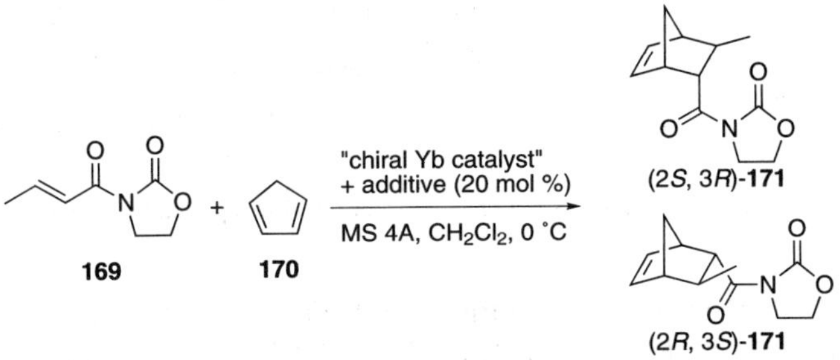

Entry	Additive	Yield (%)	*endo*/*exo*	2*S*,3*R*-**171**/2*R*,3*S*-**171**	(ee (%))
1	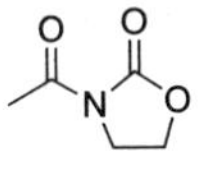	77	89/11	96.5/3.5	(93)
2	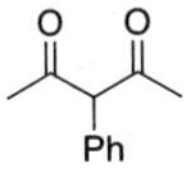	83	93/7	9.5/90.5	(81)

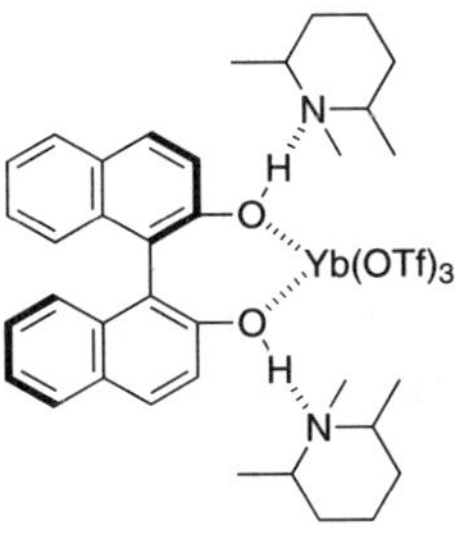

Figure 39. Proposed structure of the chiral Yb-catalyst.

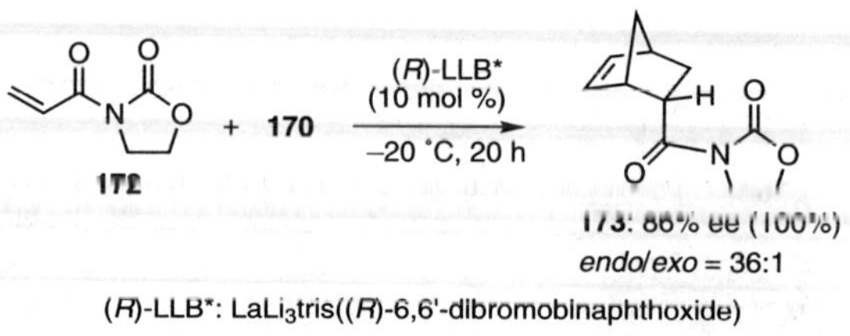

Figure 40. Catalytic, asymmetric Diels–Alder reaction promoted by La–Li–BINOL derivative complexes.

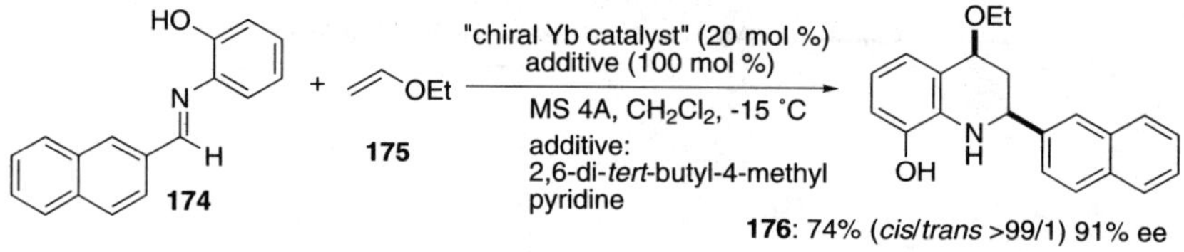

Figure 41. Catalytic, asymmetric aza-Diels–Alder reaction reported by Kobayashi et al.

derivative **177** with phenyl vinyl sulfide (**178**) (Figure 42).[96, 97] In addition to the above-mentioned Yb–BINOL catalysis, Sc–BINOL catalysis, a group 3 metal catalysis, is quite effective for asymmetric Diels–Alder reactions.[93]

Inanaga et al. have reported the preparation of the new chiral Yb(III) phosphate complex **180** (Figure 43), which is an efficient catalyst for asymmetric hetero-Diels–Alder reactions of **181** with aromatic aldehydes and/or α,β-unsaturated aldehydes. 2,6-Lutidine is found to be an effective additive to obtain high enantioselectivities.[98] Another asymmetric catalysis for hetero-Diels–Alder reaction was reported by Mikami et al.[99] An illustrative example is shown in Figure 44, wherein the interesting effects of water as an additive were observed to increase not only the enantioselectivity but also the chemical yield.

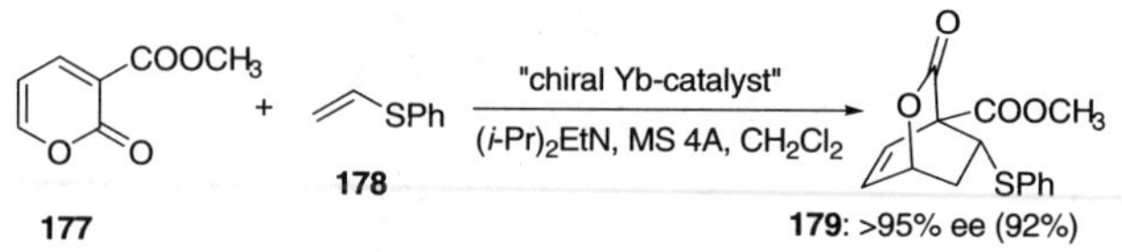

Figure 42. Catalytic, asymmetric Diels–Alder reaction of 2-pyrone derivatives.

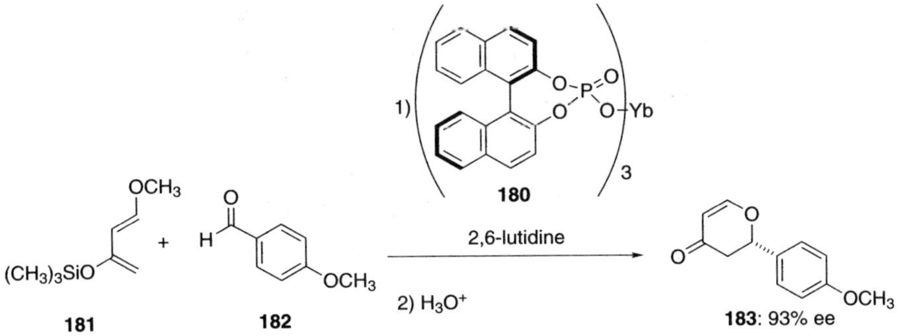

Figure 43. Catalytic, asymmetric hetero-Diels–Alder reaction reported by Inanaga et al.

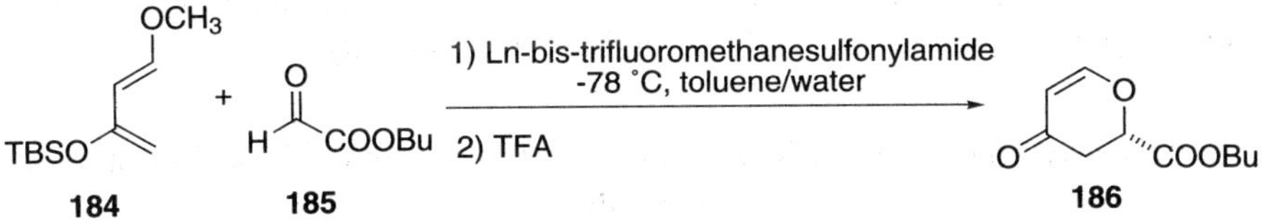

Figure 44. Catalytic, asymmetric hetero-Diels–Alder reaction reported by Mikami et al.

B. Catalytic, Asymmetric Mukaiyama Aldol Reactions

As discussed in Section III.J, in general, catalytic asymmetric aldol reactions have been studied using enol silyl ethers, enol methyl ethers, or ketene silyl acetals as a starting material. So far several types of chiral catalysis have been reported.[75–85] The chiral lanthanoid complex prepared from $Ln(OTf)_3$ and a chiral sulfonamide ligand was effective in promoting an asymmetric Mukaiyama aldol reaction with a ketene silyl acetal.[86] The preparation of the catalyst and a representative reaction are shown in Figure 45.

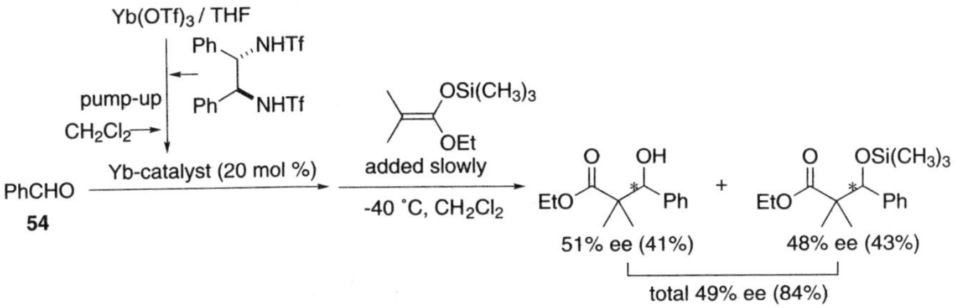

Figure 45. Catalytic, asymmetric Mukaiyama aldol reaction promoted by the chiral Yb complex.

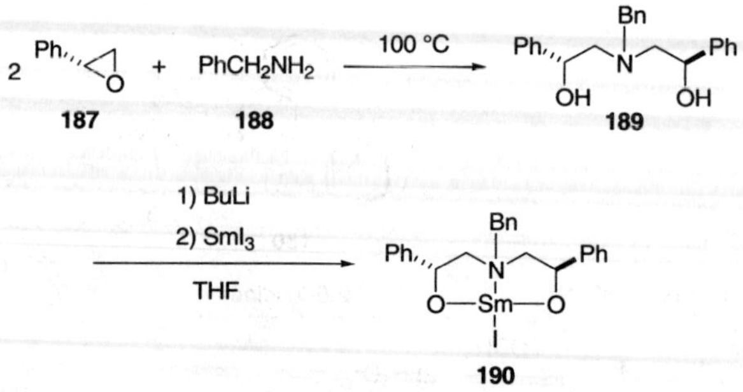

Figure 46. Catalyst preparation for an asymmetric Meerwein–Pondorf–Verley reduction.

Fujiwara has reported a unique chiral lanthanoid(II) alkoxide–promoted asymmetric Mukaiyama aldol reaction.[38] Stoichiometric amounts of the chiral alkoxide, however, were required for good enantioselectivity.

C. Catalytic, Asymmetric Reduction of Ketones with a Chiral Lanthanoid Complex

An intriguing chiral samarium complex for the Meerwein–Pondorf–Verley (MPV) reduction of ketones has been reported by Evans.[100] The soluble catalyst, prepared as indicated in Figure 46, promoted the asymmetric MPV reduction of aryl methyl ketones in up to 97% ee with as little as 5 mol % loading (Figure 47).

A new type of lanthanoid complex, prepared from BINOL and $SmCl_3$, served as an asymmetric catalyst for MPV reduction of aryl methyl ketones in the presence of molecular sieves. Moderate enantioselectivity was obtained.[101]

A chiral lanthanoid complex, which was prepared similarly to La–(*R*)-**17**,[7,23] is an effective catalyst for asymmetric reduction of ketones.[102] With 10 mol % of the catalyst, borane reduction of ketones proceeds very smoothly to give alcohols in up to 62% ee (Figure 48).

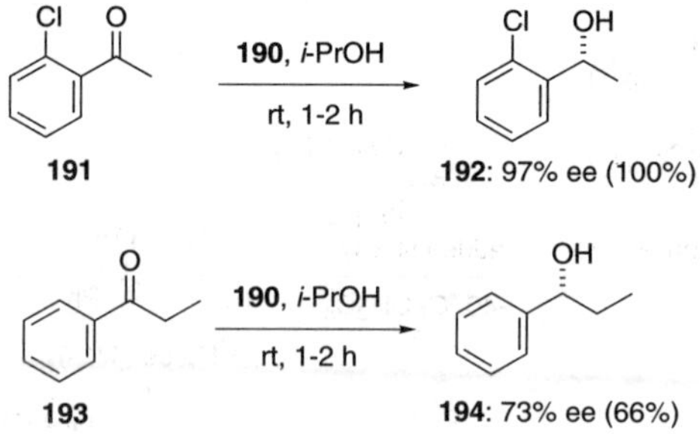

Figure 47. Catalytic, asymmetric MPV reduction.

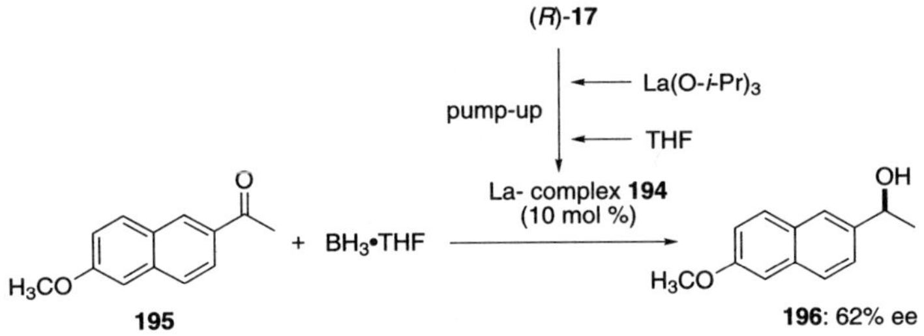

Figure 48. Catalytic, asymmetric borane reduction with the new chiral La complex.

D. Chiral Organolanthanoid Complexes for Olefin Hydrogenation, Hydroamination/Cyclization, and Hydrosilylation

A series of fascinating asymmetric organolanthanoid catalysts for enantioselective additions to olefins have been developed by Marks. Their first enantioselective hydrogenation in the presence of precatalyst **197** gave the desired alkanes in good to excellent enantiomeric excesses with high turnover frequencies (Figure 49).[103] This catalytic system has been extended to asymmetric hydroamination/cyclization of amino olefins and olefin hydrosilylation.[104–106] The mechanism for the catalytic asymmetric hydroamination/cyclization has been proposed as shown in Figure 50.

E. Other Catalytic Asymmetric Reactions Using Chiral Lanthanoid Complexes

Diastereotopic differentiation of two leaving groups has been achieved for the first time by the reaction of (dichloromethyl)borates with BuLi in the presence of $Yb(OTf)_3$ and the chiral bisoxazoline ligand **203**.[108] As shown in Figure 51, pinacol dichloromethyl boronate (**204**) gave **205** in up to 88% ee. The reaction can be said to be catalytic, as the use of 0.5 equiv. of chiral ligand **203** and 0.2 equiv. of $Yb(OTf)_3$ provided **205** in 55% ee and 86% yield.

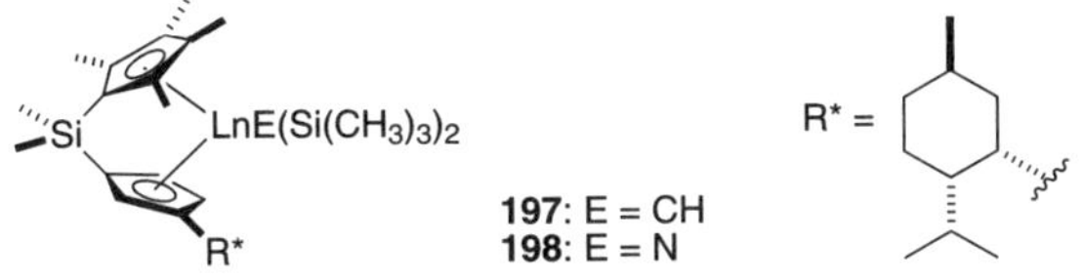

Figure 49. Chiral organolanthanoid complexes.

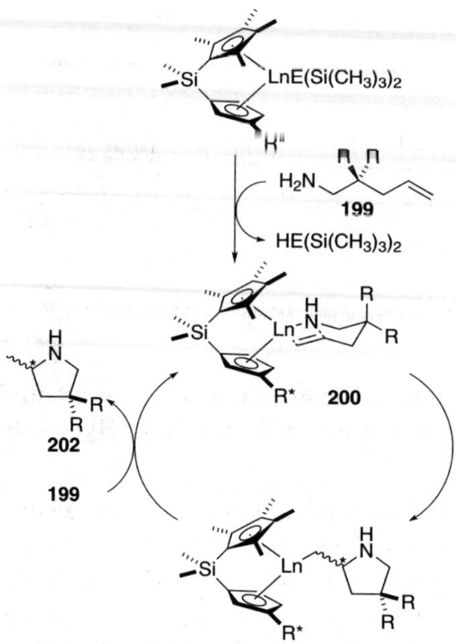

Figure 50. Proposed mechanism for catalytic asymmetric hydroamination/cyclization.

V. SUMMARY

Conceptually new multifunctional asymmetric two-center catalysts, such as the Ln–BINOL derivative, LnMB, AMB, and GaMB have been developed. These catalysts function both as Brønsted bases and as Lewis acids, making possible various catalytic, asymmetric reactions in a manner analogous to enzyme catalysis. Several such catalytic asymmetric reactions are now being investigated for potential industrial applications. Recently, the catalytic enantioselective opening of meso epoxides with thiols in the presence of a heterobimetallic complex has

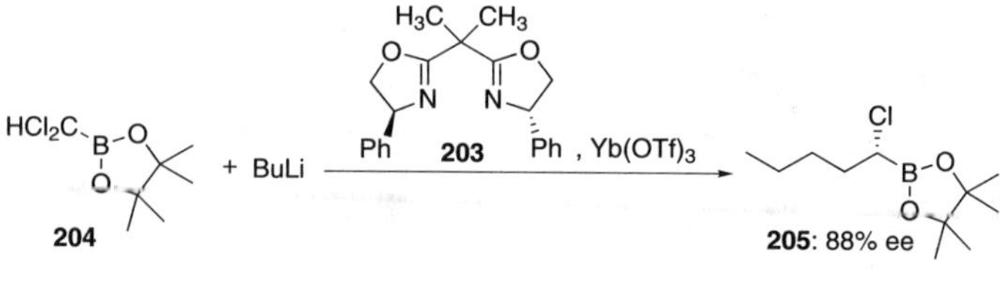

Figure 51. Enantiotopic differentiation of a prochiral compound by chiral Lewis acids.

been reported. This new complex, a gallium derivative GaLibis(binaphthoxide) (GaLB), shows excellent catalytic activity and high enantiomeric excesses for a range of cyclic and acyclic meso epoxides.[109] The successful development of the heterobimetallic concept has opened up a new field in asymmetric catalysis. In addition, recent progress in chiral lanthanoid catalysis has allowed for the development of a number of asymmetric reactions, in which lanthanoid complexes act primarily as Lewis acids.

REFERENCES

1. Shibasaki, M.; Sasai, H.; Arai, T. *Angew. Chem. Int. Ed. Engl.* **1997**, *36*, 1236–1256.
2. Steinhagen, H.; Helmchen, G. *Angew. Chem. Int. Ed. Engl.* **1996**, *35*, 2339–2342.
3. Sasai, H.; Kirio, Y.; Shibasaki, M. *J. Org. Chem.* **1990**, *55*, 5306–5308.
4. Beck, A. K.; Bastani, B.; Plattner, D. A.; Petter, W.; Seebach, D.; Braunschweiger, H.; Gysi, P.; Vecchia, L. L. *Chimia* **1991**, *45*, 238–244.
5. Sasai, H.; Suzuki, T.; Arai, S.; Arai, T.; Shibasaki, M. *J. Am. Chem. Soc.* **1992**, *114*, 4418–4420.
6. Henry, L. *C. R. Hebd. Seances Acad. Sci.* **1895**, *120*, 1265.
7. Sasai, H.; Arai, T.; Shibasaki, M. *J. Am. Chem. Soc.* **1994**, *116*, 1571–1572.
8. Bonadies, F.; Lattanzi, A.; Orelli, L. R.; Pesci, S.; Scettri, A. *Tetrahedron Lett.* **1993**, *34*, 7649–7650.
9. Hanson, R. M.; Sharpless, K. B. *J. Org. Chem.* **1986**, *51*, 1922–1925.
10. Zhang, W.; Loebach, J. L.; Wilson, S. R.; Jacobsen, E. N. *J. Am. Chem. Soc.* **1990**, *112*, 2801–2803.
11. Irie, R.; Noda, K.; Ito, Y.; Matsumoto, N.; Katsuki, T. *Tetrahedron Lett.* **1990**, *31*, 7345–7348.
12. Yamada, T.; Imagawa, K.; Nagata, T.; Mukaiyama, T. *Chem. Lett.* **1992**, 2231–2234.
13. Juliá, S.; Masana, J.; Vega, J. C. *Angew. Chem. Int. Ed. Engl.* **1980**, *19*, 929–931.
14. Kroutil, W.; Mayon, P.; Lasterra-Sánchez, M. E.; Maddrell, S. J.; Roberts, S. M.; Thornton, S. R.; Todd, C. J.; Tüter, M. *Chem. Commun.* **1996**, 845–846 and references cited therein.
15. Helder, R.; Hummelen, J. C.; Laane, R. W. P. M.; Wiering, J. S.; Wynberg, H. *Tetrahedron Lett.* **1976**, 1831–1834.
16. Colonna, S.; Gaggero, N.; Manfredi, A.; Spadoni, M.; Casella, L.; Carrea, G.; Pasta, P. *Tetrahedron* **1988**, *44*, 5169–5178.
17. Colonna, S.; Manfredi, A.; Annunziata, R.; Gaggero, N. *J. Org. Chem.* **1990**, *55*, 5862–5866.
18. Baccin, C.; Gusso, A.; Pinna, F.; Strukul, G. *Organometallics* **1995**, *14*, 1161–1167.
19. Kumar, A.; Bhakuni, V. *Tetrahedron Lett.* **1996**, *37*, 4751–4754.

20. Elston, C. L.; Jackson, R. F. W.; MacDonald, S. J. F.; Murray, P. J. *Angew. Chem. Int. Ed. Engl.* **1997**, *36*, 410–412.
21. Jacobsen, E. N.; Deng, L.; Furukawa, Y.; Martinez, L. E. *Tetrahedron* **1994**, *50*, 4323–4334.
22. Enders, D.; Zhu, J.; Raabe, G. *Angew. Chem. Int. Ed. Engl.* **1996**, *35*, 1725–1728.
23. Bougauchi, M.; Watanabe, S.; Arai, T.; Sasai, H.; Shibasaki, M. *J. Am. Chem. Soc.* **1997**, *119*, 2329–2330.
24. Sasai, H.; Suzuki, T.; Itoh, N.; Tanaka, K.; Date, T.; Okamura, K.; Shibasaki, M. *J. Am. Chem. Soc.* **1993**, *115*, 10372–10373.
25. Sasai, H.; Arai, T.; Satow, Y.; Houk, K. N.; Shibasaki, M. *J. Am. Chem. Soc.* **1995**, *117*, 6194–6198.
26. Sasai, H.; Suzuki, T.; Itoh, N.; Shibasaki, M. *Tetrahedron Lett.* **1993**, *34*, 851–854.
27. Sasai, H.; Watanabe, S.; Shibasaki, M. *Enantiomer* **1997**, *2*, 267–271.
28. Takaoka, E.; Yoshikawa, N.; Yamada, Y. M. A.; Sasai, H.; Shibasaki, M. *Heterocycles* **1997**, *46*, 157–163.
29. Sasai, H.; Suzuki, T.; Itoh, N.; Arai, S.; Shibasaki, M. *Tetrahedron Lett.* **1993**, *34*, 2657–2660.
30. Seebach, D.; Colvin, E. W.; Lehr, F.; Weller, T. *Chimia* **1979**, *33*, 1–18.
31. Rosini, G. In *Comprehensive Organic Synthesis*; Pergamon: Oxford, 1991; Vol. 2, pp. 321–340.
32. Sasai, H.; Itoh, N.; Suzuki, T.; Shibasaki, M. *Tetrahedron Lett.* **1993**, *34*, 855–858.
33. Sasai, H.; Yamada, Y. M. A.; Suzuki, T.; Shibasaki, M. *Tetrahedron* **1994**, *50*, 12313–12318.
34. Sasai, H.; Suzuki, T.; Itoh, N.; Shibasaki, M. *Applied Organometallic Chemistry* **1995**, *9*, 421–426.
35. Sasai, H.; Kim, W.-S.; Suzuki, T.; Mitsuda, M.; Ohashi, T.; Shibasaki, M. *Tetrahedron Lett.* **1994**, *35*, 6123–6126.
36. Mimoto, T.; Imai, J.; Kisanuki, S.; Enomoto, H.; Hattori, N.; Akaji, K.; Kiso, Y. *Chem. Pharm. Bull.* **1992**, *40*, 2251–2253.
37. Kageyama, S.; Mitsuto, T.; Murakawa, Y.; Nomizu, M.; Ford, H., Jr.; Shirasaka, T.; Gulnik, S.; Erickson, J.; Takada, K.; Hayashi, H.; Broder, S.; Kiso, Y.; Mitsuya, H. *Antimicrob. Agents Chemother.* **1993**, *37*, 810–817.
38. Makioka, Y.; Nakagawa, I.; Taniguchi, Y.; Takaki, K.; Fujiwara, Y. *J. Org. Chem.* **1993**, *58*, 4771–4774.
39. Maruoka, K.; Itoh, T.; Araki, Y.; Shirasaka, T.; Yamamoto, H. *Bull. Chem. Soc. Jpn.* **1988**, *61*, 2975–2976.
40. Yamamoto, K.; Noda, K.; Okamoto, Y. *J. Chem. Soc. Chem. Commun.* **1985**, 1065–1066.
41. Sasai, H.; Tokunaga, T.; Watanabe, S.; Suzuki, T.; Itoh, N.; Shibasaki, M. *J. Org. Chem.* **1995**, *60*, 7388–7389.
42. Seebach, D. *Angew. Chem. Int. Ed. Engl.* **1990**, *29*, 1320–1367

43. Iseki, K.; Oishi, S.; Sasai, H.; Shibasaki, M. *Tetrahedron Lett.* **1996**, *37*, 9081–9084.
44. Schwartz, G. K.; Jiang, J.; Kelsen, D.; Albino, A. P. *J. Nat. Cancer Inst.* **1993**, *85*, 402–407.
45. Arai, T.; Yamada, Y. M. A.; Yamamoto, N.; Sasai, H.; Shibasaki, M. *Chem. Eur. J.* **1996**, *2*, 1368–1372.
46. Young, M.; Pan, W.; Wiesner, J.; Bullough, D.; Browne, G.; Balow, G.; Potter, S.; Metzner, K.; Mullane, K. *Drug Dev. Res.* **1994**, *32*, 19–28.
47. Hammond, H. K.; McKirnan, M. D. *J. Am. Coll. Cardiol.* **1994**, *23*, 475–482.
48. Ho, T.-L. *Tandem Organic Reactions*; Wiley: New York, 1992.
49. *Cascade Reactions*; Grigg, R., Ed.; *Tetrahedron Symposium*; **1997**, *52(35)*.
50. Tietze, L. F.; Beifuss, U. *Angew. Chem. Int. Ed. Engl.* **1993**, *32*, 131.
51. Sasai, H.; Hiroi, M.; Yamada, Y. M. A.; Shibasaki, M. *Tetrahedron Lett.* **1997**, *38*, 6031.
52. Schick, H.; Roatsch, B.; Schwarz, H.; Hauser, A.; Schwarz, S. *Liebigs Ann. Chem.* **1992**, 419.
53. Oare, D.; Heathcock, C. H. In *Topics in Stereochemistry*; Eliel, E. L.; Wilen, S. H., Eds.; Wiley: New York, 1989; Vol. 19, pp. 227–407.
54. Oare, D.; Heathcock, C. H. In *Topics in Stereochemistry*; Eliel, E. L.; Wilen, S. H., Eds.; Wiley: New York, 1991; Vol. 20, pp. 87–170.
55. Rappé, A. K.; Casewit, C. J.; Colwell, K. S.; Goddard, I. W. A.; Skiff, W. M. *J. Am. Chem. Soc.* **1992**, *114*, 10024–10035.
56. Casewit, C. J.; Colwell, K. S.; Rappé, A. K. *J. Chem. Soc.* **1992**, *114*, 10035–10046.
57. Casewit, C. J.; Colwell, K. S.; Rappé, A. K. *J. Am. Chem. Soc.* **1992**, *114*, 10046–10053.
58. Rappé, A. K.; Colwell, K. S.; Casewit, C. J. *Inorg. Chem.* **1993**, *32*, 3438–3450.
59. Sasai, H.; Emori, E.; Arai, T.; Shibasaki, M. *Tetrahedron Lett.* **1996**, *37*, 5561–5564.
60. Arai, T.; Sasai, H.; Aoe, K.-I.; Okamura, K.; Date, T.; Shibasaki, M. *Angew. Chem. Int. Ed. Engl.* **1996**, *35*, 104–106.
61. Yokomatsu, T.; Yamagishi, T.; Shibuya, S. *Tetrahedron: Asymmetry* **1993**, *4*, 1783–1784.
62. Yokomatsu, T.; Yamagishi, T.; Shibuya, S. *J. Chem. Soc., Perkin Trans. 1* **1997**, 1527–1533.
63. Rath, N. P.; Spilling, C. D. *Tetrahedron Lett.* **1994**, *35*, 227–230.
64. Sasai, H.; Bougauchi, M.; Arai, T.; Shibasaki, M. *Tetrahedron Lett.* **1997**, *38*, 2717–2720.
65. Arai, T.; Bougauchi, M.; Sasai, H.; Shibasaki, M. *J. Org. Chem.* **1996**, *61*, 2926–2927.
66. Burke, T. R.; Barchi, J. J.; George, C.; Wolf, G.; Shoelson, S. E.; Yan, X. *J. Med. Chem.* **1995**, *38*, 1386–1396.
67. Kafarski, P.; Lejczak, B. *Phosphorus, Sulfur Silicon Relat. Elem.* **1991**, *63*, 193–215.
68. Sasai, H.; Arai, S.; Tabara, Y.; Shibasaki, M. *J. Org. Chem.* **1995**, *60*, 6656–6657.

69. Gröger, H.; Martens, J.; Sasai, H.; Saida, Y.; Shibasaki, M. *Tetrahedron Lett.* **1996**, *37*, 9291–9292.
70. Nelson, S. G. *Tetrahedron: Asymmetry* **1997**, *9*, 357–389.
71. Braun, M. In *Stereoselective Synthesis, Methods of Organic Chemistry (Houben-Weyl)*, E21 ed.; Helmchen, G.; Hoffmann, R., Mulzer, J.; Schaumann, E., Eds.; Thieme: Stuttgart, 1996; Vol. 3, pp. 1730–1736.
72. Franklin, A. S.; Paterson, I. *Contemporary Org. Syn.* **1994**, *1*, 317–416.
73. Sawamura, M.; Ito, Y. In *Catalytic Asymmetric Synthesis* ; VCH: New York, 1993, pp. 367–388.
74. Bach, T. *Angew. Chem. Int. Ed. Engl.* **1994**, *33*, 417.
75. Denmark, S. E.; Winter, S. B. D.; Su, X.; Wong, K.-T. *J. Am. Chem. Soc.* **1996**, *118*, 7404–7405.
76. Evans, D. A.; Murry, J. A.; Kozlowski, M. C. *J. Am. Chem. Soc.* **1996**, *118*, 5814–5815.
77. Singer, R. A.; Carreira, E. M. *J. Am. Chem. Soc.* **1995**, *117*, 12360–12361.
78. Keck, G. E.; Krishnamurthy, D. *J. Am. Chem. Soc.* **1995,** *117*, 2363–2364.
79. Sodeoka, M.; Ohrai, K.; Shibasaki, M. *J. Org. Chem.* **1995**, *60*, 2648–2649.
80. Mikami, K.; Matsukawa, S. *J. Am. Chem. Soc.* **1994**, *116*, 4077–4078.
81. Kobayashi, S.; Uchiro, H.; Shiina, I.; Mukaiyama, T. *Tetrahedron* **1993**, *49*, 1761–1772.
82. Corey, E. J.; Cywin, C. L.; Roper, T. D. *Tetrahedron Lett.* **1992**, *33*, 6907–6910.
83. Parmee, E. R.; Tempkin, O.; Masamune, S.; Abiko, A. *J. Am. Chem. Soc.* **1991**, *113*, 9365–9366.
84. Kiyooka, S.; Kaneko, Y.; Kume, K. *Tetrahedron Lett.* **1992**, *33*, 4927–4930.
85. Furuta, K.; Maruyama, T.; Yamamoto, H. *J. Am. Chem. Soc.* **1991**, *113*, 1041–1042.
86. Uotsu, K.; Sasai, H.; Shibasaki, M. *Tetrahedron: Asymmetry* **1995**, *6*, 71–74.
87. Noyori, R. *Asymmetric Catalysis in Organic Synthesis*; John Wiley & Sons, Inc.: New York, 1994.
88. Fessner, W.-D.; Schneider, A.; Held, H.; Sinerius, G.; Walter, C.; Hixon, M.; Schloss, J. V. *Angew. Chem. Int. Ed. Engl.* **1996**, *35*, 2219–2221.
89. Yamada, Y. M. A.; Yoshikawa, N.; Sasai, H.; Shibasaki, M. *Angew. Chem. Int. Ed. Engl.* **1997**, *36*, 1871–1873.
90. Bednarski, M.; Danishefsky, S. *J. Am. Chem. Soc.* **1983**, *105*, 3716–3717.
91. Kobayashi, S.; Ishitani, H. *J. Am. Chem. Soc.* **1994**, *116*, 4083–4084.
92. Kobayashi, S.; Ishitani, H.; Hachiya, I.; Araki, M. *Tetrahedron* **1994**, *50*, 11623–11636.
93. Kobayashi, S.; Araki, M.; Hachiya, I. *J. Org. Chem.* **1994**, *59*, 3758–3759.
94. Morita, T.; Arai, T., Sasai, H.; Shibasaki, M. *Tetrahedron: Asymmetry* **1998**, *9*, 1445–1450.
95. Ishitani, H.; Kobayashi, S. *Tetrahedron Lett.* **1996**, *37*, 7357–7360.

96. Markó, I. E.; Evans, G. R.; Declercq, J.-P. *Tetrahedron* **1994**, *50*, 4557–4574.
97. Markó, I. E.; Chellé-Regnaut, I.; Leroy, B.; Warriner, S. L. *Tetrahedron Lett.* **1997**, *38*, 4269–4272.
98. Hanamoto, T.; Furuno, H.; Sugimoto, Y.; Inanaga, J. *Synlett* **1997**, 79–80.
99. Mikami, K.; Kotera, O.; Motoyama, Y.; Sakaguchi, H. *Synlett* **1995**, 975–977.
100. Evans, D. A.; Nelson, S. G.; Gagné, M. R.; Muci, A. R. *J. Am. Chem. Soc.* **1993**, *115*, 9800–9801.
101. Ding, Z.-B.; Cheng, K.-J.; Wu, S.-H. *Chin. J. Chem.* **1996**, *14*, 561–564.
102. Zhang, F.-Y.; Yip, C.-W.; Chan, A. S. C. *Tetrahedron: Asymmetry* **1996**, *7*, 2463–2466.
103. Conticello, V. P.; Brard, L.; Giardello, M. A.; Tsuji, Y.; Sabat, M.; Stern, C. L.; Marks, T. J. *J. Am. Chem. Soc.* **1992**, *114*, 2761–2762.
104. Gagné, M. R.; Stern, C.; Marks, T. J. *J. Am. Chem. Soc.* **1992**, *114*, 275–294.
105. Gagné, M. R.; Brard, L.; Conticello, V. P.; Giardello, M. A.; Stern, C. L.; Marks, T. J. *Organometallics* **1992**, *11*, 2003–2005.
106. Giardello, M. A.; Conticello, V. P.; Brard, L.; Gagné, M. R.; Marks, T. J. *J. Am. Chem. Soc.* **1994**, *116*, 10241–10254.
107. Fu, P.-F.; Brard, L.; Li, Y.; Marks, T. J. *J. Am. Chem. Soc.* **1995**, *117*, 7157–7168.
108. Jadhav, P. K.; Man, H.-W. *J. Am. Chem. Soc.* **1997**, *119*, 846–847.
109. Iida, T.; Yamamoto, N.; Sasai, H.; Shibasaki, M. *J. Am. Chem. Soc.* **1997**, *119*, 4783–4784.

[illegible] Science 1999 [illegible]

[illegible]

Asymmetric Amplification

DAVID R. FENWICK† AND HENRI B. KAGAN*

Institut de Chimie Moléculaire d'Orsay,
Université Paris-Sud, 91405 Orsay, France

I. INTRODUCTION

The development of modern organic chemistry increasingly involves single enantiomers of chiral compounds as starting materials for synthesis or as the desired target. Pharmaceuticals, agrochemicals, fragrances, and materials are the main areas involved with the chemistry of enantiopure compounds. A fascinating

†Present address: Pfizer, Sandwich, Kent, CT13 9NJ (UK)

Topics in Stereochemistry, Volume 22, Edited by Scott E. Denmark.
ISBN 0-471-25316-2

question, however, remains open to speculation—how and why did biomolecules evolve as single enantiomers of a defined absolute configuration? Enantiopure compounds may, in principle, be obtained by the use of a chiral auxiliary (in stoichiometric or catalytic amounts) that will control the resolution of a racemic mixture or the transformation of achiral compounds. Classically the chiral auxiliary is provided by the "chiral pool" as an enantiopure compound [alkaloids, α-amino acids, (−)-menthol, *etc.*] although some natural products may sometimes be of various enantiomeric excesses (*e.g.*, α-pinene 70–90% ee according to its origin). In this chapter will be considered some cases where enantiomerically impure chiral auxiliaries are involved in asymmetric synthesis or kinetic resolution. One expects a degradation in the stereoselectivity of the chiral auxiliary because its enantiomeric excess (ee_{aux}) is inferior to 100%. However, the results are not always as expected—sometimes the ee of the product can be higher or lower than that anticipated from using a chiral auxiliary of given ee. If the ee of the product of an asymmetric synthesis ($ee_{product}$) or ee of recovered substrate in a kinetic resolution (ee_s) is higher than expected (based on the ee_{aux}), one says that there is an *asymmetric amplification*.[1, 2] If the ee is lower than expected, this deviation is termed an *asymmetric diminution* (or *depletion*).[2] In this chapter we will concentrate on *asymmetric amplification*, an expression that needs to be properly defined (vide infra).

II. DEFINITION OF ASYMMETRIC AMPLIFICATION

To amplify is defined in dictionaries as *to increase* or *to enlarge*. Its meaning is not so different from *multiplication*, which is defined as "the action of increasing the number or the amount of things belonging to the same species."[3] The term *asymmetric amplification* (or *amplification of chirality*) has been used in the literature in various ways (for some examples see Refs. 5–8), usually to describe processes giving rise to an increase in the amount of enantioenriched material, or to the enhancement of enantiomeric excess (or enantiomeric ratio),[9] or to a combination of the both. Here we will consider that there is an amplification of chirality when a reaction provides a compound with an enantiomeric excess higher than expected. The convenient but less precise expression *asymmetric amplification* is now widely used and will also be used in this chapter.[12]

The size of an asymmetric amplification is difficult to quantify. The most convenient way is to compare the enantiomeric excess (ee) or enantiomeric ratio (er) of a product generated in an experiment to the ee (or er) expected on the basis of classical laws. Thus if a product of 40% ee is expected from an asymmetric synthesis but a product with 90% ee is obtained, the amplification factor is 2.5. If this calculation is based upon enantiomeric ratio (2.33 and 8.26 for 40% ee and 90% ee, respectively), the amplification factor is 3.54. The latter calculation based upon er is more convenient because it gives an expanded scale for highly enantioenriched products. Comparing ees of 95% and 99%, for example, suggests a weak amplification (1.04), while comparison of the corresponding ers (39 and 199) gives the more informative amplification factor of 5.1.[14]

III. ENANTIOENRICHED VERSUS ENANTIOPURE AUXILIARIES

As mentioned in the introduction, a chiral auxiliary for asymmetric synthesis or kinetic resolution may not necessarily be enantiomerically pure. Even natural products considered to be composed of a single enantiomer may later be proven to be contaminated by minor amounts of the other enantiomer. Currently available techniques are estimated to allow the measurement of ees up to 99.99% (er ≈ 20,000).[16, 17] These numbers define, for the time being, the experimental border between enantiopure and enantioimpure compounds.

Many asymmetric syntheses have been performed with partially resolved chiral auxiliaries (enantiomeric excess $ee_{aux} < 100\%$). The enantiomeric excess of the product ($ee_{product}$) of the reaction is lower than that obtained (ee_o) in the reaction with enantiopure chiral auxiliary. The maximum value ee_o was traditionally calculated by "correcting" $ee_{product}$ with ee_{aux}:

$$ee_o(\%) = \frac{ee_{product}}{ee_{aux}} \times 100 \tag{1}$$

This assumption (to our knowledge there is no demonstration of this calculation) is easily established by considering the case of a simplified model of asymmetric catalysis (Scheme 1). In process I the asymmetric reaction is controlled by a chiral catalyst, cat*R*, while in process II it is the enantiomeric catalyst, cat*S*, that is working. The two systems are mirror images and will generate products of opposite absolute configuration but with the same enantiomeric excess (absolute value).

It is possible to calculate the ee of the product ($ee_{product} = ee_o$) in system I using the following equations (2)–(6).

$$\frac{d[R]}{dt} = k_1 \, \mathrm{cat}R[A]\,[B] \tag{2}$$

$$\frac{d[S]}{dt} = k_2 \, \mathrm{cat}R[A]\,[B] \tag{3}$$

Hence,

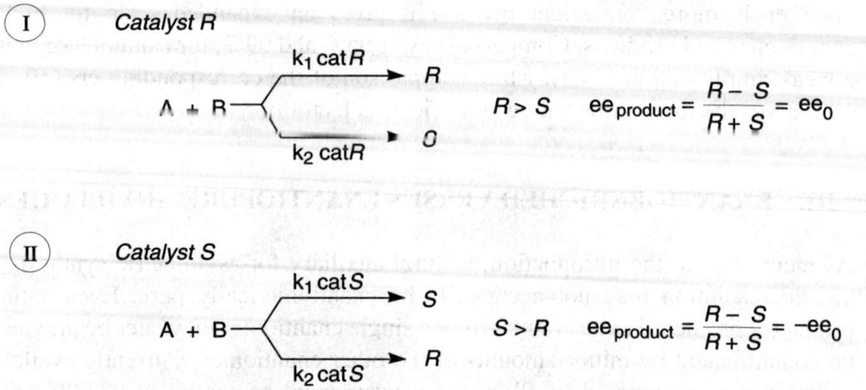

Scheme 1. Asymmetric catalysis from achiral reactants *A* and *B* under the influence of a chiral catalyst (cat*R* or cat*S*).

$$\frac{d[R]}{d[S]} = \frac{k_1}{k_2} \tag{4}$$

which after integration gives

$$\frac{[R]}{[S]} = \frac{k_1}{k_2} = s \tag{5}$$

Thus the ratio of the enantiomeric products is *independent* of time. The enantiomeric excess ee_o is given by

$$ee_o = \frac{[R] - [S]}{[R] + [S]} = \frac{s - 1}{s + 1} \tag{6}$$

The same calculations can be applied to system II by exchanging k_1 and k_2, giving

$$\frac{[S]}{[R]} = \frac{k_1}{k_2} = s \tag{7}$$

The enantiomeric excess for this system is $-ee_o$.

If the catalyst is a mixture of enantiomers (cat*R* and cat*S*, with cat*R* being the major enantiomer), one can calculate $ee_{product}$ by combining the equations describing systems I and II (which lead to products of opposite absolute configuration). The ee of the catalyst, ee_{aux}, can be expressed in terms of the relative amounts of its two enantiomers cat*R* and cat*S* (with cat*R* predominating)

$$ee_{aux} = \frac{\text{cat}R - \text{cat}S}{\text{cat}R + \text{cat}S} \quad (8)$$

The product will be the result of parallel reactions through systems I and II giving rise to the *R* enantiomer (ee_o in system I) and *S* enantiomer ($-ee_o$ in system II). The overall result is equivalent to mixing two samples with different ees giving rise to one mole of product with $ee_{product}$. This ee may be calculated by applying equation (9).[10, 18]

$$ee_{product} = (R_1 + S_1)ee_1 + (R_{II} + S_{II})ee_{II} \quad (9)$$

In this formula $(R_I + S_I)$ and $(R_{II} + S_{II})$ are the fractions of products obtained through systems I and II giving 1 mole of product. The corresponding ees are ee_I, ee_{II}, and $ee_{product}$ (with absolute values ≤ 1 for simplicity).

Applying the above definitions gives

$$\frac{R_I + S_I}{R_{II} + S_{II}} = \frac{\text{cat}R}{\text{cat}S} = \frac{1 + ee_{aux}}{1 - ee_{aux}} \quad (10)$$

$$R_I + S_I + R_{II} + S_{II} = 1 \quad (11)$$

Thus

$$R_I + S_I = \frac{a}{1+a} \quad \text{and} \quad R_{II} + S_{II} = \frac{1}{1+a}, \quad \text{where } a = \frac{1 + ee_{aux}}{1 - ee_{aux}} \quad (12)$$

Replacing these values for $(R_I + S_I)$ and $(R_{II} + S_{II})$ in Eq. (9), and taking $ee_I = ee_o$ and $ee_{II} = -ee_o$, gives

$$ee_{product} = ee_o \frac{a-1}{a+1} \quad (13)$$

Using the definition of *a* in Eq. (12) gives

$$ee_{product} = ee_o ee_{aux} \quad (14)$$

This relationship shows that there is a *linear correlation* between the *enantiomeric excess of the product* of an asymmetric synthesis and the *enantiomeric excess of the chiral auxiliary*. As pointed out in Ref. 10, such a simple relationship does not exist if enantiomeric ratios (er) are used to describe the enantiomer distributions. It was classically Eq. (14), or the equivalent Eq. (1), which was implicitly used to extrapolate the enantiomeric excess of the product (ee_o) prepared under the influence of enantiopure chiral auxiliary.

The main assumption behind the calculations leading to Eq. 14 is that the catalysts are working independently from each other, with catalysts of opposite absolute configuration giving rise to enantiomeric products. In other words, the mixing of enantiomeric catalysts has been treated as an *ideal mixing* of compounds, without perturbation terms. This is also reflected in the use of the simple law of mixing for chiral compounds (Eq. 11). Is this assumption valid, however?

There were various indications in the literature that a *nonideal behavior* of mixtures of enantiomers could sometimes occur. Let us consider the specific rotations associated with various mixtures of enantiomers.

The optical purity (op) of a compound has been calculated, since the time of Pasteur and Biot, by

$$\text{op} = \frac{[\alpha]_{\text{exp}}}{[\alpha]_{\text{max}}} \tag{15}$$

where $[\alpha]_{\text{exp}}$ and $[\alpha]_{\text{max}}$ are the values for the specific rotations of a partially resolved sample and the enantiopure sample, respectively. The optical purity was assumed to be equivalent to the ee, given by the ratio

$$\text{ee} = \frac{[R] - [S]}{[R] + [S]} \tag{16}$$

if the *R* enantiomer is in excess. For a long time polarimetry was the only tool for measuring the relative amounts of each enantiomer in a mixture. The equation op = ee, however, is not always valid. The first breakdown of this law was observed by Horeau in 1969 when studying the optical rotation of samples of 1-ethyl-1-methylsuccinic acid of various ees in chloroform (Scheme 2).[19] It has been interpreted by intermolecular interactions (H bonds) between the enantiomers of the acid affording diastereomeric aggregates.[19, 20] Methanol, however, effectively solvates the acid and disrupts this aggregation, restoring the additivity rule based upon the contributions of the enantiomeric species.

Interactions between enantiomers in solution have also been observed in NMR spectroscopy. Uskokovic et al. discovered that the ^1H NMR spectra of dihydroquinine of various ees are different.[21] Chromatography of enantioenriched

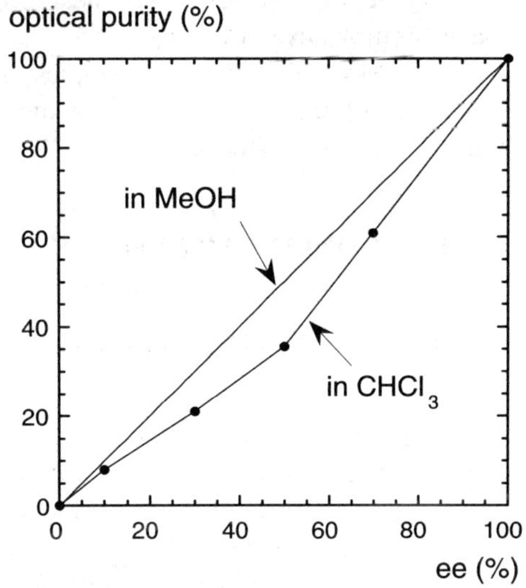

Scheme 2. Nonequivalence of optical purity (op) and enantiomeric excess (ee) for 1-ethyl-1-methylsuccinic acid in chloroform.[19]

mixtures under achiral conditions may give a fractionation of the enantiomeric excess.[22] This has been interpreted as the result of diastereomeric interactions in the mobile or stationary phases. Finally, the fractionation of ee by distillation of a partially resolved sample of isopropyl trifluorolactate is evidence for the importance of diastereomeric associations.[23]

The potential perturbations afforded by diastereomeric interactions between enantiomers during reactions were pointed out by Wynberg and Feringa in 1976.[24] They stated that "when a chiral substance undergoes a reaction, the reaction rate and the product ratio will depend—inter alia—upon the enantiomeric excess present in the starting material."

The authors provided some experimental support for their hypothesis by performing various diastereoselective reactions on racemic or optically active compounds. Indeed, they found some differences in stereoselectivity between the two cases. Reduction of DL-camphor or D-camphor with lithium aluminum hydride, for example, gave ratios of isoborneol to borneol of 7.93 and 9.20, respectively.

Izumi and Tai discussed in their book published in 1976 the case of asymmetric syntheses performed with partially resolved reagents or catalysts. They concluded that some complications may occur for catalysts bearing several chiral ligands.[25]

The first experimental studies of asymmetric catalysis with enantiomerically impure ligands were disclosed in 1986 by one of us in collaboration with C. Agami.[2] The deviations observed with respect to Eq. (14), which defines a straight line in the graph $ee_{product} = f(ee_{aux})$, were called nonlinear effects (NLE). We later gave approximate models for the analysis of NLEs and for making predictions with regard to their behavior.[26] Since then many examples of NLEs have been discovered in a diverse range of reaction systems.[27] In the next section we will present the basic features of nonlinear effects.

IV. NONLINEAR EFFECTS

A. Graphical Representation

Equation (14) plotted on a graph of $ee_{product}$ versus ee_{aux} gives a straight line (**1** in Scheme 3). On this graph, and those that follow, ee_{aux} and $ee_{product}$ are considered, for clarity, as taking values between 0 and 100% (by multiplying the values of ee_{aux} and $ee_{product}$ in the various formulas by 100). When for a given ee_{aux} the corresponding value of $ee_{product}$ is always found experimentally to lie above line **1**, as for curve **2**, this phenomenon is termed a *positive nonlinear effect*, (+)-NLE. If the experimental values of $ee_{product}$ are below line **1**, as for curve **3**, there is a negative nonlinear effect, (–)-NLE. The denominations "positive" or "negative" were suggested in order to recall that the deviation from linearity occurs above or below line **1**.

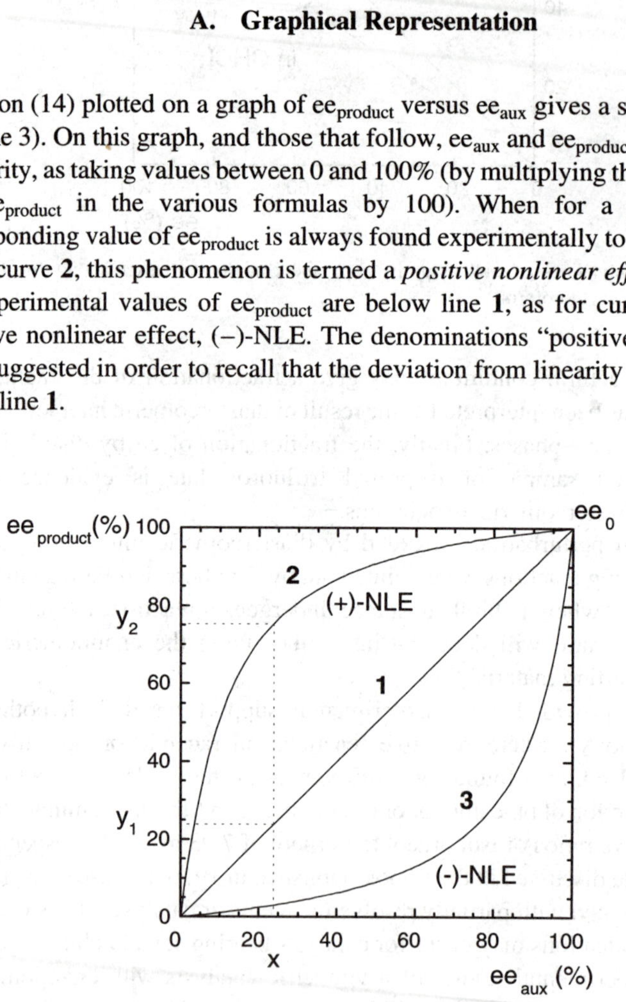

Scheme 3. Graphical representation of nonlinear effects.

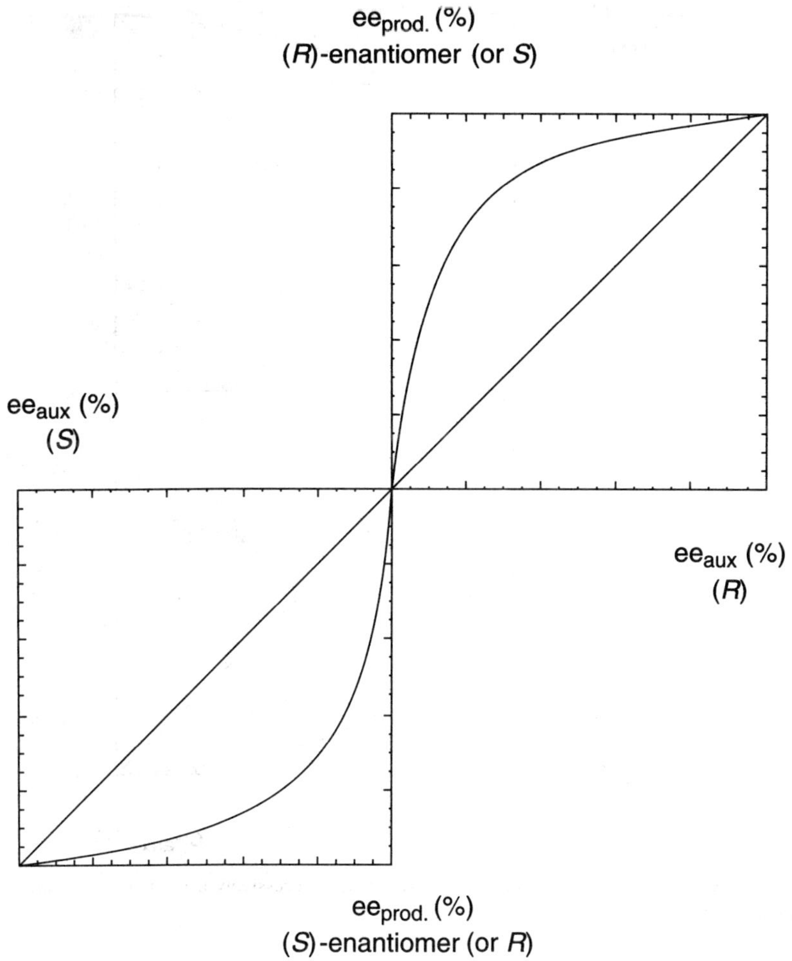

Scheme 4. Graphical representation of a (+)-NLE when successively using both enantiomers of a chiral auxiliary.

Usually NLEs are studied for a chiral auxiliary of a given absolute configuration by variation of its ee between 0 and 100%. The graph is presented in a standard way in Scheme 3, which is irrespective of the absolute configuration of the chiral auxiliary and product. Sometimes experiments are performed with both enantiomers of the chiral auxiliary. In this case the results may be plotted as in Scheme 4 or 5. Symmetry relationships relate the curves in Schemes 4 and 5, since experiments performed with enantiomeric catalysts are strictly mirror image.

Let us compare curves **2** and **3** in Scheme 3. At a given value (x) of ee$_{aux}$ there are two corresponding values of ee$_{product}$, y_1 and y_2, on curves **1** and **2**, respectively.

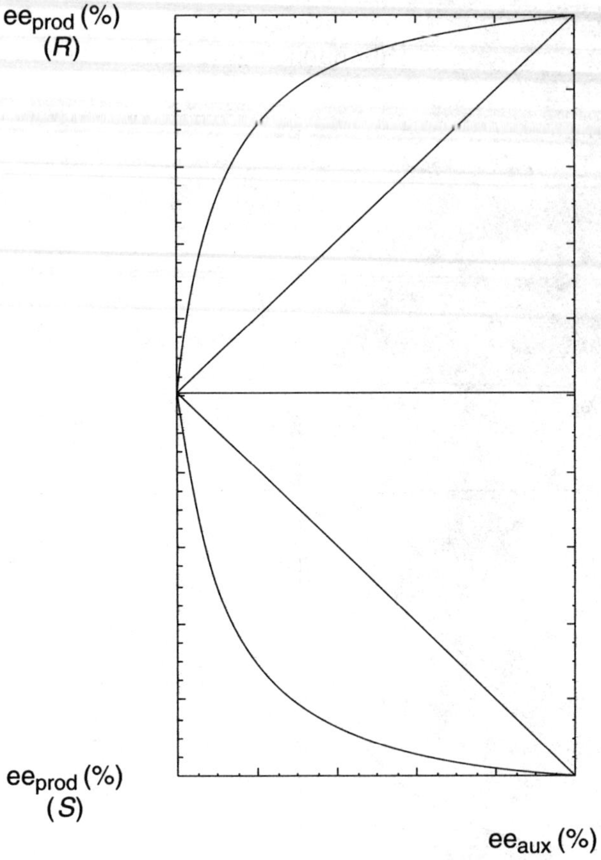

Scheme 5. Graphical representation of a (+)-NLE when successively using both enantiomers of a chiral auxiliary.

The value y_2 is much higher than the "normal" value y_1 obtained in the linear correlation, hence the term (+)-NLE. Oguni et al. introduced in 1988 the expression *asymmetric amplification* as being synonymous with (+)-NLE in order to describe these effects. Although this expression is now widely used in the context of nonlinear effects, its validity is questionable (vide supra).

B. Early Examples of Asymmetric Amplification

We showed in 1986 that the Sharpless epoxidation of geraniol gives a well-defined (+)-NLE, although the amplification is moderate (Scheme 6).[2] Curiously, however,

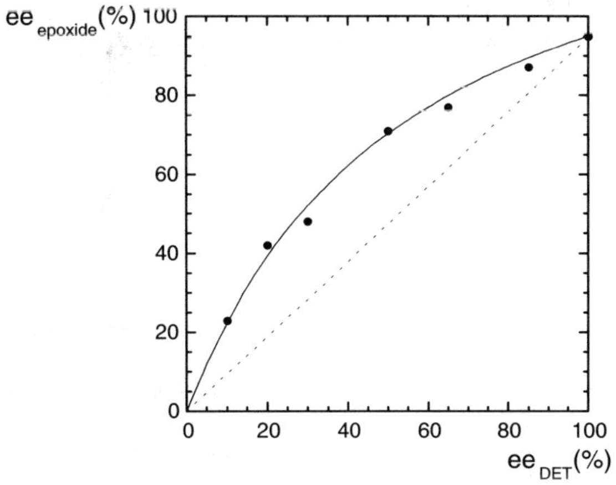

Scheme 6. (+)-NLE in the Sharpless epoxidation of geraniol.[2]

asymmetric oxidation of methyl *p*-tolyl sulfide in the presence of a water-modified Sharpless reagent provided a sulfoxide with a (–)-NLE.[2]

Very strong asymmetric amplifications have been observed in the addition of diethylzinc to benzaldehyde catalyzed by various β-amino alcohols. The first experiments were published by Oguni et al. in 1988[1] and by Noyori et al. in 1989.[32] The experimental results are shown in Scheme 7. Using (–)-DAIB **2** of 15% ee, for example, gave the addition product in 92% ee, while using enantiopure (–)-DAIB gave the product in 98% ee.

C. Index of Amplification

In reactions exhibiting a (+)-NLE, some authors have quantified the magnitude of the amplification by comparing the ee of the product obtained with that expected if there were a linear correlation between catalyst and product ees. However, as previously mentioned, the ratio of ees gives a less informative index of amplification than when compared with ers. Thus we propose to define routinely the size of an amplification (the amplification index I) as the ratio of the enantiomeric ratio (er) of the product experimentally observed to the er of the product expected for a linear relationship. This amplification index I has been plotted against ee_{aux} in the case of the diethylzinc addition to benzaldehyde catalyzed by (–)-DAIB **2**, as described by Noyori et al. (Scheme 8, top). Here the maximum amplification (with $I = 32$) occurs when DAIB has an ee_{aux} of approximately 20%. Expressing the amplification in terms of the ratio of ees,

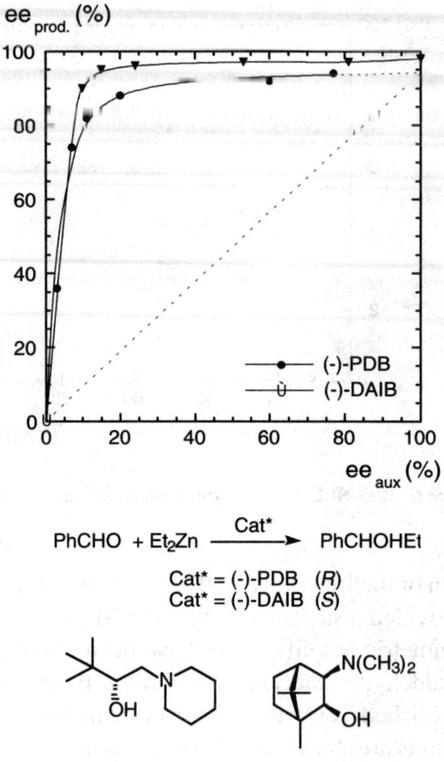

(-)-PDB
1

(-)-DAIB
2

Scheme 7. (+)-NLEs in the PDB,[1] and DAIB[32] catalyzed addition of diethylzinc to benzaldehyde.

however, gives a misleading view of the size of the amplification (Scheme 8, bottom).

The ene reaction between 1-methylstyrene and methyl glyoxylate, catalyzed by a BINOL–titanium complex, gives a considerable amplification, as reported by Nakai and Mikami (Scheme 9).[33] Many reports of asymmetric amplifications have subsequently been published for a wide range of reactions (vide infra).

D. Mechanisms for Asymmetric Amplification

We have described two fundamental models for explaining the phenomenon of asymmetric amplification. In the *kinetic model* we proposed that the unusual

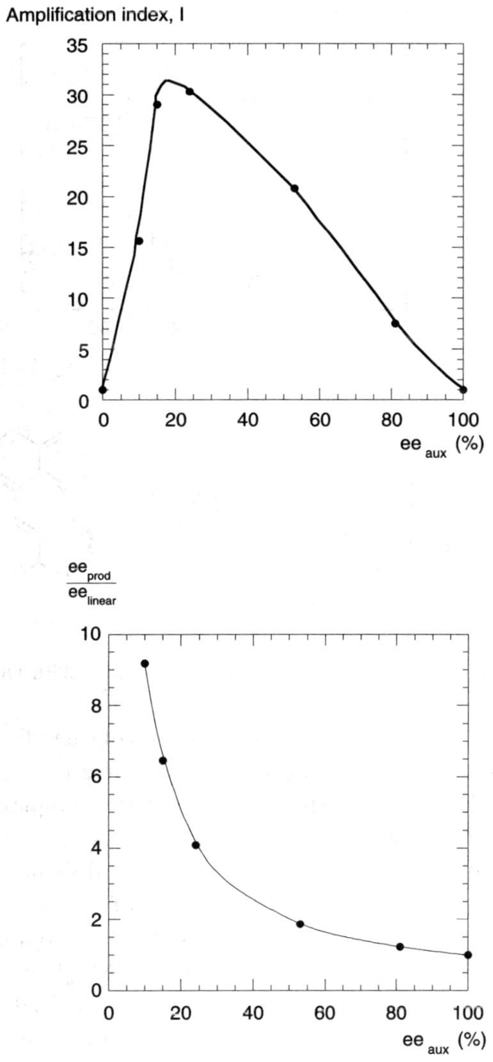

Scheme 8. Amplification index *I* (top) and ratio of product ee to expected ee for a system with no NLE (bottom) versus enantiomeric excess of auxiliary (ee_{aux}) in the (–)-DAIB catalyzed addition of diethyl zinc to benzaldehyde.

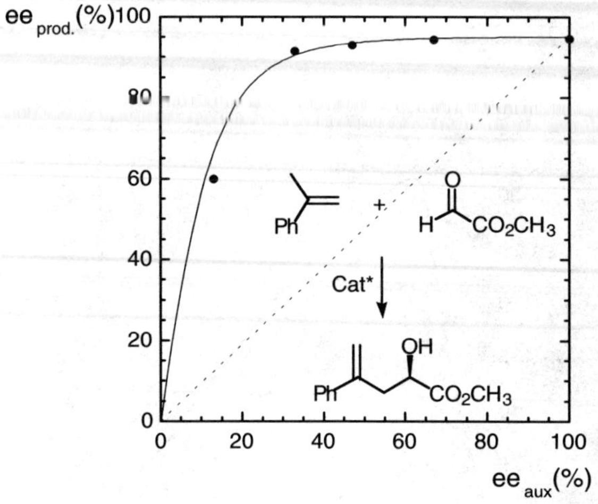

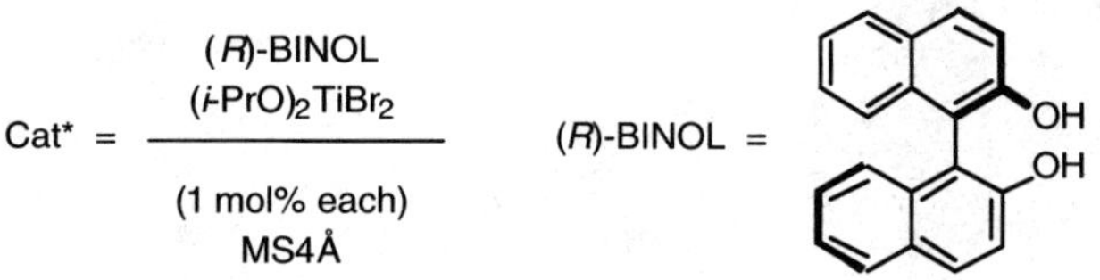

Scheme 9. (+)-NLE in the glyoxylate-ene reaction catalyzed by a BINOL–Ti reagent.[33]

behavior of the chiral auxiliary reflects the formation of catalytic species involving several chiral ligands.[2] If L_R and L_S are the two enantiomeric ligands of a metal (M) bearing n ligands (L), one may simply represent the complexes as ML_n. ML_2 complexes, or the dimers $(ML)_2$, are the simplest ways of assembling two chiral ligands in a molecular species. This model, however, does not take into account other co-ligands or additional possibilities of stereoisomerism. Three stereoisomeric complexes are generated, namely, two homochiral complexes (ML_RL_R and ML_SL_S) and one heterochiral complex (ML_RL_S), assumed to have a *meso* structure. The chiral products P_R and P_S are obtained from these three catalysts (Scheme 10). The *meso* catalyst will give a racemic mixture of products, whereas the two homochiral catalysts will give products of opposite absolute configuration with enantiomeric excess $ee_{product} = ee_o$. The overall enantiomeric excess of the product will depend on the relative amounts of the three catalysts present (x, y, and z) and on their relative reactivities ($k_{homo} = k_{RR} = k_{SS}$, $k_{meso} = k_{RS}$).

A simple calculation has been performed[2] to express $ee_{product}$ as a function of the relative amount of *meso* catalyst present β and of its relative reactivity g:

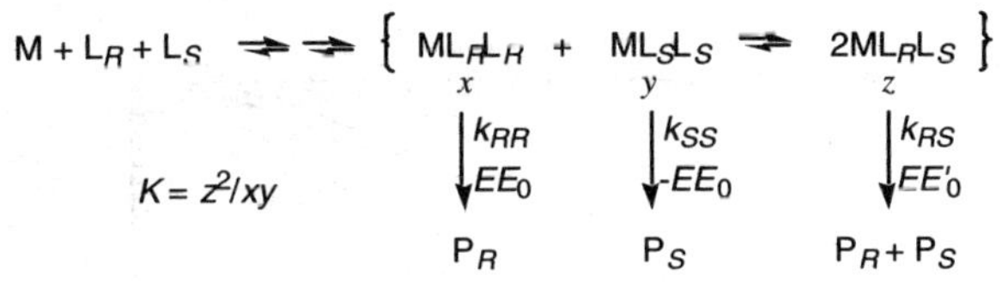

Scheme 10. Complexes formed between the two enantiomers of a chiral ligand and a metal.

$$\beta = \frac{z}{x+y} \quad \text{and} \quad g = \frac{k_{RS}}{k_{RR}} \tag{17}$$

An additional parameter K is needed to calculate the distribution of the three complexes. It is defined as the equilibrium constant of interconversion between the three complexes and is related to the relative amount (β) of *meso* catalyst. The kinetic calculations give equation (18), where ees are defined as being ≤ 1.

$$ee_{product} = ee_o ee_{aux}\, f, \quad f = \frac{1+\beta}{1+g\beta} \tag{18}$$

In order to achieve an amplification of chirality, it requires that $f > 1$. If $\beta = 0$ (no *meso* catalyst) or $g = 1$ (same reactivity of *meso* and homochiral catalysts), then $f = 1$. The condition $f > 1$ is achieved for $1 + \beta > 1 + g\beta$, or $g < 1$. Thus the necessary condition for asymmetric amplification in the above model is for the heterochiral or *meso* catalyst to be less reactive than the homochiral catalyst. If the *meso* catalyst is more reactive, then $f < 1$, and hence a negative nonlinear effect is observed. The size of the asymmetric amplification is regulated by the value of f, which increases as K does. The more *meso* catalyst (of the lowest possible reactivity) there is, the higher will be $ee_{product}$. This is well illustrated by computed curves in Scheme 11. The variation of $ee_{product}$ with ee_{aux} is represented for various values of g (the relative reactivity of the *meso* complex) with $K = 4$ (corresponding to a statistical distribution of ligands; Scheme 11, top). The variation in the relative amounts of the three complexes with ee_{aux} is also represented for a statistical distribution of ligands (Scheme 11, bottom).

The asymmetric amplification is a consequence of an in situ increase in the ee of the active catalyst, since racemic ligand is trapped in the unreactive or weakly reactive *meso* catalyst. In the *reservoir effect* a similar phenomenon occurs outside the catalytic cycle. Let us assume that part of the initial chiral ligand, characterized by ee_{aux}, is diverted into a set of catalytically inactive complexes (Scheme 12).

If this reservoir traps some of the initial ligand with an enantiomeric excess (ee_{res}) different from ee_{aux}, then the remaining ligand will have an effective

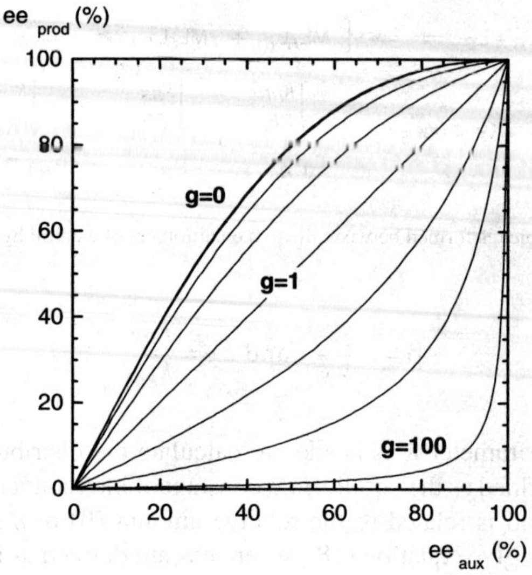

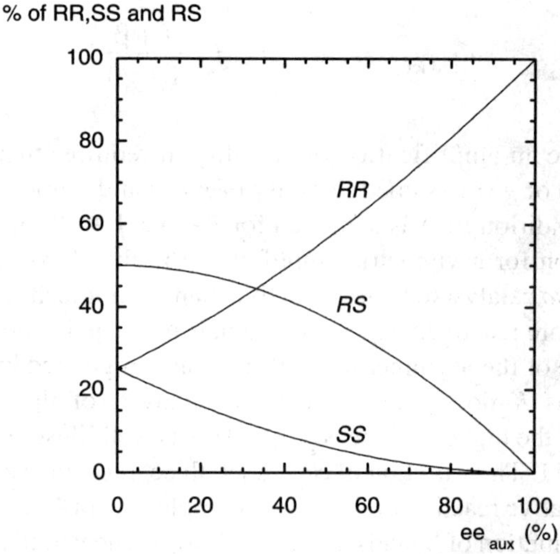

Scheme 11. Variation of $ee_{product}$ with ee_{aux} in ML_2 model with various values of *g* (top) and the dependency of the relative amounts of homochiral and heterochiral catalysts on the ee_{aux} (bottom).

enantiomeric excess (ee_{eff}). The usual case is the formation of inert species of racemic composition by, for example, autoassociation of a complex into a *meso* dimer. The available ligand for catalysis will then have an ee given by Eq. (19), where α is the molar fraction (<1) of the initial ligand that exists as a racemic mixture in the reservoir

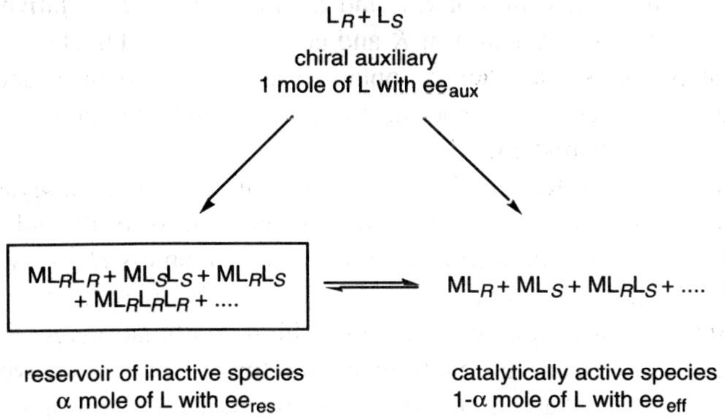

Scheme 12. Reservoir effect.

$$ee_{eff} = \frac{ee_{aux}}{1 - \alpha} \tag{19}$$

The classical equation for the correlation of $ee_{product}$ and ee_{aux}, Eq. (14), can now be used but replacing ee_{aux} by ee_{eff}. The size of the reservoir is crucial for the value of the amplification. Equation (19), shows that a ligand of $ee_{aux} = 20\%$ will have an $ee_{eff} = 80\%$ if $\alpha = 0.75$ (75% of storage). The value α is a function of the initial enantiomeric excess of the ligand (ee_{aux}). A common scenario is a pre-equilibrium between monomers and dimers (inactive) with an equilibrium constant K, which fixes the size of the reservoir as a function of ee_{aux}. Using these assumptions leads to

$$ee_{eff} = \frac{K\, ee_{aux}}{\sqrt{1 + K(K\, ee_{aux}^2 + 2)} - 1} \tag{20}$$

Strong amplifications at low ee_{aux} may be achieved using suitable values of K.

In conclusion, asymmetric amplification is essentially generated by the "annihilation" of two ligands of opposite configuration in unreactive species (inside or outside the catalytic cycle).[26] A direct consequence of the reservoir effect or of the lower reactivity of *meso* species is the enhancement of the enantiomeric excess of the reactive species. This beneficial effect, however, is balanced by a decrease in the reaction rate, since a smaller amount of active catalyst is available for reaction. The lowering of reaction rate will be especially strong for low ee_{aux} due to the greater formation of *meso* species. This effect has been experimentally observed[34] and recently quantitatively discussed by Blackmond in connection with ML_n models.[35] In the ML_2 model (Scheme 10 and Eq. 17), for example, it is possible to calculate the values of x, y, and z, which define the relative amounts of the three catalysts (as a function of K and ee_{aux}, Scheme 11). The overall rate of formation of products may then be obtained as a function of K and ee_{aux}. The comparison of calculated and experimental rates should provide a simple tool for the support of a mechanistic model.

These mechanisms for the in situ enhancement of the catalyst ee bear a similarity to the "Horeau duplication," which can be used for the enhancement of ee.[36] In this approach a mixture of enantiomeric monomers *R* and *S* are coupled with an achiral bifunctional linker to give a mixture of dimers *R–R*, *S–S*, and *R–S*. The *R–S* dimer (diastereomeric to the homochiral combinations *R–R* and *S–S*) is separated and the remaining enantiomeric mixture of dimers cleaved to give a mixture of monomers *R* and *S* of enhanced ee (even if the coupling reaction gives a statistical distribution of products). A generalization of the Horeau duplication principle has been discussed by Rautenstrauch in cases where a chiral catalyst acts on a substrate bearing several prochiral sites, giving an apparent amplification of the stereoselectivity.[15] A symmetrical diketone, for example, may be reduced to a chiral diol of much higher ee than expected from the ee of the monoreduction. The amplification arises from the formation and separation of the *meso* diol.

V. CATALYTIC REACTIONS WITH ASYMMETRIC AMPLIFICATION

A. Organozinc Additions

Asymmetric amplification in the 1,2- or 1,4-addition of dialkylzincs is very commonly observed. As previously mentioned, the expression *asymmetric amplification* originated with experiments performed by Oguni et al. in the reaction of diethylzinc with benzaldehyde catalyzed by chiral β-amino alcohols.[1]

The most fully understood system in this class of reactions, however, is the DAIB-catalyzed addition of diethylzinc to aldehydes, due to the very detailed mechanistic studies performed by Noyori et al.[32, 34, 37] They were able to determine the structure of several intermediates involved in the reaction and established the kinetic law. Part of the catalytic cycle is depicted in Scheme 13. The origin of the asymmetric amplification lies in the formation of dimers of DAIB–zinc alkoxides. The heterochiral dimer is quite stable in the concentration range of the experiment (2×10^{-1} to 5×10^{-1} M in toluene for DAIB), whereas the homodimers are prone to dissociation and react further with diethylzinc to give a di-zinc complex that is the active species in the catalytic cycle. They react with benzaldehyde and give rise to the asymmetric transfer of the ethyl group. The final product, as a zinc alkoxide, does not interfere with the reaction (and hence there is no autoinduction), since it

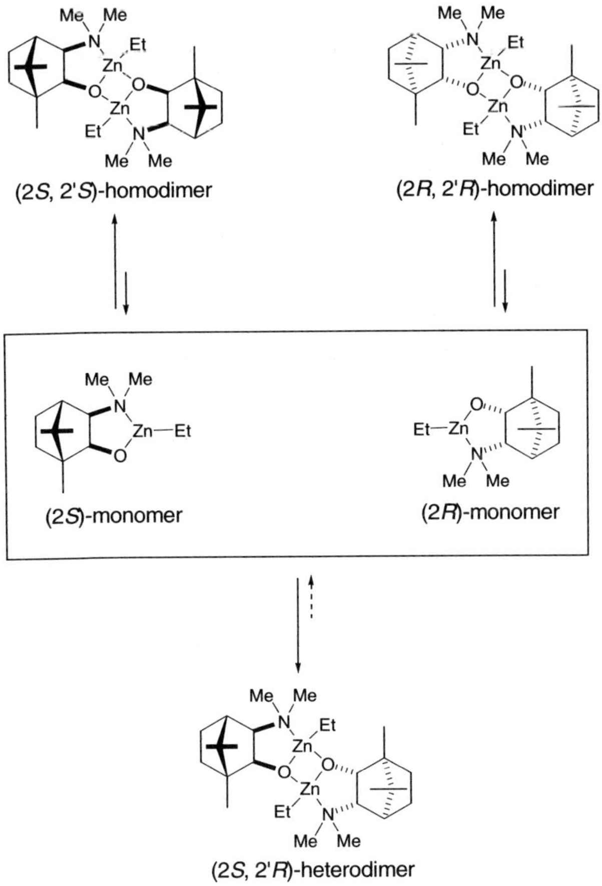

Scheme 13. Reservoir effect in the DAIB-catalyzed addition of diethylzinc to benzaldehyde.[32]

autoassociates into a stable tetramer. From the experimental data given by the authors, one can model the system (as far as DAIB is concerned) as a reservoir connected to a catalytic cycle. For $ee_{DAIB} = 20\%$, which gives $ee_{product} = 90\%$, one calculates $\alpha = 0.77$, which means that 77% of the ligand is stored in the reservoir. It was shown, however, that the *meso* dimer can react under some experimental conditions, leading to a minor amount of racemic product.

Asymmetric amplification in the diethylzinc addition to aldehydes has been observed with many β-amino alcohols as catalyst, presumably because of a reservoir effect similar to that discovered by Noyori et al. Asymmetric amplification has also been found for other classes of chiral catalysts—diamines, diols, titanium complexes, etc. The various examples are collected in Table 1. The

Table 1
Asymmetric Amplification in the 1,2-Addition of Diethylzinc to Benzaldehyde

Ligand			Addition product		Amplification index I[a]	Ref.
Structure	mol %	ee (%)	Yield (%)	ee (%)		
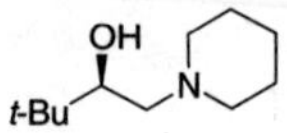	2	20.5	96	88 (*R*)	10.5	1
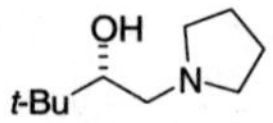	2	20	92	73 (*S*)	4.6	1
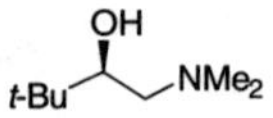	2	19.8	82	60 (*R*)	3.0	1
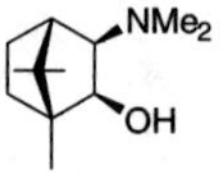	8	24	—[b]	96 (*S*)	30.3	32
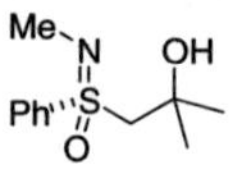	10	23	70	64 (*S*)	3.1	38
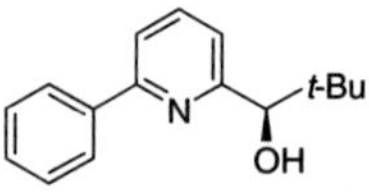	5	14	49	87 (*R*)	11.2	39
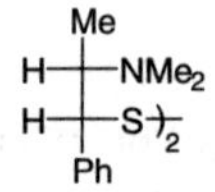	5	50	—[b]	81[b]	3.4	40
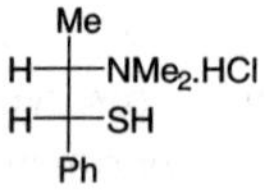	5	50	—[b]	81[b]	3.4	40
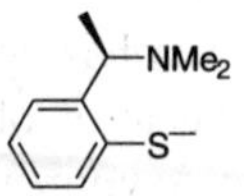	2	60	—[b]	84 (*S*)	3.2	41

[a]Either maximum value calculated directly from the experimental data or estimated from the graph showing the (+)-NLE in the paper.
[b]Not determined.

Table 2
Asymmetric Amplification in the 1,4-Addition of Diethylzinc to Chalcone

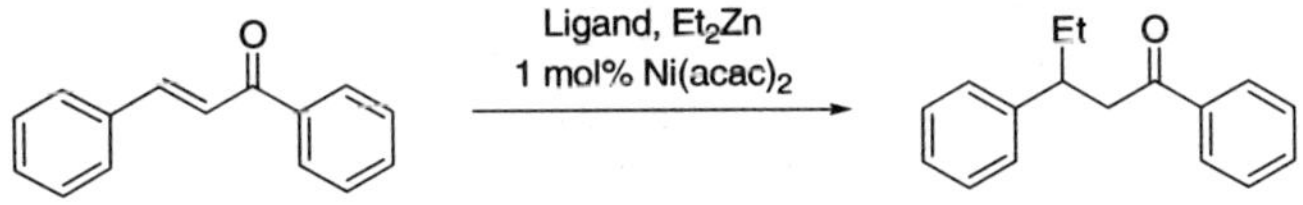

Ligand			Addition product		Amplification index I^a	Ref.
Structure	mol %	ee (%)	Yield (%)	ee (%)		
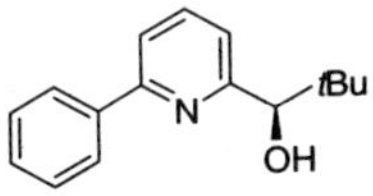	20	32	77	66 (*R*)	2.9	42
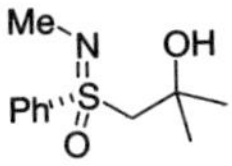	20	18	—[b]	22 (*R*)	1.3	43
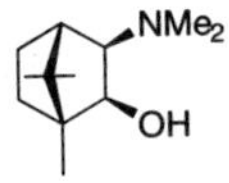	16	38	—[b]	45 (*R*)	1.5	44

[a]Either maximum value calculated directly from the experimental data or estimated from the graph showing the (+)-NLE in the paper.

[b]Not determined.

size of the maximum amplification has also been given by calculating the amplification index I:

$$I = \frac{er_{product}}{er_{linear}} \tag{21}$$

where er_{linear} is the expected enantiomeric ratio for a linear correlation, calculated from Eq. (14), while $er_{product}$ is the experimental value.

The 1,4-addition of diethylzinc to α,β-unsaturated ketones is catalyzed by nickel (II) complexes. Bolm showed in 1991 that a chiral amino alcohol derived from pyridine and $Ni(acac)_2$ catalyzed the addition of diethylzinc to chalcone with a moderate asymmetric amplification.[42] ML_2-type species are assumed, with the *meso* complex being the less reactive. Other examples of this reaction are collected in Table 2.

Table 3 contains the two known examples of systems exhibiting (+)-NLEs in the 1,4-addition of organocuprates to α,β-unsaturated ketones.

Table 3
Asymmetric Amplification in the 1,4-Addition of Organocuprates to α,β-Unsaturated Ketones

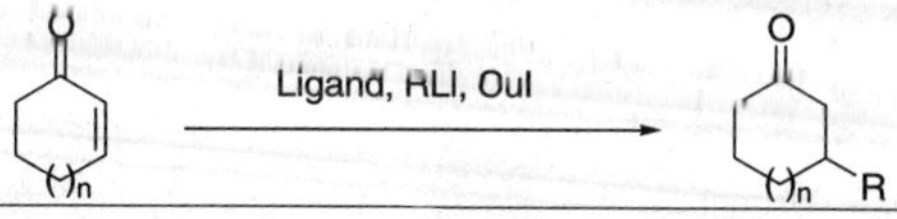

Ligand			Addition product		Amplification index I^a	Ref.
Structure	mol %	ee (%)	Yield (%)	ee (%)		
			$n = 10$, R = Me			
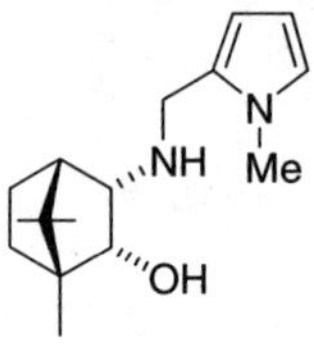	367	80	75	93 (*R*)	3.1	45
			$n = 2$, R = Bu			
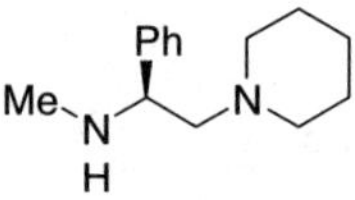	125	84	—[b]	94 (*S*)	3.5	46

[a]Either maximum value calculated directly from the expeimental data or estimated from the graph showing the (+)-NLE in the paper.
[b]Not determined.

B. Oxidations

Sharpless asymmetric epoxidation of allylic alcohols, asymmetric epoxidation of conjugated ketones, asymmetric sulfoxidations catalyzed, or mediated, by chiral titanium complexes, and allylic oxidations are the main classes of oxidation where asymmetric amplification effects have been discovered. The various references are listed in Table 4 with the maximum amplification index observed.

C. Ene Reactions

This was one of the first classes of catalyzed reactions that were shown to give rise to strong asymmetric amplifications (vide supra, Scheme 9).[33] Many subsequent studies have been performed with variations in the structure of the chiral catalyst, generated from BINOL and titanium salts.[50–52] Asymmetric amplification is frequently observed with these catalysts in the glyoxylate ene reaction (Table 5).

Table 4
Asymmetric Amplification in Oxidation

Ligand		Oxidation product			Amplification index I^a	Ref.
Structure	mol %	ee (%)	Yield (%)	ee (%)		
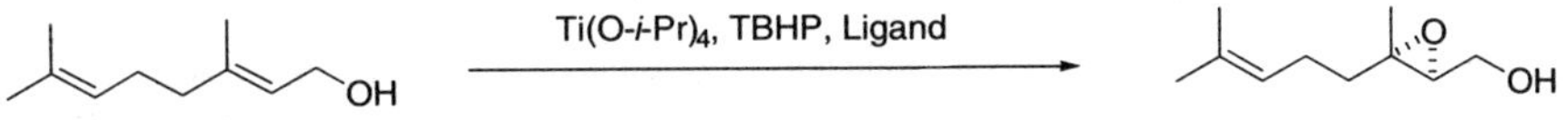						
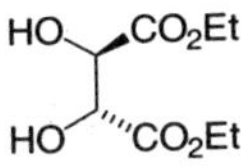	5	50	—[b]	71 (2*R*, 3*R*)	2.1	2
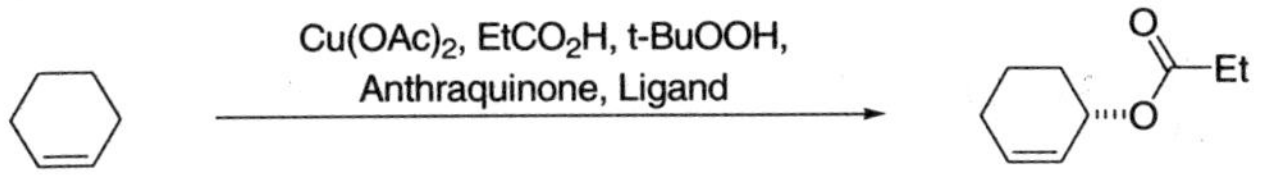						
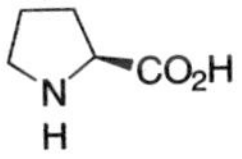	7.5	20	—[b]	29 (*S*)	1.4	47

(*R*)-BINOL	5	64	—[b]	75 (α*S*, β*R*)	2.0	48
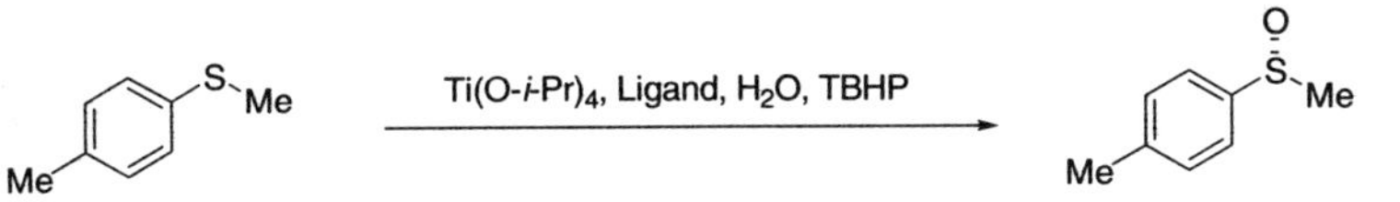						
(*R*)-BINOL	20	50	—[b]	75 (*R*)	2.6	49

[a]Either maximum value calculated directly from the experimental data or estimated from the graph showing the (+)-NLE in the paper.

[b]Not determined.

D. Cycloadditions

Many Diels–Alder or hetero-Diels–Alder reactions (catalyzed by chiral Lewis acids) have been described in the literature. The first example involving an asymmetric amplification was disclosed by Narasaka et al. in 1989.[53]

These authors used titanium taddolate as a chiral catalyst, under heterogeneous conditions, giving partial precipitation of an inactive *meso* dimer. Titanium

Table 5
Asymmetric Amplification in the Glyoxylate-Ene Reaction

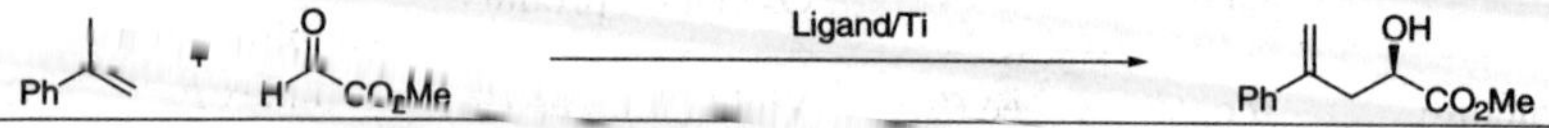

Ligand			Addition product			
Structure	mol %	ee (%)	Yield (%)	ee (%)	Amplification index I^a	Ref.
(*R*)-BINOL + $TiBr_2(O\text{-}i\text{-}Pr)_2$ + MS 4Å	1	33	92	91.4 (*R*)	11.7	50
(BINOL)Ti(μ-O)$_2$Ti(BINOL)	0.2	45	—[b]	90 (*R*)	7.3	52
(BINOL)Ti(μ-O)$_2$Ti(BINOL)	5	25	—[b]	85 (*R*)	7.5	57
(BINOL)Ti(μ-O)$_2$Ti(BINOL)	20	56	—[b]	91 (*R*)	6.3	57

[a]Either maximum value calculated directly from the experimental data or estimated from the graph showing the (+)-NLE in the paper.

[b]Not determined.

binolate has also been successfully used as a chiral catalyst in Diels–Alder reactions.[51] It is interesting to note that according to the preparation of the Ti-binolate catalyst there can be either asymmetric amplification or complete linearity.[51, 52] The treatment of BINOL (of various ees) with $Ti(O\text{–}i\text{–}Pr)_2Cl_2$ in the presence of molecular sieves (MS4Å), followed by filtration of the sieves, gives a catalyst that exhibits asymmetric amplification. The preparation of the two enantiopure catalysts, under the same conditions as above, followed by mixing of the two catalysts in various amounts, gives a system devoid of an NLE. The explanation lies in the formation of dimeric Ti complexes (with Cl or oxygen bridges). In the first procedure diastereomeric dimers are generated, leading to asymmetric amplification. In the second method there is only a mixture of enantiomeric dimeric catalyst, and hence normal behavior is observed. However, the addition of molecular sieves to this mixture catalyzes the interconversion between titanium complexes, with the concomitant formation of diastereomeric complexes and the subsequent appearance of an asymmetric amplification.

Table 6
Asymmetric Amplification in Cycloadditions

Ligand			Addition product			
Structure	mol %	ee (%)	Yield (%)	ee (%)	Amplification index I^a	Ref.
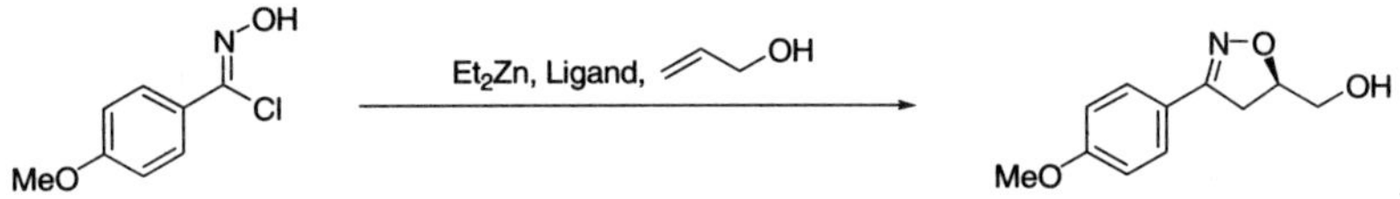						
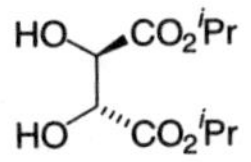	20	75	—[b]	80 (*R*)	1.7	54
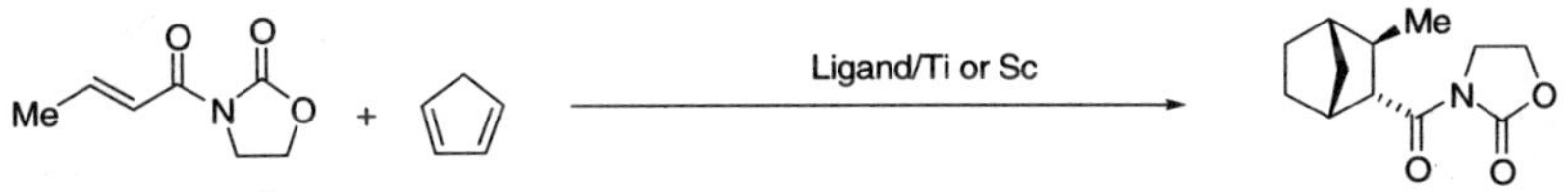						
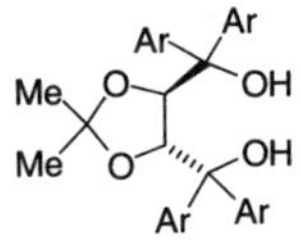 + $TiCl_2$(O-*i*-Pr)$_2$ + MS 4Å Ar = 2-Naphthyl	10	50	—[b]	68 (2*S*, 3*R*)	2.1	55
(*R*)-BINOL + $Sc(OTf)_3$ + MS 4Å + TMP	20	80	—[b]	90 (2*S*, 3*R*)	3.1	56
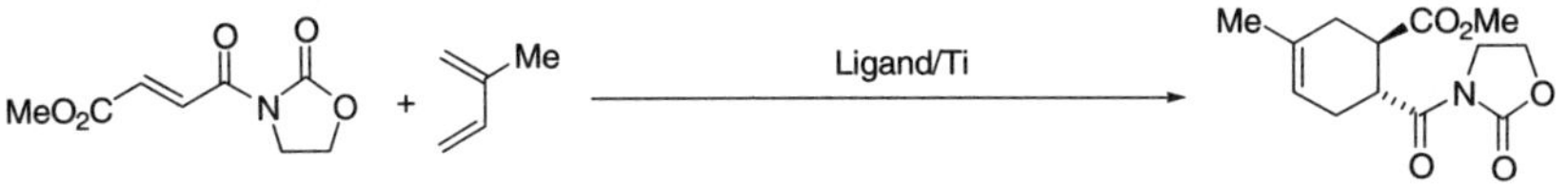						
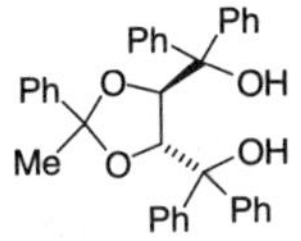 + $TiCl_2$(O-*i*-Pr)$_2$ + MS 4Å	200	25	—[b]	83 (1*R*, 2*R*)	7.1	53
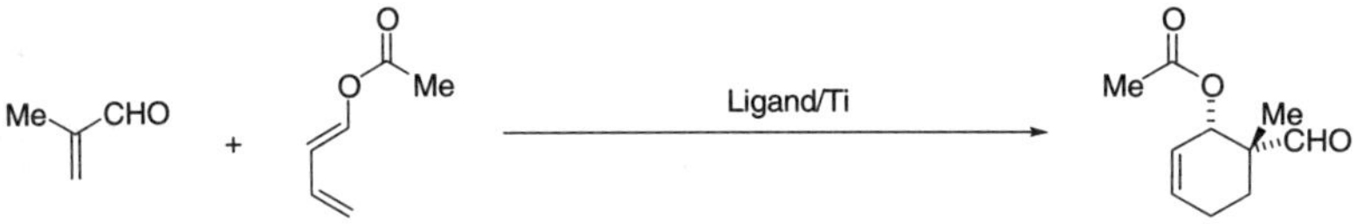						
(*R*)-BINOL + $TiCl_2$(O-*i*-Pr)$_2$	10	50	50	74 (1*R*, 2*S*)	2.4	51

[a]Either maximum value calculated directly from the experimental data or estimated from the graph showing the (+)-NLE in the paper.

[b]Not determined.

Table 7
Miscellaneous Examples of Asymmetric Amplification

Ligand			Addition product		Amplification index I^a	Ref.
Structure	mol %	ee (%)	Yield (%)	ee (%)		
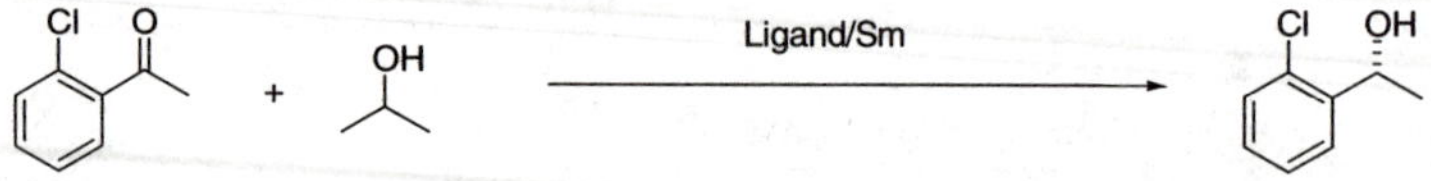						
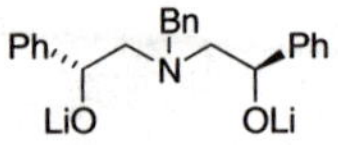 + SmI_3	5	80	—[b]	80 (*R*)	1.2	58
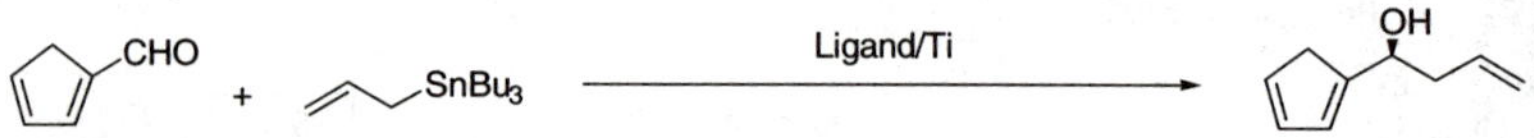						
(*R*)-BINOL + $Ti(O\text{-}i\text{-}Pr)_4$ + MS 4Å	10	50	—[b]	88 (*S*)	5.3	59
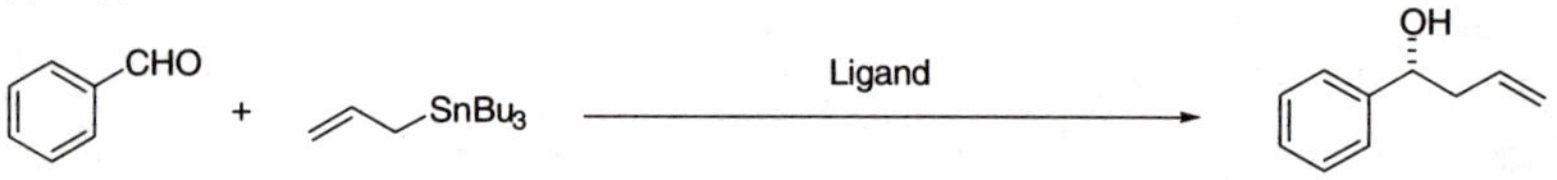						
(*R*)-BINOL + $Ti(O\text{-}i\text{-}Pr)_4$ + MS 4Å	30	70	>95	>95 (*R*)	9.6	60
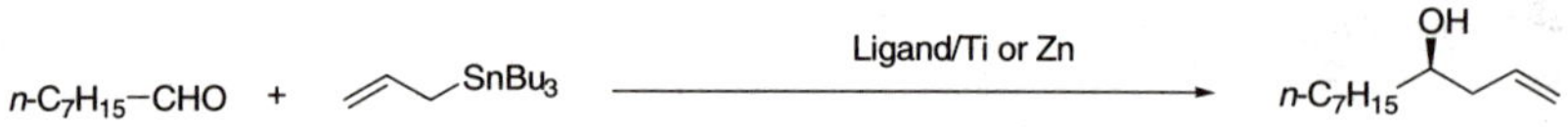						
(*S*)-BINOL + $TiCl_2(O\text{-}i\text{-}Pr)_2$ + MS 4Å	20	70	—[b]	92 (*R*)	4.9	61
(*S*)-BINOL + $Zr(O\text{-}i\text{-}Pr)_4.i\text{-}PrOH$ + MS 4Å	20	40	—[b]	52 (*R*)	1.5	61
(*S*)-BINOL + $Zr(O\text{-}i\text{-}Pr)_4.i\text{-}PrOH$	20	70	—[b]	73 (*R*)	1.6	61
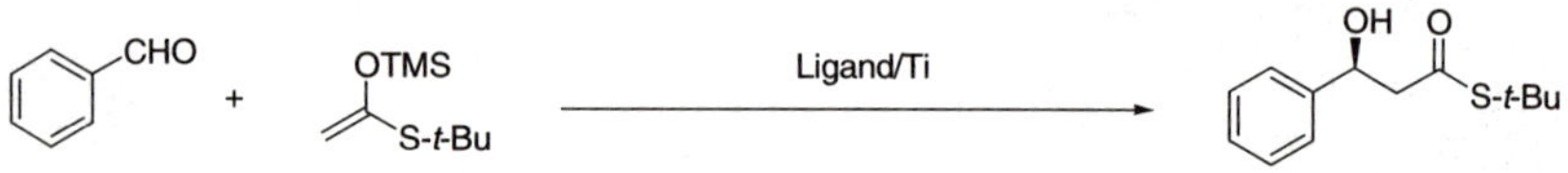						
(*R*)-BINOL + $Ti(O\text{-}i\text{-}Pr)_4$ + MS 4Å	20	50	50	91 (*S*)	7.4	62
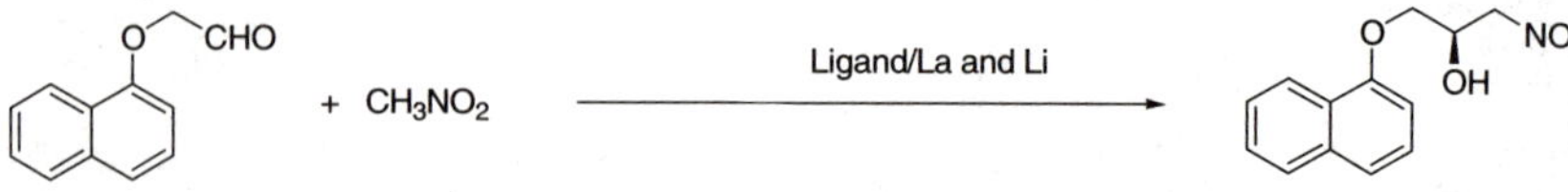						
(*S*)-BINOL.Li + $LaCl_3.7H_2O$ + NaOH	10	56	65	68 (*R*)	1.7	63

(*continued*)

Table 7
(*Continued*)

Ligand			Addition product		Amplification index I^a	Ref.
Structure	mol %	ee (%)	Yield (%)	ee (%)		

(*S*)-BINOL + TiF_4 + CH_3CN	10	50	—[b]	82 (*R*)	3.7	64
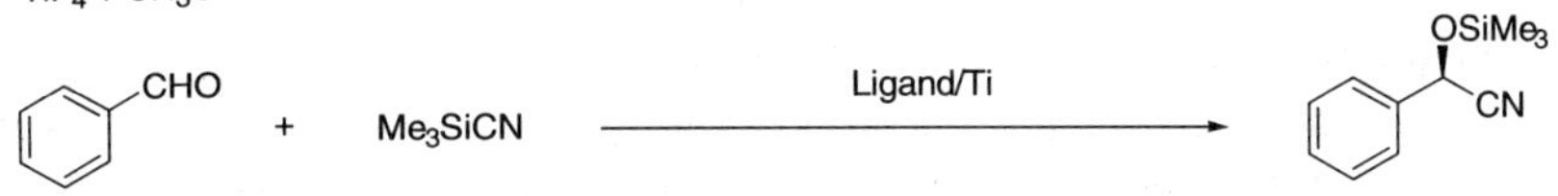						
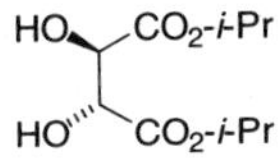 + Ti(O-*i*-Pr)$_4$	100	50	83	70 (*R*)	2.3	65
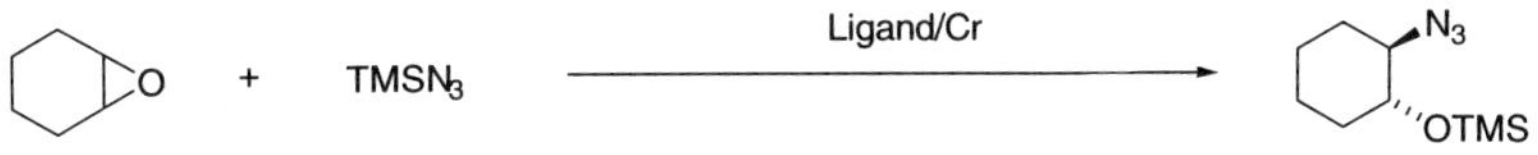						
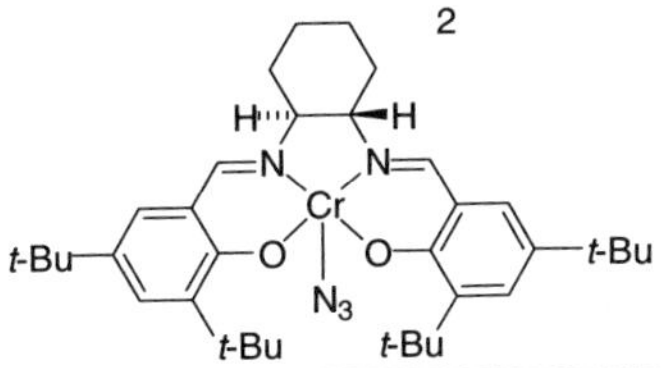	2	70	—[b]	77 (1*R*, 2*R*)	3.1	66

[a]Either maximum value calculated directly from the expeimental data or estimated from the graph showing the (+)-NLE in the paper.

[b]Not determined.

Examples of asymmetric amplification in Diels–Alder or hetero-Diels–Alder reactions are listed in Table 6 (see page 281), together with an example of amplification during a 1,3-dipolar cycloaddition.

E. Miscellaneous Reactions Exhibiting Asymmetric Amplification

Asymmetric amplification has been observed in organotin-catalyzed C–C bond formation. Keck, for example, discovered that allylation of aldehydes with allyltri-*n*-butylstannane catalyzed by titanium binolate catalysts gives a strong (+)-NLE.[59] The size of the asymmetric amplification is very sensitive to the solvent and the concentration.[62] The more concentrated the Ti catalyst, the greater the amplification, presumably due to increased dimerization into *meso* complexes.

Table 7 (see page 283) collects all the examples of asymmetric amplifications not quoted in the previous sections.

VI. AUTOINDUCTION OR AUTOCATALYSIS WITH ASYMMETRIC AMPLIFICATION

Autoinduction has been defined by Albert and Wynberg as the case where the product of the reaction modifies the further course of the reaction by changing the nature of the reagent or the catalyst.[67, 68] In catalytic reactions (the only type considered here) the autoinduction can involve the formation of product different from, or identical to, the catalyst. This last situation has been termed *autocatalysis*. These two cases are illustrated in Scheme 14.

$$A + B \xrightarrow{\text{Cat}} P \quad \text{then} \quad A + B \xrightarrow{\text{Cat—P}} P \qquad \textit{Autoinduction}$$

$$A + B \xrightarrow{\text{Cat = P}} P \qquad \textit{Autocatalysis}$$

Scheme 14. Mechanisms of autoinduction and autocatalysis.

The asymmetric version of these two catalytic processes is possible, and a few examples have been described in the literature.

Danda et al. studied the formation of the cyanohydrin (*S*)-**4** from 3-phenoxybenzaldehyde **3** and HCN in the presence of catalytic amounts of the diketopiperazine (*R,R*)-**5** (Scheme 15).[69] They established that the diketopiperazine is modified by the cyanohydrin product, giving the active chiral catalyst. The diketopiperazine (*R,R*)-**5** (100% ee, 2 mol %) gave the cyanohydrin (*S*)-**4** in 92% ee. When (*R,R*)-**5** with 2% ee was used in the presence of 4 mol equiv. of (*S*)-**4** of 92.0% ee, (*S*)-**4** was generated with 81.6% ee, whereas in the presence of 4 mol equiv. of (*R*)-**4** of 84.9% ee (*R*)-**4** was generated with 74.0% ee. This shows that the initially added cyanohydrin is the stereocontrolling factor in the reaction. The authors discovered that diketopiperazine of 2% ee is almost inactive on its own, presumably due to conglomeration into an inert *meso* structure. Addition of a small amount of (*S*)-**4** dissociates this dimer to give the combination (*R,R*)-**5***(*S*)-**4**, which is more active than (*R,R*)-**5** alone. Thus the active catalyst is generated by the in situ combination of one enantiomer of the initial diketopiperazine with the major enantiomer of

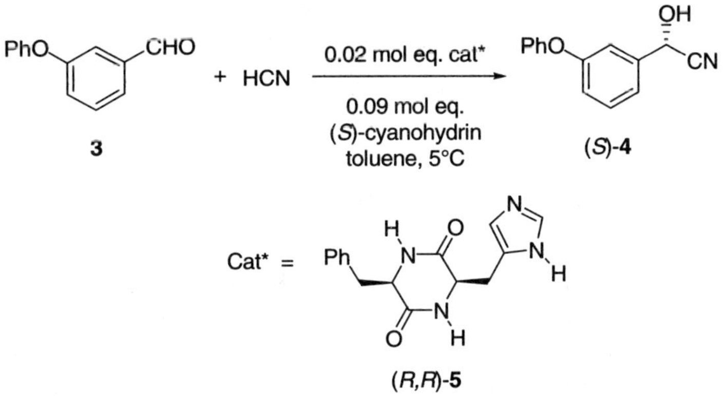

Scheme 15. Asymmetric catalytic autoinduction in the addition of HCN to benzaldehyde in the presence of the (*R*,*R*)-diketopiperazine **5**.[69]

the added cyanohydrin. Experiments established that both (*R*,*R*)-**5**(*S*)-**4** and (*S*,*S*)-**5**(*R*)-**4** are catalytically active. Asymmetric amplification may be operating in this catalytic system since $ee_{product}$ can be greater than the ee of the two initially added components. More experiments, however, are required to rigorously demonstrate the existence of an asymmetric amplification. This interesting reaction works well under heterogeneous conditions, though its exact mechanism has not yet been fully elucidated. Shvo et al. proposed that two adjacent imidazole bases function as the reactive sites within an H-bonded insoluble polymer.[70]

The self-reproduction of a chiral catalyst is called *asymmetric autocatalysis* and was proposed by Calvin as being the means by which homochirality was propagated in prebiotic times.[71] This model, however, requires that complete transfer of chirality occurs during reactions and that the ee of the products is maintained at 100%, a situation unlikely to be observed in any reaction. A minor departure from 100% ee will cause the ee of the catalyst to decrease, which will be further amplified as the reaction proceeds, ultimately giving a product close to a racemic composition. To overcome this difficulty, Frank proposed the kinetic scheme illustrated in Scheme 16.[72] This proposal is based on the autocatalytic formation of the enantiomers *R* and *S* followed by an inhibition reaction, involving the coupling of *R* and *S* to give a product *C*, which then takes no further part in the reaction scheme (by, for example, the precipitation of an insoluble racemate). In this way racemic material is continuously removed, and a small initial enantiomeric excess is increasingly enlarged, as clearly shown on computed curves. This approach recalls both the reservoir effect, since racemic material is trapped in an inactive "pool,"[26] and the Horeau duplication.[36]

Examples of asymmetric autocatalysis are difficult to find. Only quite recently have Soai et al. shown that a catalyzed diorganozinc addition to aldehydes may be

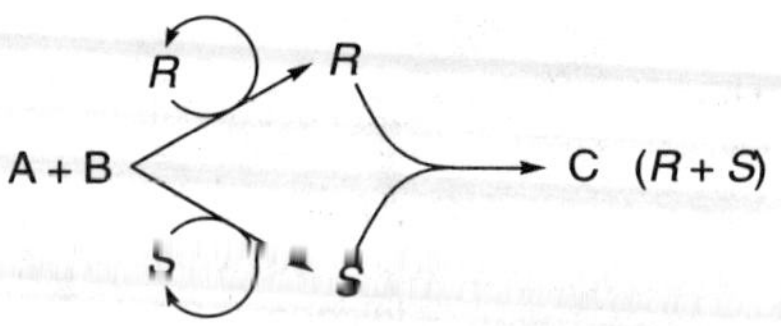

Scheme 16. Frank model of asymmetric autocatalysis.

suitably modified to lead to asymmetric autocatalysis.[73, 74] The basic features of this process are indicated in Scheme 17. This process is particularly efficient because there is a (+)-NLE that compensates for the incomplete (although high) enantioselectivity of the reaction (99% ee). The asymmetric amplification is the indispensable companion of an efficient autocatalysis starting from a catalyst of low ee. The pyrimidyl zinc alkoxide **7** (R = H or Me) acts as a catalyst for the addition of diisopropyl zinc to a pyrimidyl carboxaldehyde **6**, generating **7** in high ee and with the same configuration. The alcohol precursor of alkoxide **7** (R = H) was used in 5% ee as an asymmetric autocatalyst (20 mol %).[73] It produced an alcohol **8** (R = H) of 39% ee in 62% yield, meaning that the newly created alcohol is of 55% ee (calculated by subtracting the contribution from the initially added alcohol). In a similar reaction, a catalyst of 39% ee gave a product of 76% ee (alcohol formed with 87% ee during the reaction), and a catalyst of 89% ee provided alcohol of 89% ee (alcohol formed with 90% ee during the reaction). The curve showing the ee of the newly formed product as a function of the ee of the catalyst is typically that of (+)-NLE, that is, a reaction with a strong asymmetric amplification. The case of 5% ee giving 55% ee corresponds to an amplification index *I* of 7. In the previously described experiment where the ee of the pyrimidyl alcohol increases from 5 to 89%, one can calculate that the amount of *S*-enantiomer increases by a factor of 94, while the *R*-enantiomer increases only by a factor of 8. An improvement to the experimental procedure was subsequently discovered by Soai et al. in the case of a 2-methylpyrimidyl alcohol **8** (R = Me).[74] A one-pot reaction, employing several consecutive additions of diisopropyl zinc and

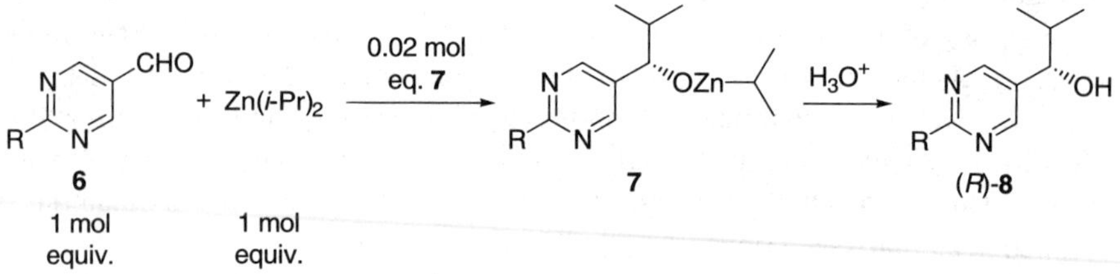

Scheme 17. Soai's autocatalysis.[73, 74]

aldehyde and with a small amount of (*R*)-**8** catalyst of low ee (6.4%) gave a large amount of (*R*)-**8** with 93.1% ee. In the case of using only 3 mg of (*R*)-**8** with 0.18% ee as catalyst, 323 mg of (*R*)-**8** with 93.1% ee was obtained. Asymmetric autocatalysis sustained by an asymmetric amplification is a fascinating class of reactions, and a few impressive examples have now been well established. It can be expected that further reports of this phenomenon, which is especially relevant to the problem of the generation and propagation of biomolecular homochirality on earth, will be seen in the near future.

VII. ASYMMETRIC AMPLIFICATION IN CHIRAL POISONING OR CHIRAL ACTIVATION OF A RACEMIC CATALYST

A chiral catalyst may be generated in situ from racemic ligand by the inclusion of controlled amounts of chiral additives. The various possibilities are presented in Scheme 18.

A chiral poison will destroy part of the racemic ligand or catalyst, in situ, by kinetic resolution. The remaining enantioenriched material will then act as the chiral catalyst. Chiral activation of a catalyst, on the other hand, generates a new

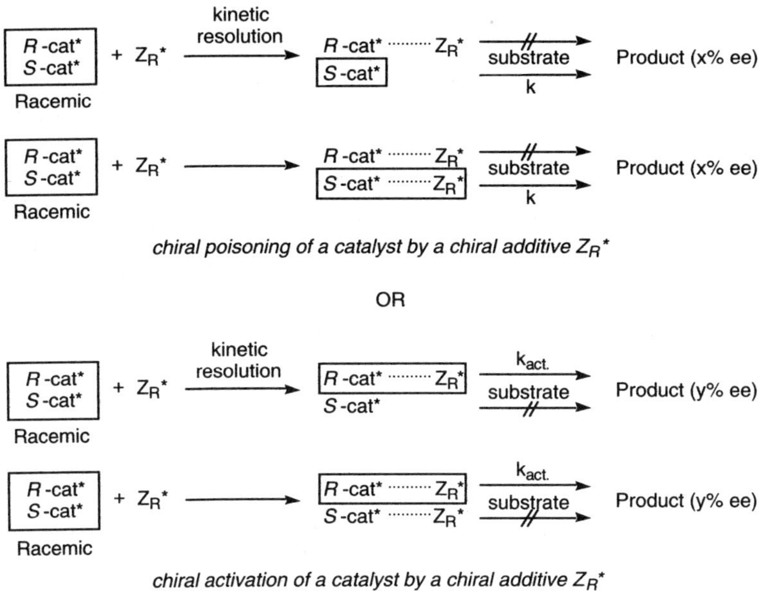

Scheme 18. Asymmetric amplification in the chiral poisoning or chiral activation of a racemic catalyst.

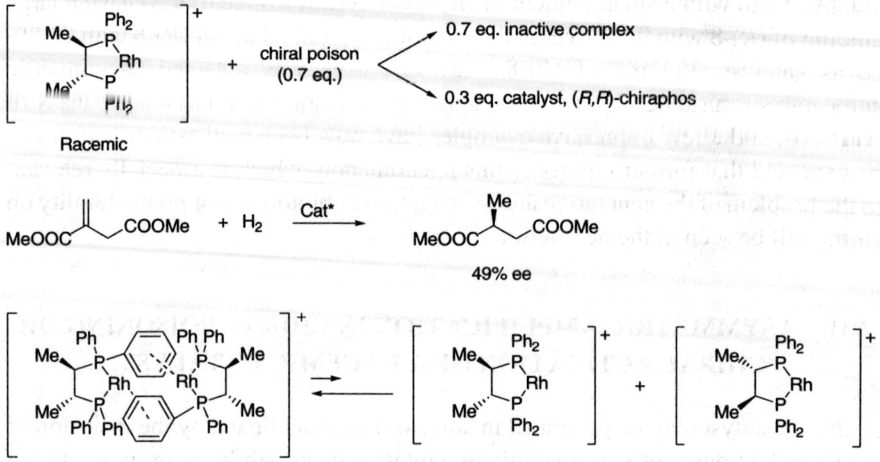

Scheme 19. Faller's chiral poisoning.[75]

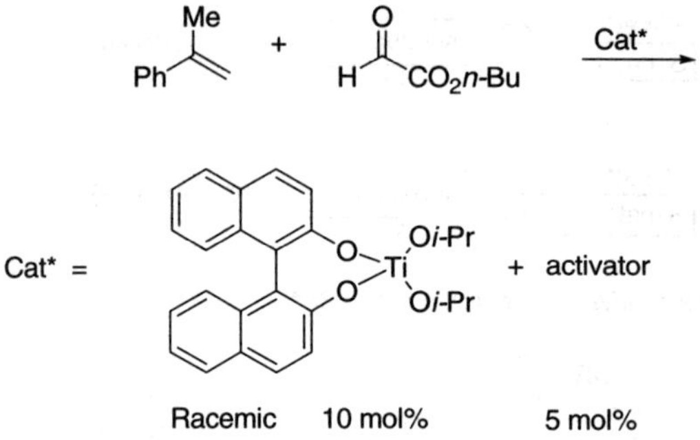

Activator	Yield (%)	ee (%)
None	5.9	0
(*R*)-BINOL	52	89.8

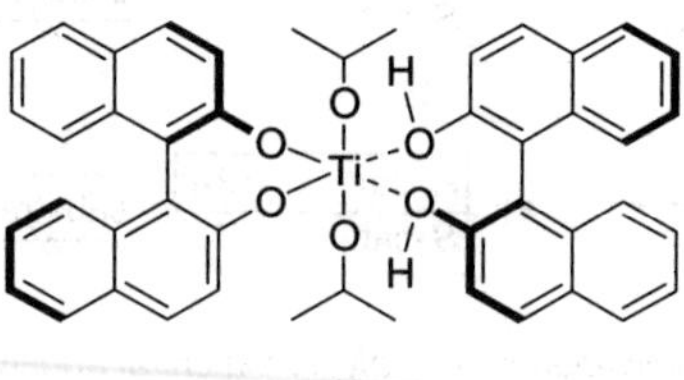

9

Scheme 20. Mikami's chiral drugging.[53]

chiral catalyst by kinetic resolution that is more active than the initial catalyst. In both cases a (+)-NLE can occur that amplifies the enantioselectivity of the active catalyst.

Faller et al. found that a racemic cationic rhodium–chiraphos complex can be poisoned by half an equivalent of (*R*)-cysteine (Scheme 19).[75] The remaining rhodium–chiraphos complex catalyzes an asymmetric hydrogenation of dimethyl itaconate with 49% ee. The ee is higher than that expected solely on the basis of a kinetic resolution. Independent experiments demonstrated that $[Rh\text{–chiraphos}]^+$ gives rise to asymmetric amplification. Chiraphos of 33% ee, for example, gives a product of 60% ee rather than the 30% ee calculated for a linear correlation. This has been explained by an equilibrium between rhodium dimers, with the hetero or *meso* dimer being of higher stability, thus removing racemic material from the reaction.

Mikami et al. studied the Diels–Alder reaction between α-methylstyrene and *n*-butyl glyoxylate catalyzed by a titanium binolate catalyst.[76–78] Addition of 0.5 equivalents of (*R*)-BINOL to 1 equivalent of the racemic catalyst accelerated the reaction and gave the product with 89.8% ee (Scheme 20). Enantiopure catalyst derived solely from (*R*)-BINOL gave the product with 94.5% ee. Here the amplification originates from the creation of a new chiral complex **9** of higher efficiency (rate and enantioselectivity) with respect to each enantiomer of the original racemic catalyst.

VIII. ASYMMETRIC AMPLIFICATION FOR CHIRAL REAGENTS

In reactions involving stoichiometric quantities of chiral auxiliary, nonlinear effects may arise if there are possibilities for the formation of diastereomeric reagents (by aggregation, dimerization, or other processes). The ee of the product will depend upon the relative amounts of reagent and substrate and upon the conversion. This is due to changes in the concentration of chiral reagents with conversion. If there are diastereomeric reagents made in situ from auxiliaries not enantiopure, the more reactive diastereoisomer will be used first by the prochiral substrate. The only case studied thus far involves the H. C. Brown reagent Ipc_2BCl. It was used to reduce ketone **10** or acetophenone **12** by a research group at Merck[79, 80] and by one of us,[81] respectively (Scheme 21). There is a strong asymmetric amplification in both cases. Using (–)-α-pinene of 70% ee, for example, provided alcohol **11** with 95% ee (against 98% ee if pinene of 99% ee is used). Similarly, carbinol **13** was obtained in 94% ee when α-pinene of only 80% ee was used (Curve B, Scheme 21). It is interesting to note that mixtures of enantiomerically pure reagents did not give rise to NLEs (Curve A, Scheme 21).[81] This suggests that, once formed, the stereoisomeric reagents do not interconvert during the reduction of ketones and that the *meso* reagent is a very slow reacting species. A practical application of asymmetric amplification is well demonstrated

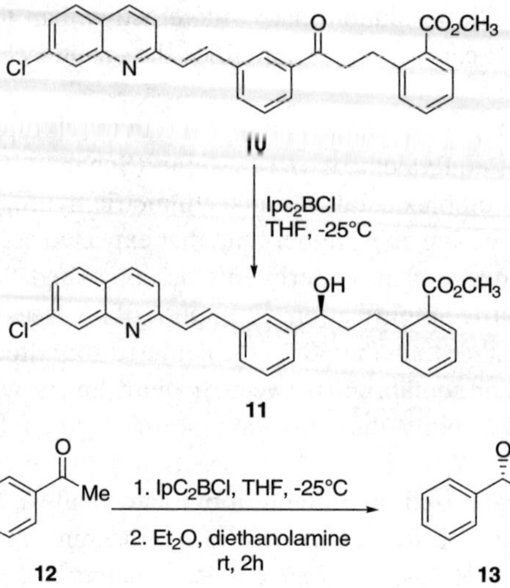

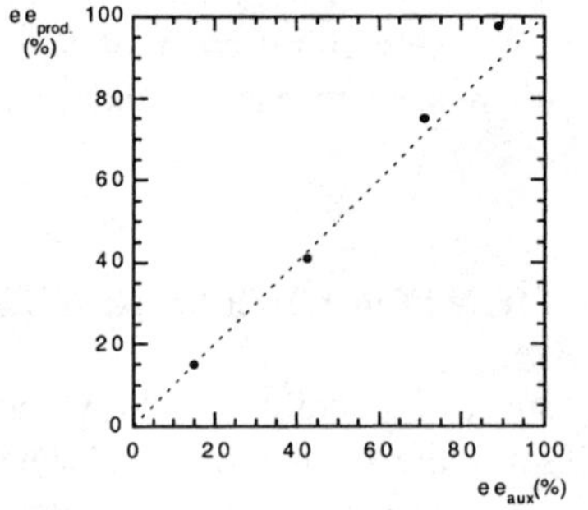

ee_{prod} versus ee_{aux} in reduction of **12** to **13** using mixtures of enantiomerically pure Ipc_2BCl

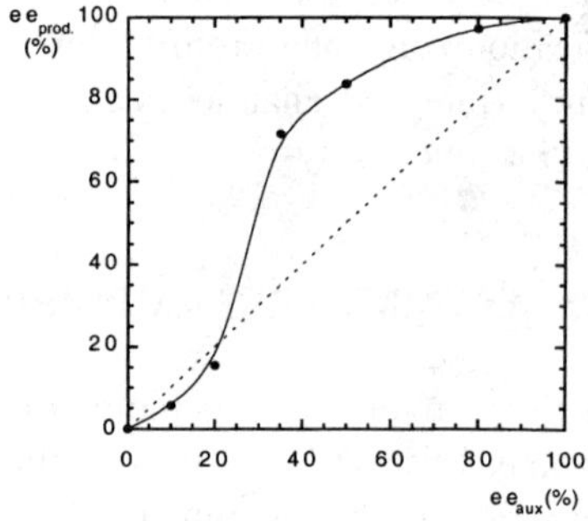

ee_{prod} versus ee_{aux} in reduction of **12** to **13** using Ipc_2BCl prepared from mixtures of enantiomerically pure (+)- and (–)-α-pinene

Various stereoisomeric Ipc_2BCl:

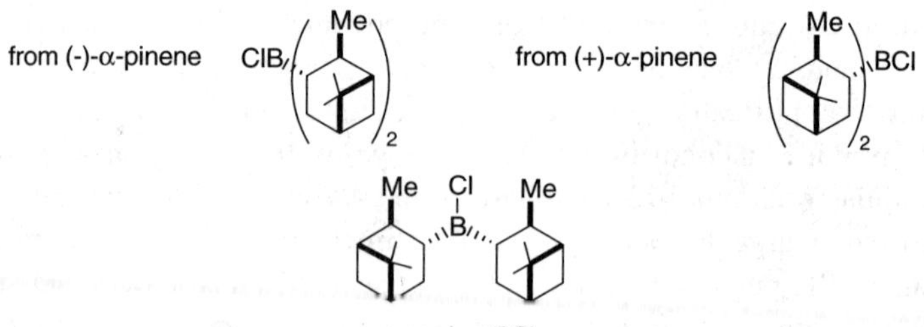

meso Ipc_2BCl

Scheme 21. Asymmetric amplification in stoichiometric reductions with Ipc_2BCl.[79–81]

here, since commercial α-pinene is usually of 70–90% ee—the improvement in ee conveniently occurs during the preparation of the reagent. The price to pay, however, is the decreased reactivity of the reagent and the necessity to use an excess of Ipc_2BCl with respect to the ketone.

IX. USE OF ASYMMETRIC AMPLIFICATION IN MECHANISTIC STUDIES

The presence of a nonlinear effect, either negative or positive, is a useful piece of information for the mechanistic study of a reaction. It implies that diastereomeric species are formed from the chiral auxiliary. If an asymmetric amplification is observed, it can be indicative of the formation of *meso* dimers (or tetramers etc.) of low reactivity. When the kinetic study of an asymmetric catalysis shows a rate second order with respect to catalyst concentration, it may be useful to investigate the possibility of nonlinear effects in the system. Jacobsen et al., for example, studied the

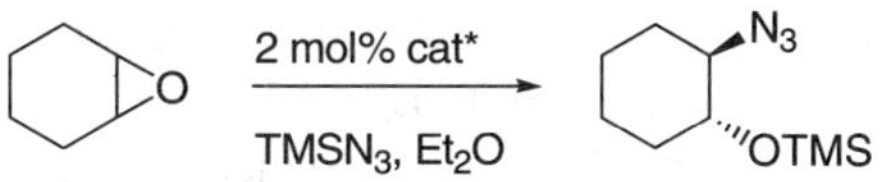

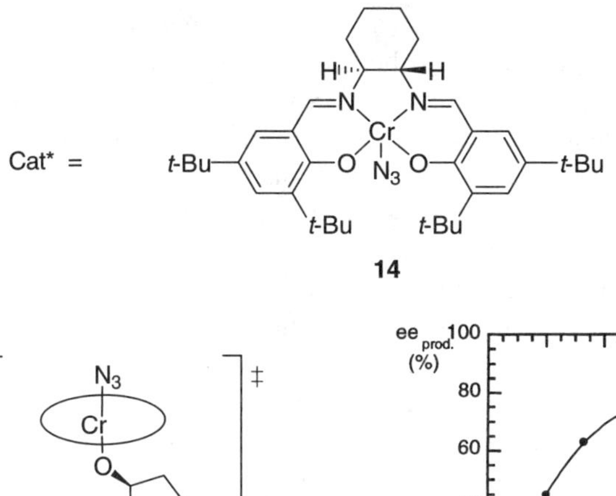

14

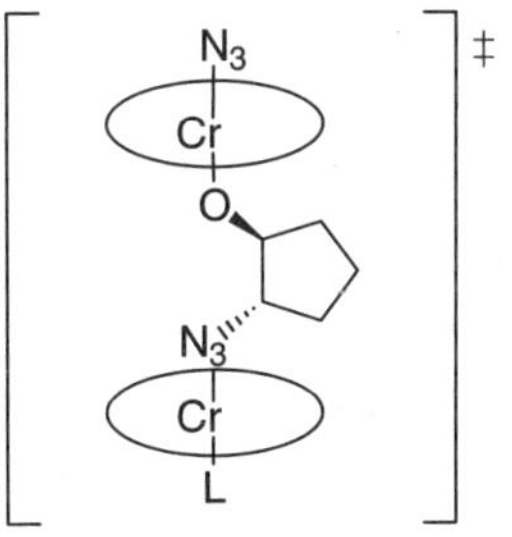

15

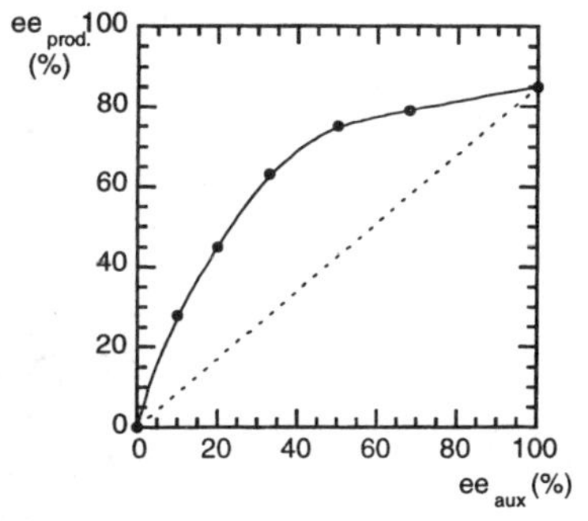

Scheme 22. Asymmetric amplification in the opening of *meso* epoxides with $TMSN_3$ catalyzed by a chiral chromium–salen complex.[66]

asymmetric opening of *meso* epoxides using $TMSN_3$ and catalyzed by a chiral chromium–salen complex.[66] The reaction was found to involve the formation of a chromium–azido complex **14** (Scheme 22). The kinetic investigation showed a second-order dependency upon the chromium–salen complex concentration. The authors then suspected the participation of two chromium–salen complexes in a bimetallic rate-determining step—one acting as a Lewis acid for epoxide activation, the other delivering the azido group. This hypothesis was nicely confirmed by an NLE study. A respectable asymmetric amplification was observed, in agreement with model **15**, where the heterochiral combination is disfavored with respect to the homochiral one, the latter giving an excellent asymmetric induction.

X. CONCLUSION

Asymmetric amplification in reactions involving partially resolved chiral auxiliaries is now a well-established phenomenon that is very attractive since it gives improved enantioselectivities with respect to expectations based upon the ee of the auxiliary. It may have practical application in that enantiomerically pure chiral auxiliaries are not always required for highly selective asymmetric synthesis. Asymmetric amplification is also of fundamental importance in order to achieve efficient asymmetric autocatalysis. Finally, evidence of an asymmetric amplification is a very useful piece of information in mechanistic studies.

Asymmetric amplification is the direct consequence of species of racemic composition being stable and of low reactivity. It should be found in all reactions dealing with chiral auxiliaries. One can expect such a phenomenon in diastereoselective reactions performed on chiral substrates (e.g., reduction of a keto steroid by a chiral catalyst) or kinetic resolution, a hypothesis that has just recently been confirmed.[82, 83]

It is important to remember that the advantage of increased enantiomeric purity gained during an asymmetric amplification is always accompanied by the disadvantage of reduced reaction rates when compared with the enantiopure chiral auxiliary.[35] Although this reduction in reaction rate may not present a problem for fast reactions, it may be necessary to increase the amount of catalyst or chiral reagent in slower reactions in order to have reasonable reaction times.

ACKNOWLEDGMENTS

One of us (D.R.F.) acknowledges the Royal Society for a European Exchange Postdoctoral Fellowship. We thank the C.N.R.S. and Institut Universitaire de France for their financial support. We also gratefully acknowledge the assistance of Timo Luukas with the preparation of the diagrams.

REFERENCES

1. Oguni, N.; Matsuda, Y.; Kaneko, T. *J. Am. Chem. Soc.* **1988**, *110*, 7877.
2. Puchot, C.; Dunach, E.; Zhao, S.; Agami, C.; Kagan, H. B. *J. Am. Chem. Soc.* **1986**, *108*, 2353.
3. The strong connection between "amplification" and "increase of material" is well illustrated in the "PCR amplification," classically used in nucleic acid chemistry to prepare a large amount of given DNA from a small amount of the same DNA.[4]
4. Mullis, K. B. *Angew. Chem. Int. Ed. Engl.* **1994**, *33*, 1209.
5. Blair, N. E.; Bonner, W. A. *Origins of Life* **1980**, *10*, 255.
6. Spach, G. *Symbioses* **1981**, *13*, 72.
7. Chen, C. S.; Liu, Y. C. *J. Org. Chem.* **1991**, *56*, 1966.
8. Faller, J. W.; Ma, Y. *Organometallics* **1992**, *11*, 2726.
9. For a discussion on the advantages and disadvantages of using enantiomeric excess (ee) or enantiomer ratio (er), see Refs. 10 and 11.
10. Kagan, H. B. *Rec. Trav. Chim. Pays-Bas* **1995**, *114*, 203.
11. Schurig, V. *Enantiomer* **1996**, *1*, 139.
12. D. Seebach et al. recently wrote[13]: "We hesitate to use the term 'chiral amplification,' because it is hard to see how 'amplification' could be non-congruent with its mirror image."
13. Seebach, D.; Marti, R. E.; Hinterman, T. *Helv. Chim. Acta* **1996**, *79*, 1725.
14. The use of enantiomeric ratio or stereoisomeric ratio is also well suited to calculations involving multiple consecutive asymmetric reactions on a molecule.[15] An apparent amplification of stereoselectivity may be observed. This topic is outside the scope of the present chapter.
15. Rautenstrauch, V. *Bull. Soc. Chim. Fr.* **1994**, *131*, 515.
16. Shi, X.; Leal, W. S.; Meinwald, J. *Bioorganic Med. Chem.* **1996**, *4*, 297.
17. Rautenstrauch, V.; Lindström, M.; Bourdin, B.; Currie, J.; Oliveros, E. *Helv. Chim. Acta* **1993**, *76*, 607.
18. El Baba, S.; Poulin, J. C.; Kagan, H. B. *Tetrahedron* **1984**, *40*, 4275.
19. Horeau, A. *Tetrahedron Lett.* **1969**, *36*, 3121.
20. Horeau, A.; Guetté, J. P. *Tetrahedron* **1974**, *30*, 1923.
21. Williams, T.; Pitcher, R. G.; Bommer, P.; Gutzwiller, J.; Uskokovic, M. *J. Am. Chem. Soc.* **1969**, *91*, 1871.
22. Diter, P.; Taudien, S.; Samuel, O.; Kagan, H. B. *J. Org. Chem.* **1994**, *59*, 370, and references cited therein.
23. Katagiri, T.; Yoda, C.; Furuhashi, K.; Ueki, K.; Kubota, T. *Chem. Lett.* **1996**, 115.
24. Wynberg, H.; Feringa, B. *Tetrahedron* **1976**, *32*, 2831.
25. Izumi, Y.; Tai, A. *Stereo-Differentiating Reactions*; Academic Press: New York, **1977**, p. 244.

26. Guillaneux, D.; Zhao, S. H.; Samuel, O.; Rainford, D.; Kagan, H. B. *J. Am. Chem. Soc.* **1994**, *116*, 9430.

27. For some reviews, see Refs. 28–31.

28. Girard, C.; Kagan, H. B. *Angew Chem. Int. Ed.* **1998**, *37*, 2922.

29. Bolm, C. In *Advanced Asymmetric Catalysis*; Stephenson, G. R., Ed., Blackie Academic and Professional: New York, 1996, pp. 9–26.

30. Noyori, R. In *Catalytic Organic Reactions*; Wiley: New York, 1994, pp. 273–297.

31. Avalos, M.; Babiano, R.; Cintas, P.; Jiménez, J. L.; Palacios, J. C. *Tetrahedron: Asymmetry* **1997**, *8*, 2997.

32. Kitamura, M.; Okada, S.; Noyori, R. *J. Am. Chem. Soc.* **1989**, *11*, 4028.

33. Terada, M.; Mikami, K.; Nakai, T. *J. Chem. Soc. Chem. Commun.* **1990**, 1623.

34. Yamakawa, M.; Noyori, R. *J. Am. Chem. Soc.* **1995**, *117*, 6327.

35. Blackmond, D. G, *J. Am. Chem. Soc.* **1997**, *119*, 12934.

36. Vigneron, J. P.; Dhaenens, M.; Horeau, A. *Tetrahedron* **1973**, *29*, 1055.

37. Noyori, R.; Kitamura, M. *Angew. Chem. Int. Ed. Engl.* **1990**, *30*, 49.

38. Bolm, C.; Müller, J.; Schlingloff, G.; Zehnder, M.; Neuburger, M. *J. Chem. Soc. Chem. Commun.* **1993**, 182.

39. Bolm, C.; Schlingloff, G.; Harms, K. *Chem. Ber.* **1992**, *125*, 1191.

40. Fitzpatrick, K.; Hulst, R.; Kellog, R. M. *Tetrahedron: Asymmetry* **1995**, *6*, 1861.

41. Rijnberg, E.; Hovestad, N. J.; Kleij, A. W.; Jastrzebski, J. T. B. H.; Boersma, J.; Janssen, M. D.; Spek, A. L.; van Koten, G. *Organometallics* **1997**, *16*, 2847.

42a. Bolm, C. *Tetrahedron: Asymmetry* **1991**, *2*, 701.

42b. Bolm, C.; Ewald, M.; Felder, M. *Chem. Ber.* **1992**, *125*, 1205.

43. Bolm, C.; Felder, M.; Müller, J. *Synlett* **1992**, 439.

44. de Vries, A. H. M.; Jansen, J. F. G. A.; Feringa, B. L. *Tetrahedron* **1994**, *50*, 4479.

45a. Tanaka, K.; Matsui, J.; Kawabata, Y.; Suzuki, H.; Watanabe, A. *J. Chem. Soc. Chem. Commun.* **1991**, 1632.

45b. Tanaka, K.; Matsui, J.; Suzuki, H. *J. Chem. Soc. Perkin Trans. 1* **1993**, 153.

46. Rossiter, B. E.; Miao G.; Swingle, N. M.; Eguchi, M.; Hernández, A. E.; Patterson, R. G. *Tetrahedron: Asymmetry* **1992**, *3*, 231.

47. Zondervan, C.; Feringa, B. L. *Tetrahedron: Asymmetry* **1996**, *7*, 1895.

48. Bougauchi, M.; Watanabe, S.; Arai, T.; Sasai, H.; Shibasaki, M. *J. Am. Chem. Soc.* **1997**, *119*, 2329.

49. Komatsu, N.; Hashizume, M.; Sugita, T.; Uemura, S. *J. Org. Chem.* **1993**, *58*, 4529.

50a. Terada, M.; Mikami, K.; Nakai, T. *J. Chem. Soc. Chem. Commun.* **1990**, 1623.

50b. Mikami, K.; Terada, *M. Tetrahedron* **1992**, *48*, 5671.

51. Mikami, K.; Motoyama, Y.; Terada, M. *J. Am. Chem. Soc.* **1994**, *116*, 2812.

52. Terada, M.; Mikami, K. *J. Chem. Soc. Chem. Commun.* **1994**, 833.

53. Iwasawa, N.; Hayashi, Y.; Sakurai, H.; Narasaka, K. *Chem. Lett.* **1989**, 1581.

54. Shimizu, M.; Ukaji, Y.; Inomata, K. *Chem. Lett.* **1996**, 455.

55. Seebach, D.; Dahinden, R.; Marti, R. E.; Beck, A. K.; Plattner, D. A.; Kühnle, F. N. M. *J. Org. Chem.* **1995**, *60*, 1788.

56. Kobayashi, S.; Ishitani, H.; Araki, M.; Hachiya, I. *Tetrahedron Lett.* **1994**, *35*, 6325.

57. Kitamoto, D.; Imma, H.; Nakai, T. *Tetrahedron Lett.* **1995**, *36*, 1861.

58. Evans, D.A.; Nelson, S. G.; Gagné, M. R.; Muci, A. R. *J. Am. Chem. Soc.* **1993**, *115*, 9800.

59. Keck, G. E.; Krishnamurthy, D.; Grier, M. C. *J. Org. Chem.* **1993**, *58*, 6543.

60. Faller, J. W.; Sams, D. W. I.; Liu, X. *J. Am. Chem. Soc.* **1996**, *118*, 1217.

61. Bedeschi, P.; Casolari, S.; Costa, A. L.; Tagliavini, E.; Umani-Ronchi, A. *Tetrahedron Lett.* **1995**, *36*, 7897.

62. Keck, G. E.; Krishnamurthy, D. *J. Am. Chem. Soc.* **1995**, *117*, 2363.

63. Sasai, H.; Suzuki, T.; Itoh, N.; Shibasaki, M. *Tetrahedron Lett.* **1993**, *34*, 851.

64. Gauthier, Jr., D. R.; Carreira, E. M. *Angew. Chem. Int. Ed. Engl.* **1996,** *35*, 2363.

65. Hayashi, M.; Matsuda, T.; Oguni, N. *J. Chem. Soc. Chem. Commun.* **1990**, 1364.

66. Hansen, K. B.; Leighton, J. L.; Jacobsen, E. N. *J. Am. Chem. Soc.* **1996**, *118*, 10924.

67a. Albert, A. H.; Wynberg, H. *J. Am. Chem. Soc.* **1989**, *111*, 7265.

67b. Wynberg, H. *Chimia*, **1989**, *43*, 150.

68. For a brief review, see: Bolm, C.; Bienewald, F.; Seger, A. *Angew. Chem. Int. Ed. Engl.* **1996**, *35*, 1657.

69. Danda, H.; Nishikawa, H.; Otaka, K. *J. Org. Chem.* **1991**, *56*, 6740.

70. Shvo, Y.; Gal, M.; Becker, Y.; Elgavi, A. *Tetrahedron: Asymmetry* **1996**, *7*, 911, and references cited therein.

71. Calvin, M. In *Chemical Evolution*; Clarendon: London, 1969, Chapter 7.

72. Frank, F. C. *Biochim. Biophys. Acta* **1953**, *11*, 459.

73. Soai, K.; Shibata, T.; Marioka, H.; Choji, K. *Nature* **1995**, *378*, 767.

74. Shibata, T.; Hayase, T.; Yamamoto, J.; Soai, K. *Tetrahedron: Asymmetry* **1997**, *8*, 1717.

75. Faller, J. W.; Parr, J. *J. Am. Chem. Soc.* **1993**, *115*, 804.

76. Mikami, K.; Matsukawa, S. *Nature* **1997**, *385*, 613.

77. Mikami, K.; Matsukawa, S. *Enantiomer* **1996**, *1*, 69.

78. Mikami, K.; Matsukawa, S.; Volk, T.; Terada, M. *Angew. Chem. Int. Ed. Eng.* **1997**, *36*, 2768.

79. Shinkai, I.; King, A. O.; Larsen, R. D. *Pure Appl. Chem.* **1994**, *66*, 7551.

80. Zhao, M.; King, A. O.; Larsen, R. D.; Verhoeven, T. R.; Reider, P. J. *Tetrahedron Lett.* **1997**, *38*, 2641.

81a. Girard, C.; Kagan, H. B. *Tetrahedron: Asymmetry* **1995**, *6*, 1881.

81b. Girard, C.; Kagan, H. B. *Tetrahedron: Asymmetry* **1997**, *8*, 3851.

82. Komatsu, N.; Hashizume, M.; Suguta, T.; Uemura, S. *J. Org. Chem.* **1993**, *58*, 7624.

83. For a general theory of kinetic resolution with a partially resolved chiral auxiliary, see Ref. 84.
84. Kagan, H. B.; Girard, C.; Luukas, T.; Fenwick, D. R. manuscript in preparation.

CUMULATIVE TITLE INDEX

SUBJECT INDEX